Reaction Mechanisms of Inorganic and Organometallic Systems

TOPICS IN INORGANIC CHEMISTRY
A Series of Advanced Textbooks in Inorganic Chemistry

Series Editor
Peter C. Ford, University of California, Santa Barbara

Chemical Bonding in Solids, J. Burdett

Reaction Mechanisms of Inorganic and Organometallic Systems, 2nd Edition, R. Jordan

Reaction Mechanisms of Inorganic and Organometallic Systems

Second Edition

Robert B. Jordan

New York   Oxford
OXFORD UNIVERSITY PRESS
1998

Oxford University Press

Oxford   New York
Athens   Auckland   Bangkok   Bogota   Bombay   Buenos Aires
Calcutta   Cape Town   Dar es Salaam   Delhi   Florence   Hong Kong
Istanbul   Karachi   Kuala Lumpur   Madras   Madrid   Melbourne
Mexico City   Nairobi   Paris   Singapore   Taipei   Tokyo   Toronto   Warsaw

and associated companies in
Berlin   Ibadan

Published by Oxford University Press, Inc.,
198 Madison Avenue, New York, New York 10016

Oxford is a registered trademark of Oxford University Press

Library of Congress Cataloging-in-Publication Data

Jordan, Robert B.
   Reaction mechanisms of inorganic and organometallic systems /
   Robert B. Jordan — 2nd ed.
        p.   cm. — (Topics in inorganic chemistry)
   Includes bibliographical references and index.
   ISBN 0-19-511555-4 (acid-free paper)
   1. Chemical reaction, Conditions and laws of.   2. Organometallic
compounds.   3. Inorganic compounds.   I. Title.   II. Series.
QD502.J67   1997   97–31222
541.3'9—dc21   CIP

9 8 7 6 5 4 3 2 1

Printed in the United States of America
on acid-free paper

Preface

This book evolved from the lecture notes of the author for a one-semester course given to senior undergraduates and graduate students over the past 20 years. In this second edition, the organization of the first edition has been retained. The first two chapters have been modified with the intention of making them more palatable to students with a minimum of previous exposure to the basics of kinetics. This material includes new sections on the qualitative relationship of rate law and mechanism and on the conditions for application of the steady state and rapid equilibrium assumptions. There are new sections on square planar isomerization in Chapter 4 and on the bio-inorganic chemistry of nitric oxide in Chapter 8. A new Chapter 9 on experimental methods along with examples of their applications, and some new problems for each chapter have been added. The material generally has been updated to developments through 1996, and the overall result is that this edition is 30 percent larger than the first.

There is more material than can be covered in depth in one semester, but the organization allows the lecturer to give less coverage to certain areas without jeopardizing an understanding of other areas. It is assumed that the students are familiar with elementary crystal field theory and its applications to electronic spectroscopy and energetics, and concepts of organometallic chemistry, such as the 18-electron rule, π bonding, and coordinative unsaturation. The terminology and developments of elementary kinetics are given in the first two chapters; some background from a physical chemistry course would be useful, and familiarity with simple differential and integral calculus is assumed.

It is expected that students will consult the original literature to obtain further information and to gain a feeling for the excitement in the field. This experience also should enhance their ability to critically evaluate such work. Sample problems for each chapter are given at the end of the book. Many problems are taken from the literature, and original references are given; outlines of answers to the problems will be supplied to instructors who request them from the author.

The issue of units continues to be a vexing one in this area. A major goal of this course has been to provide students with sufficient background so that they can read and analyze current research papers. To do this and be able to compare results, the reader must be vigilant about the units used by different authors. There is now general agreement that the time unit for rate constants is s^{-1}, and this has been adopted throughout the text. Energy units are a different matter. Since both joules and calories are in common usage, both units have been retained in the text, with the choice made on the basis of the units in the original work as much as possible. However, within individual sections the text attempts to use one energy unit. In spectroscopic areas, cm^{-1} is dominant, but spectra are often given in nanometers, and the conversion between these units and calories and joules is given when it seems important to understanding the issue being covered. Bond lengths are given in angstroms, which are still commonly quoted for crystal structures, and in picometers, which are used in theoretical calculations. The formulas for such calculations are given in the original or most common format, and units for the various quantities are always specified.

The author is greatly indebted to all those whose research efforts have provided the core of the material for this book. The author is pleased to acknowledge those who have provided the inspiration for this book: first, my parents, who contributed the early atmosphere and encouragement; second, Henry Taube, whose intellectual and experimental guidance ensured my continuing enthusiasm for mechanistic studies. Finally and foremost, Anna has been a vital force in the creation of this book through her understanding of the time commitment, her force of will to see the project completed, and her invaluable comments, criticisms, and assistance in producing the manuscript.

R. B. J.

Edmonton, Alberta
November 1997

Contents

Reaction Mechanisms of
Inorganic and Organometallic
Systems

1

Tools of the Trade

This chapter covers the basic terminology related to the types of studies that are commonly used to provide information about a reaction mechanism. More background material is available from general physical chemistry texts[1,2] and books devoted to kinetics.[3,4,5] The reader also is referred to the initial volumes of the series edited by Bamford and Tipper.[6] Experimental techniques that are commonly used in inorganic kinetic studies are discussed in Chapter 9.

1.1 BASIC TERMINOLOGY

As with most fields, the study of reaction kinetics has some terminology with which one must be familiar in order to understand advanced books and research papers in the area. The following is a summary of some of these basic terms and definitions. Many of these may be known from previous studies in introductory and physical chemistry, and further background can be obtained from textbooks devoted to the physical chemistry aspects of reaction kinetics.

Rate
For the general reaction

$$a A + b B \ldots \longrightarrow m M + n N \ldots \qquad (1.1)$$

the reaction rate and the rate of disappearance of reactants and rate of formation of products are related by

$$\text{Rate} = -\frac{1}{a}\frac{d\,[A]}{d\,t} = -\frac{1}{b}\frac{d\,[B]}{d\,t} = \frac{1}{m}\frac{d\,[M]}{d\,t} = \frac{1}{n}\frac{d\,[N]}{d\,t} \qquad (1.2)$$

In practice, it is not uncommon to define the rate only in terms of the species whose concentration is being monitored. The consequences that can result from different definitions of the rate in relation to the stoichiometry are described below under the definition of the rate constant.

Rate Law

The rate law is the experimentally determined dependence of the reaction rate on reagent concentrations. It has the following general form:

$$\text{Rate} = k\,[A]^m\,[B]^n \dots \tag{1.3}$$

where k is a proportionality constant called the rate constant. The exponents m and n are determined experimentally from the kinetic study. The exponents in the rate law have no necessary relationship to the stoichiometric coefficients in the balanced chemical reaction. The rate law may contain species that do not appear in the balanced reaction and may be the sum of several terms for different reaction pathways.

The rate law is an essential piece of mechanistic information because it contains the concentrations of species necessary to get from the reactant to the product by the lowest energy pathway. A fundamental requirement of an acceptable mechanism is that it must predict a rate law consistent with the experimental rate law.

Order of the Rate Law

The order of the rate law is the sum of the exponents in the rate law. For example, if m = 1 and n = −2 in Eq. (1.3), the rate law has an overall order of −1. However, except in the simplest cases, it is best to describe the order with respect to individual reagents; in this example, first order in [A] and inverse second order in [B].

Rate Constant

The rate constant, k, is the proportionality constant that relates the rate to the reagent concentrations (or activities or pressures, for example), as shown in Eq. (1.3). The units of k depend on the rate law and must give the right-hand side of Eq. (1.3) the same units as the left-hand side.

A simple example of the need to define the rate in order to give the meaning of the rate constant is shown for the reaction

$$2\,A \xrightarrow{\ k\ } B \tag{1.4}$$

Following Eq. (1.2), and assuming the rate is second order in [A], then

$$\text{Rate} = -\frac{1}{2}\frac{d\,[A]}{d\,t} = \frac{d\,[B]}{d\,t} = k\,[A]^2 \tag{1.5}$$

If the experiment were done by following the disappearance of A, then the experimental rate constant would be 2k, and it must be divided by 2 to get the numerical value of k as defined by Eq. (1.5). However, if the formation of B was followed, then k would be determined directly from the experiment.

Half-Time

The half-time, $t_{1/2}$, is the time required for a reactant concentration to change by half of its total change. This term is used to convey a qualitative idea of the time scale and has a quantitative relationship to the rate constant in simple cases. In complex systems, the half-time may be different for different reagents, and one should specify the reagent to which the $t_{1/2}$ refers.

Lifetime

For a particular species, the lifetime, τ, is the concentration of that species divided by its rate of disappearance. This term is commonly used in so-called lifetime methods, such as NMR, and in relaxation methods, such as temperature jump.

1.2 ANALYSIS OF RATE DATA

In general, a kinetic study begins with the collection of data of concentration versus time of a reactant or product. As will be seen later, this can also be accomplished by determining the time dependence of some variable that is proportional to concentration, such as absorbance or NMR peak intensity. The next step is to fit the concentration–time data to some model that will allow one to determine the rate constant if the data fit the model.

The following section develops some integrated rate laws for the models most commonly encountered in inorganic kinetics. This is essentially a mathematical problem; given a particular rate law as a differential equation, the equation must be reduced to one concentration variable and then integrated. The integration can be done by standard methods or by reference to integration tables. Many more complex examples are given in advanced textbooks on kinetics.

1.2.a Zero-Order Reaction

A zero-order reaction is rare for inorganic reactions in solution but is included for completeness. For the general reaction

$$A \xrightarrow{\ k\ } B \qquad\qquad (1.6)$$

the zero-order rate law is given by

$$\frac{d\,[B]}{d\,t} = k \qquad\qquad (1.7)$$

and integration over the limits $[B] = [B]_0$ to $[B]$ and $t = 0$ to t yields

$$[B] - [B]_0 = k t \qquad (1.8)$$

This predicts that a plot of [B] or [B] − [B]$_0$ versus t should be linear with a slope of k.

1.2.b First-Order Irreversible System

Strictly speaking, there is no such thing as an irreversible reaction. It is just a system in which the rate constant in the forward direction is much larger than that in the reverse direction. The kinetic analysis of the irreversible system is just a special case of the reversible system that is described in the next section.

For the representative irreversible reaction

$$A \xrightarrow{\ k_1\ } B \qquad (1.9)$$

the rate of disappearance of A and appearance of B are given by

$$-\frac{d\,[A]}{d\,t} = \frac{d\,[B]}{d\,t} = k_1\,[A] \qquad (1.10)$$

The problem, in general, is to convert this differential equation to a form with only one concentration variable, either [A] or [B], and then integrate the equation to obtain the *integrated rate law*. The choice of the variable to retain will depend on what has actually been measured experimentally. The elimination of one concentration is done by considering the reaction stoichiometry and the initial conditions. The most general conditions are that both A and B are present initially at concentrations [A]$_0$ and [B]$_0$, respectively, and the concentrations at any time are defined as [A] and [B].

For this simple case, the rate law in terms of [A] can be obtained by simple rearrangement to give

$$-\frac{d\,[A]}{[A]} = k_1\,d\,t \qquad (1.11)$$

Then, integration over the limits [A] = [A]$_0$ to [A] and t = 0 to t gives

$$\ln\,[A] - \ln\,[A]_0 = -k_1\,t \qquad (1.12)$$

and predicts that a plot of ln [A] versus t should be linear with a slope of −k$_1$. The linearity of such plots often is taken as evidence of a first-order rate law. Since the assessment of linearity is somewhat subjective, it is better to show that the slope of such plots is the same for different initial concentrations of A.

The equivalent exponential form of Eq. (1.12) is

$$[A] = [A]_0 e^{-k_1 t} \tag{1.13}$$

and it is now common to fit data to this equation by nonlinear least squares to obtain k_1.

In order to obtain the integrated form in terms of B, it is necessary to use the mass balance conditions. For 1:1 stoichiometry, the changes in concentration are related by

$$[A]_0 - [A] = [B] - [B]_0 \tag{1.14}$$

At the end of the reaction, $[A] = 0$ and $[B] = [B]_\infty$, and substitution of these values into Eq. (1.14) gives

$$[A]_0 = [B]_\infty - [B]_0 \tag{1.15}$$

After rearrangement of Eq. (1.14) and substitution from Eq. (1.15), one obtains

$$[A] = [B]_\infty - [B] \tag{1.16}$$

Then, substitution for [A] from Eq. (1.16) into Eq. (1.10) gives an equation that can be integrated over the limits $[B] = [B]_0$ to $[B]$ and $t = 0$ to t, to obtain

$$\ln([B]_\infty - [B]) - \ln([B]_\infty - [B]_0) = -k_1 t \tag{1.17}$$

Alternatively, this equation can be obtained by substitution for $[A]_0$ and $[A]$ from Eqs. (1.15) and (1.16) into Eq. (1.12). This predicts that a plot of $\ln([B]_\infty - [B])$ versus t should be linear with a slope of $-k_1$.

At the half-time, $t = t_{1/2}$, $[A] = [A]_0/2$, $[B] = [B]_0 + ([B]_\infty - [B]_0)/2$, and substitution into Eq. (1.12) or Eq. (1.17) gives

$$t_{1/2} = \frac{\ln 2}{k_1} = \frac{0.693}{k_1} \tag{1.18}$$

Therefore, *the half-time is independent of the initial concentrations.*

A very important practical advantage of the first-order system is that *the analysis can be done without any need to know the initial concentrations.* This means that the collection of concentration–time data can be started at any time arbitrarily defined as $t = 0$. This is a significant difference from the second-order case that is described later in this chapter.

1.2.c First-Order System Coming to Equilibrium

For a system coming to equilibrium, both the forward and reverse reactions must be included in the kinetic analysis, and one must take into account that significant concentrations of both reactants and products will be present at the end of the reaction. A first-order system coming to equilibrium may be represented by

$$A \; \underset{k_{-1}}{\overset{k_1}{\rightleftharpoons}} \; B \qquad\qquad (1.19)$$

The rate of disappearance of A equals the rate of appearance of B, and these are given by

$$-\frac{d\,[A]}{d\,t} = \frac{d\,[B]}{d\,t} = k_1\,[A] - k_{-1}\,[B] \qquad\qquad (1.20)$$

Just as with the irreversible system, the problem is to convert this equation to a form with only one concentration variable, either [A] or [B], and then integrate the equation to obtain the *integrated rate law*. The initial concentrations are defined as $[A]_0$ and $[B]_0$, and those at any time as [A] and [B]. The final concentrations at equilibrium are $[A]_e$ and $[B]_e$. Then, mass balance gives

$$[A]_0 + [B]_0 = [A] + [B] = [A]_e + [B]_e \qquad\qquad (1.21)$$

To obtain the rate law in terms of B, Eq. (1.21) can be rearranged to obtain the following expressions for [A] and $[A]_0$:

$$[A] = [A]_0 + [B]_0 - [B] \quad \text{and} \quad [A]_0 = [A]_e + [B]_e - [B]_0 \quad (1.22)$$

so that

$$[A] = [A]_e + [B]_e - [B]_0 + [B]_0 - [B] = [A]_e + [B]_e - [B] \quad (1.23)$$

Substitution for [A] from Eq. (1.23) into Eq. (1.20) gives

$$\frac{d\,[B]}{d\,t} = k_1 \left([A]_e + [B]_e \right) - \left(k_1 + k_{-1} \right) [B] \qquad\qquad (1.24)$$

Note that the initial concentrations have been eliminated.

Since Eq. (1.24) contains only one concentration variable, [B], it can be integrated directly. However, it is convenient in the end to eliminate $[A]_e$ by noting that, at equilibrium, the rate in the forward direction must be equal to the rate in the reverse direction:

$$k_1 [A]_e = k_{-1} [B]_e \qquad (1.25)$$

and substitution for $k_1[A]_e$ into Eq. (1.24) gives

$$\frac{d [B]}{dt} = k_{-1} [B]_e + k_1 [B]_e - (k_1 + k_{-1}) [B]$$

$$= (k_1 + k_{-1})([B]_e - [B]) \qquad (1.26)$$

This equation can be rearranged and integrated over the limits [B] = $[B]_0$ to [B] and t = 0 to t to obtain

$$\ln([B]_e - [B]) - \ln([B]_e - [B]_0) = -(k_1 + k_{-1})t \qquad (1.27)$$

Therefore, a plot of $\ln([B]_e - [B])$ versus t should be linear with a slope of $-(k_1 + k_{-1})$ and the kinetic study yields the sum of the forward and reverse rate constants. If the equilibrium constant, K, is known, then k_1 and k_{-1} can be calculated since $K = k_1/k_{-1}$.

Just as in the irreversible first-order system, *the analysis can be done without any need to know the initial concentrations*, and the collection of concentration–time data can be started at any time defined as t = 0.

At the reaction half-time, $t = t_{1/2}$, $[B] = [B]_0 + ([B]_e - [B]_0)/2$, and substitution into Eq. (1.27) gives

$$t_{1/2} = \frac{\ln 2}{k_1 + k_{-1}} = \frac{0.693}{k_1 + k_{-1}} \qquad (1.28)$$

It should be noted that the irreversible first-order system is a special case of the reversible system. For the irreversible system, $k_1 \gg k_{-1}$, so that $(k_1 + k_{-1}) = k_1$.

1.2.d Second-Order System Coming to Equilibrium

The second-order reversible system will be described next, and the simpler irreversible system will be developed later as a special case of the reversible one. This reversible system can be described by

$$A + B \underset{k_{-2}}{\overset{k_2}{\rightleftharpoons}} C \qquad (1.29)$$

and the rate of formation of C is given by

$$\frac{d [C]}{dt} = k_2 [A] [B] - k_{-2} [C] \qquad (1.30)$$

To simplify the development, one can assume that there is no C present initially and that the stoichiometry is 1:1:1, so that mass balance gives initially

$$[A] = [A]_0 - [C] \quad \text{and} \quad [B] = [B]_0 - [C] \tag{1.31}$$

and at equilibrium

$$[A]_e = [A]_0 - [C]_e \quad \text{and} \quad [B]_e = [B]_0 - [C]_e \tag{1.32}$$

If the stoichiometry is other than 1:1:1, then the appropriate coefficients must be used in the mass balance conditions.

Substitution for [A] and [B] from Eq. (1.31) into Eq. (1.30) gives an equation that can be integrated because [C] is the only concentration variable. However, it is convenient to eliminate k_{-2} from the equation before integrating by noting that the forward and reverse rates are equal at equilibrium:

$$k_2 ([A]_0 - [C]_e) ([B]_0 - [C]_e) = k_{-2} [C]_e \tag{1.33}$$

so that

$$k_{-2} = \frac{k_2 ([A]_0 - [C]_e) ([B]_0 - [C]_e)}{[C]_e} \tag{1.34}$$

and substitution into Eq. (1.30) yields

$$\frac{d [C]}{dt} = k_2 \frac{([C]_e - [C]) ([A]_0 [B]_0 - [C]_e [C])}{[C]_e} \tag{1.35}$$

This equation can be rearranged and integrated over the limits [C] = 0 to [C] and t = 0 to t to give the following solution:

$$\ln \left(\frac{[A]_0 [B]_0 - [C]_e [C]}{[C]_e ([C]_e - [C])} \right) + \ln \left(\frac{[C]_e^2}{[A]_0 [B]_0} \right) = \left(\frac{([A]_0 [B]_0 - [C]_e^2) k_2}{[C]_e} \right) t \tag{1.36}$$

A plot of the first term on the left-hand side of Eq. (1.36) versus t should be linear with a slope related to k_2, as indicated by the right-hand side of Eq. (1.36). It is apparent that *one must know the initial concentrations, [A]$_0$ and [B]$_0$, and the final concentration, [C]$_e$, in order to do the analysis and to determine the value of k_2 from the slope.* These requirements make this an unpopular and uncommon situation for experimental studies.

1.2.e Second-Order Irreversible System

This system can be obtained as a special case of the reversible system by simple consideration of the stoichiometry conditions. If $[A]_0 < [B]_0$ and reaction (1.29) goes essentially to completion, then $[C]_e = [A]_0$, and substitution of this condition into Eq. (1.36) gives

$$\ln\left(\frac{[B]_0 - [C]}{[A]_0 - [C]} \right) + \ln\left(\frac{[A]_0}{[B]_0} \right) = \left([B]_0 - [A]_0 \right) k_2 t \qquad (1.37)$$

In this case, the initial concentrations of both reactants are required in order to plot the first term on the left versus t and to determine k_2 from the slope. These conditions are not as restrictive as those for the reversible second-order system, but they are still worse than those for the first-order system.

At the half-time for this second-order reaction, $[C] = [A]_0/2$, and substitution into Eq. (1.37) shows that

$$t_{1/2} = \frac{1}{\left([B]_0 - [A]_0 \right) k_2} \ln\left(\frac{2 [B]_0 - [A]_0}{[B]_0} \right) \qquad (1.38)$$

1.2.f Pseudo-First-Order Reaction Conditions

The pseudo-first-order reaction condition is very widely used, but it is seldom mentioned in textbooks. Although many reactions have second-order or more complex rate laws, the experimental kineticist wishes to optimize experiments by taking advantage of the first-order rate law since it imposes the fewest restrictions on the conditions required to determine a reliable rate constant. The trick is to use the pseudo-first-order condition.

The *pseudo-first-order condition* requires that the concentration of the reactant whose concentration is monitored is at least 10 times smaller than that of all the other reactants, so that the concentrations of all the latter remain essentially constant during the reaction. Under this condition, the rate law usually simplifies to a first-order form, and one gains the advantage of not needing to know the initial concentration of the deficient reagent.

In the preceding irreversible second-order example, if it is assumed that the conditions have been set so that $[B]_0 \gg [A]_0$, then $[B]_0 \gg [C]$. In addition, the concentration of B will remain constant at $[B]_0$, and the final concentration of C is $[C]_\infty = [A]_0$ if the reaction is irreversible and has a 1:1 stoichiometry. Substitution of these conditions into Eq. (1.37) gives

$$\ln\left(\frac{[C]_\infty}{[C]_\infty - [C]} \right) = [B]_0 k_2 t \qquad (1.39)$$

This equation predicts that a plot of $\ln([C]_\infty - [C])$ versus t should be linear with a slope of $-k_2[B]_0$. This is identical in form to the first-order rate law except that k_1 is replaced by $k_2[B]_0$. The latter constant is often called the observed, k_{obsd}, or experimental, k_{exp}, rate constant. Since $[B]_0$ is known, it is possible to calculate k_2.

In a more general case, if the rate of disappearance of reactant A is a function of the concentrations of other species X, Y, and Z, then the rate of disappearance of A may be given by

$$-\frac{d[A]}{dt} = k[X]^x[Y]^y[Z]^z[A] \qquad (1.40)$$

If the conditions are such that $[X]_0, [Y]_0, [Z]_0 \gg [A]$, so that [X], [Y], and [Z] remain essentially constant, then

$$-\frac{d[A]}{dt} = k[X]_0^x[Y]_0^y[Z]_0^z[A] = k_{exp}[A] \qquad (1.41)$$

and the rate law has the first-order form.

1.2.g Comparison of First-Order and Second-Order Conditions

The differences between second-order and first-order kinetic behavior and the transition from second-order to pseudo-first-order conditions are illustrated by the curves in Figure 1.1. These curves are based on the system

$$A + B \xrightarrow{\ k_2\ } C \qquad (1.42)$$

and show the time dependence of the formation of C. The deficient reagent is A, with $[A]_0 = 0.10$ M, different initial concentrations of B of 0.50, 0.70, 0.85, and 1.0 M are used, and $k_2 = 3 \times 10^{-3}$ M^{-1} s^{-1}. The lower curve, with $[B]_0 = 0.50$ M, represents a typical time dependence for second-order conditions. Note that this time dependence has a more gradual approach to the final value than the others in Figure 1.1. Thus, a qualitative examination can provide an indication of the second-order nature of the reaction.

As $[B]_0$ is increased, C is formed more rapidly, and the curves approach a first-order shape. When $[B]_0 = 1.0$ M, $[B]_0 \geq 10 \times [A]_0$ and the pseudo-first-order condition has been reached. Then, the curve calculated from Eq. (1.37) for second-order conditions is almost indistinguishable from the line calculated from Eq. (1.39) for pseudo-first-order conditions. The correspondence of these curves is the rationale for the general rule that *pseudo-first-order conditions require at least a ten-fold excess of the reagents whose concentrations are to remain constant.*

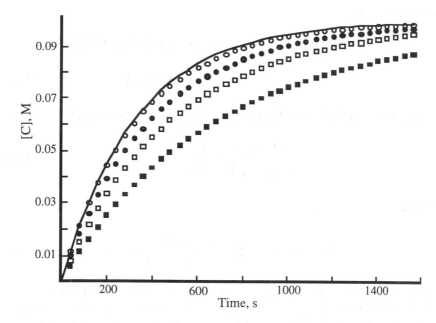

Figure 1.1. The time dependence for formation of product C, with the initial deficient reagent $[A]_0 = 0.10$ M. Results are calculated for pseudo-first-order (—) conditions and concentrations (M) of excess reagent B of 0.50 (■), 0.70 (◘), 0.85 (●), and 1.0 (○).

1.3 CONCENTRATION VARIABLES AND FIRST-ORDER RATES

The rate laws have been developed in terms of concentrations, but in many cases it is not practical or possible to determine actual molar concentrations as a function of time. However, it is easy to measure some property that is known to be directly proportional to molar concentration, such as absorbance, NMR integrated peak intensity, conductance, or refractive index. In such cases, the first-order system can still be analyzed to determine the rate constant.

The following development assumes an irreversible system but can easily be expanded to the more general reversible case. For the system in (1.9) with the integrated rate law given by Eq. (1.13), the property being observed is designated as I and its proportionality constant with concentration is ε. Then, initially

$$I_0 = \varepsilon_A [A]_0 \tag{1.43}$$

at the end

$$I_\infty = \varepsilon_B [B]_\infty = \varepsilon_B [A]_0 \tag{1.44}$$

and at any time

$$I = \varepsilon_A [A] + \varepsilon_B [B] = \varepsilon_A [A] + \varepsilon_B ([A]_0 - [A])$$

$$= (\varepsilon_A - \varepsilon_B)[A] + I_\infty \qquad (1.45)$$

Substitution for [A] and $[A]_0$ in terms of I and I_∞ into Eq. (1.13) gives

$$\frac{(I - I_\infty)}{(\varepsilon_A - \varepsilon_B)} = \frac{I_\infty}{\varepsilon_B} e^{-k_1 t} \qquad (1.46)$$

so that

$$\ln(I - I_\infty) = \ln\left(\frac{(\varepsilon_A - \varepsilon_B)I_\infty}{\varepsilon_B}\right) - k_1 t \qquad (1.47)$$

and a plot of $\ln(I - I_\infty)$ versus t should be linear with a slope of $- k_1$.

It is not necessary to know the concentration of A or any values of ε in order to determine the rate constant, but one does need I_∞. Sometimes it is impossible to measure I_∞ because of secondary reactions, or it is inconvenient because the reaction is slow. Such systems can be analyzed by nonlinear least-squares fitting of the data over as much of the reaction as possible, or in a more classical way by the Guggenheim method described in more detail by Moore and Pearson[3] (p. 71), Mangelsdorf,[7] and Espenson[8] (p. 25).

1.4 COMPLEX RATE LAWS

It is not unusual for a rate law to be more complex than the simple zero-, first-, or second-order cases we have considered. In general, the rate law has the following form:

$$-\frac{d[A]}{dt} = f([X], [Y], [Z])[A] = k_{exp}[A] \qquad (1.48)$$

where f([X], [Y], [Z]) is a function of the concentrations of X, Y, and Z.

In inorganic systems, the concentration dependence of k_{exp} can be quite complex, but some common forms of k_{exp} are

$$k_{exp} = k'[X] \qquad\qquad k_{exp} = k' + k''[X]$$

$$\qquad\qquad\qquad\qquad\qquad\qquad\qquad\qquad (1.49)$$

$$k_{exp} = \frac{k'[X]}{k'' + [Y]} \qquad\qquad k_{exp} = \frac{k'[Y] + k''}{k''' + [Z]}$$

The dependence of k_{exp} on [X], [Y], and [Z] is determined from a series of kinetic experiments under pseudo-first-order conditions, keeping [Y] and [Z] constant and changing [X] to determine the dependence of k_{exp} on [X], and then repeating the process for Y and Z.

In many elementary considerations of kinetic data, it is suggested that the order of a reaction with respect to a particular reagent can be determined from a plot of the logarithm of the rate constant versus the logarithm of the reagent concentration. This procedure is only appropriate for simple forms such as the first example in Eq. (1.49). Although log–log plots may appear linear for the more complex forms, the plots will yield meaningless fractional orders and should be avoided. Unfortunately, there is no truly general method of analysis to yield the reaction order, but this is seldom a serious problem when the reagent concentrations have been varied over a reasonable concentration range.

1.5 COMPLEX KINETIC SYSTEMS

Sometimes, even under pseudo-first-order conditions, the kinetic observations do not obey the first-order integrated rate law. This may indicate a number of chemical problems, such as impurities, a nonlinear analytical method, or precipitate formation. However, it is also possible that the system is more complex, with parallel and/or successive reactions, as shown in the following system:

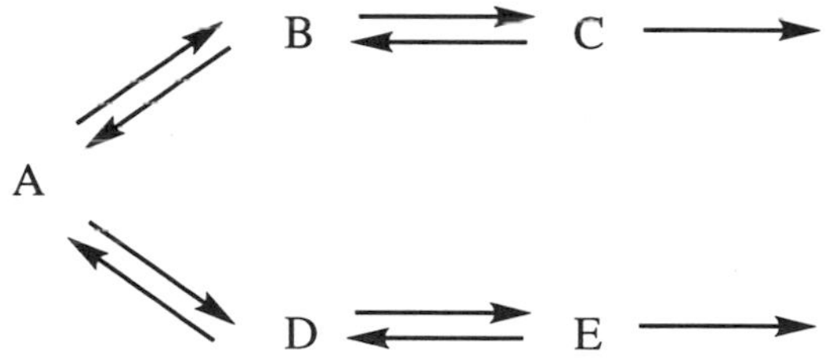

$$
\begin{array}{ccc}
& B \rightleftharpoons C \longrightarrow \\
\nearrow & & \\
A & & \quad\quad\quad\quad (1.50) \\
\searrow & & \\
& D \rightleftharpoons E \longrightarrow
\end{array}
$$

If pseudo-first-order conditions are maintained, it is always possible to solve the differential equations to determine the integrated rate law.[9,10] The solution has the general form

$$[\text{Product}] = M\,e^{-\gamma_1 t} + N\,e^{-\gamma_2 t} + \cdots \quad\quad (1.51)$$

where M, N, . . . and γ_1, γ_2, . . . are constants that depend on the rate constants for the individual steps. The number of exponential terms equals the number of steps in the reaction network. For such systems, the time dependence of the concentration variable is usually fitted by nonlinear least squares to determine the constants. In practice, exceptionally good data are required to extract more than two γ values.

The following system of successive reactions is often encountered:

$$A \underset{\beta_1}{\overset{k_1}{\rightleftarrows}} B \underset{\beta_2}{\overset{k_2}{\rightleftarrows}} C \qquad (1.52)$$

If one measures some property I that is directly proportional to concentration, then the integrated rate law predicting the time dependence of I is given by

$$I = I_\infty + \left(\frac{1}{\gamma_2 - \gamma_1} \right) \times$$

$$\{ [k_1 (I_B - I_A) - \gamma_2 (I_\infty - I_A)] e^{-\gamma_1 t} - [k_1 (I_B - I_A) - \gamma_1 (I_\infty - I_A)] e^{-\gamma_2 t} \} \quad (1.53)$$

where I_A and I_B are the values of the property I for species A and B, respectively, I_∞ is the final value of I at "infinite time" and γ_1 and γ_2 are the apparent rate constants. The latter are related to the specific rate constants in (1.52) by

$$\gamma_{1,2} = \frac{k_1 + \beta_1 + k_2 + \beta_2}{2} \pm$$

$$\frac{\sqrt{(k_1 + \beta_1 + k_2 + \beta_2)^2 - 4(k_1 k_2 + k_1 \beta_2 + \beta_1 \beta_2)}}{2} \qquad (1.54)$$

The form of Eq. (1.53) is useful for computer fitting procedures because one usually has some idea of reasonable values of I_A and I_B and an experimental value for I_∞. The equation for simpler schemes in which either or both of the reverse rate constants are zero can be obtained by setting the appropriate terms equal to zero in Eq. (1.54).

A somewhat simpler example of a system that proceeds by parallel paths to give different products is shown by the following:

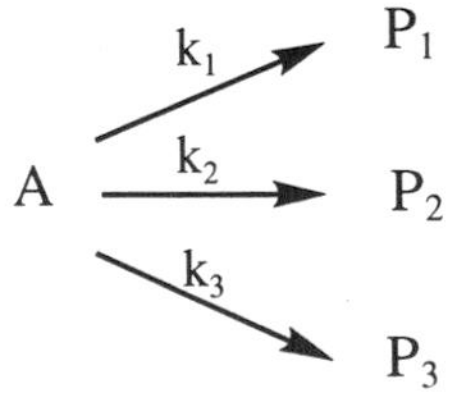

$$A \begin{array}{c} \overset{k_1}{\nearrow} P_1 \\ \overset{k_2}{\longrightarrow} P_2 \\ \underset{k_3}{\searrow} P_3 \end{array} \qquad (1.55)$$

The rate and integrated rate law are given by

$$-\frac{d[A]}{dt} = (k_1 + k_2 + k_3)[A] \qquad (1.56)$$

and

$$[A] = [A]_0 e^{-(k_1 + k_2 + k_3)t} \qquad (1.57)$$

Therefore, the kinetics will give $k_{exp} = (k_1 + k_2 + k_3)$. In order to evaluate the individual k_i values, it is necessary to determine the final product amounts, $[P_i]_\infty$, and use the relationship

$$[P_i]_\infty = \frac{k_i [A]_0}{k_1 + k_2 + k_3} \qquad (1.58)$$

Such systems are discussed in more detail by Espenson[8] (pp. 55–56).

1.6 TEMPERATURE DEPENDENCE OF RATE CONSTANTS

To obtain information about the energetics of a reaction, the temperature dependence of the rate constant is determined. For complex rate laws, this will also involve a study of the concentration dependence of the rate at different temperatures, in order to determine the temperature dependence of the various terms contributing to the rate law. Once the experimental information is available for the specific rate constants, it is usually analyzed in terms of one of the following formalisms.

1.6.a Arrhenius Equation

Arrhenius seems to have been the first to find empirically that rate constants have a temperature dependence analogous to that of equilibrium constants, as given by the following exponential or logarithmic forms:

$$k = A \exp\left(-\frac{E_a}{R T}\right) \qquad (1.59)$$

$$\ln k = \ln A - \frac{E_a}{R}\left(\frac{1}{T}\right) \qquad (1.60)$$

where A is called the Arrhenius pre-exponential factor, E_a is the Arrhenius activation energy, R is the gas constant, and T is the temperature in Kelvin. The units of A will be the same as those of k, and E_a will be in cal mol^{-1} or J mol^{-1}, depending on the units chosen for R (1.987 cal mol^{-1} K^{-1} or 8.314 J mol^{-1} K^{-1}). The k is determined by appropriate experiments at several temperatures, and a plot of ln k versus T^{-1} has a slope of $- E_a/R$. The rate constant is always predicted to increase with increasing temperature. Typical values are 10 to 30 kcal mol^{-1} for E_a and, for a first-order rate constant, 10^{10} to 10^{14} s^{-1} for A.

The Arrhenius equation is still widely used in certain areas of kinetics and for complex systems where the measured rate constant is thought to be a complex composite of specific rate constants.

1.6.b Transition-State Theory

This theory was developed originally for a simple dissociation process in the gas phase, and it assumes that the reaction can be described by the following sequence:

$$A—B \; \underset{}{\overset{K^*}{\rightleftharpoons}} \; \{ A\text{----}B \}^* \; \xrightarrow{k_3} \; A + B \qquad (1.61)$$

Activated complex
or transition state

The theory proposes that the activated complex or transition state will proceed to products when the bond has a thermal energy kT, so that its vibrational frequency is $v = kT/h$ s^{-1} ($k \equiv$ Boltzmann's constant, 1.38×10^{-16} erg K^{-1}; h $\equiv$ Planck's constant, 6.622×10^{-27} erg s), which will equal the rate constant k_3. Furthermore, it is assumed that the activated complex is always in equilibrium with the reactant with a normal equilibrium constant, $K^* = [A\text{----}B]^*/[A—B]$, so that

$$\frac{d\,[B]}{d\,t} = k_3\,[A\text{----}B]^* = k_3\,K^*\,[A—B] = \frac{kT}{h}\,K^*\,[A—B] \quad (1.62)$$

and

$$\ln K^* = -\frac{\Delta G^{o*}}{RT} = -\frac{\Delta H^{o*}}{RT} + \frac{\Delta S^{o*}}{R} \qquad (1.63)$$

where ΔG^{o*}, ΔH^{o*}, and ΔS^{o*} are the standard molar free energy, enthalpy, and entropy differences, respectively, between the activated complex and the reactants.

If the first-order rate expression, $d[B]/dt = k_{exp}\,[A—B]$, is compared to Eq. (1.62), then substitution for K^* from Eq. (1.63) shows that

$$k_{exp} = \frac{kT}{h}\,K^* = \frac{kT}{h}\,\exp\!\left(-\frac{\Delta G^{o*}}{RT}\right) \qquad (1.64)$$

This expression can be rearranged, expanded by substitution from Eq. (1.63), and put into logarithmic form to give

$$\ln\!\left(\frac{k_{exp}}{T}\right) = \ln\!\left(\frac{k}{h}\right) - \frac{\Delta H^{o*}}{RT} + \frac{\Delta S^{o*}}{R} \qquad (1.65)$$

Therefore, a plot of $\ln(k_{exp}/T)$ versus T^{-1} should be linear with a slope of $-\Delta H^{o*}/R$. The value of ΔS^{o*}, in cal mol^{-1} K^{-1}, can be calculated from a known value of k_{exp} at a particular T:

$$\Delta S^{o*} = 4.576 \left[\log\left(\frac{k_{exp}}{T}\right) - 10.319 \right] + \frac{\Delta H^{o*}}{T} \qquad (1.66)$$

In common usage, the standard state designations are dropped and ΔH^{*} and ΔS^{*} are called the activation enthalpy and activation entropy, respectively, for the reaction.

The Arrhenius parameters are related to ΔH^{*} and ΔS^{*} by the following relationships:

$$\Delta H^{*} = E_a - RT \quad \text{and} \quad \Delta S^{*} = 4.58 \left(\log A - 13.2 \right) \quad (1.67)$$

Some typical data[11] for the temperature dependence for the linkage isomerization of $(H_3N)_5Co—ONO^{2+}$ to form $(H_3N)_5Co—NO_2^{2+}$ are shown in Figure 1.2. The rate law for the reaction in alkaline solution under pseudo-first-order conditions shows a spontaneous path (k_S) and an OH$^-$ catalyzed path (k_{OH}). The ΔH^{*} is somewhat higher for the k_{OH} path, as indicated by the steeper slope in Figure 1.2.

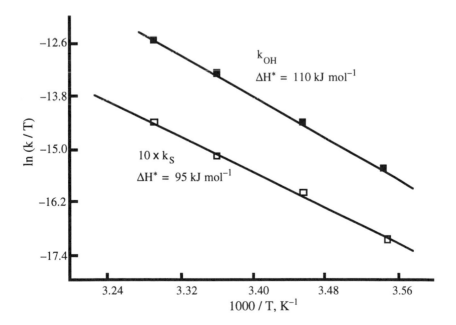

Figure 1.2. The temperature dependence of the linkage isomerization of $(H_3N)_5Co—ONO^{2+}$.

The transition-state theory parameters are used commonly to describe the temperature dependence of k_{exp} even for processes in solution that are far more complex than assumed in the original formulation of the theory. Therefore, the ΔH^* and ΔS^* are best considered as experimental parameters that are useful for calculating k_{exp} at other temperatures and for comparison to other closely related systems and processes.

Within transition-state theory, reactions are often depicted in terms of "reaction coordinate diagrams". These are plots of the energy of the system versus the "reaction coordinate", which is an ambiguous measure of the extent to which reactant has been converted to product. Examples of such diagrams are shown in Figure 1.3.

The *transition state* (or activated complex) is the species at the highest energy point on the reaction coordinate diagram. An *intermediate* is the species present at any valley on a reaction coordinate diagram. When a valley is shallow, it can be ambiguous whether or not an intermediate really is formed. In chemistry, an intermediate is expected to have a lifetime longer than a few vibrational lifetimes ($>10^{-13}$ s) and the valley should be deeper than the thermal energy ($RT = 2.5$ kJ mol^{-1} at 25°C).

1.7 PRESSURE DEPENDENCE OF RATE CONSTANTS

It is possible to measure the rate of a reaction at various applied pressures and determine the variation of k with P. In recent years, such measurements have become increasingly widespread for a variety of inorganic reactions, and the interpretation of the pressure dependence adds a further tool to the arsenal of parameters available for mechanistic interpretations. In this area, pressure now is given in megapascals, MPa, and this is related to earlier units by 1 atm = 0.101 MPa = 1.01 bar.

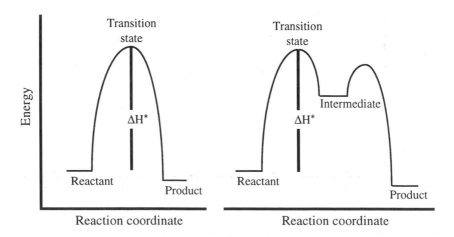

Figure 1.3. Reaction coordinate diagrams.

The variation of K with P is given by the van't Hoff equation:

$$\left(\frac{\delta \ln K}{\delta P}\right)_T = -\frac{\Delta V^o}{RT} \tag{1.68}$$

where ΔV^o is the difference in the partial molar volume between the products and reactants. Since $K = k_1/k_{-1}$, it is logical to express the variation of k_1 with P by

$$\left(\frac{\delta \ln k_1}{\delta P}\right)_T = -\frac{\Delta V_1^*}{RT} \tag{1.69}$$

where ΔV_1^* is defined as the volume of activation for the forward step and is equal to the partial molar volume of the transition state minus the partial molar volume of the reactant(s). If ΔV_1^* is independent of pressure, then integration of Eq. (1.69) at constant temperature over the limits $P = 0$ to P and $k_1 = (k_1)_0$ to k_1 gives

$$\ln k_1 = \ln (k_1)_0 - \frac{\Delta V_1^* P}{RT} \tag{1.70}$$

and a plot of $\ln k_1$ versus P should be linear with a slope of $-\Delta V_1^*/RT$. Since these studies usually cover pressures up to several thousand atmospheres, $\ln(k_1)_0$ is taken as the value at ambient pressure. If the plot is not linear, it is assumed that the reactant and/or transition state may be compressible and their volumes as a function of pressure can be described by

$$\Delta V_1^* = \left(\Delta V_1^*\right)_0 - \Delta \beta^* P \tag{1.71}$$

where $\Delta \beta^*$ represents the compressibility of the system. Substitution of Eq. (1.71) into Eq. (1.69) and integration over the same limits yields

$$\ln k_1 = \ln (k_1)_0 - \frac{\left(\Delta V_1^*\right)_0 P}{RT} + \Delta \beta^* \left(\frac{P^2}{2RT}\right) \tag{1.72}$$

More complex explanations of nonlinearity also are possible.[12]

The pressure dependence for the isomerization of $(H_3N)_5Co\!-\!ONO^{2+}$ is shown in Figure 1.4. For the pathway catalyzed by OH^-, with the rate constant k_{OH}, the authors[11] have ascribed the slight curvature of the plot to the effect of compressibility with $\Delta \beta^* = 5$ cm^3 $kbar^{-1}$ mol^{-1}. The difference between the upper straight line and the dashed curve shows the extent of the deviation from linearity.

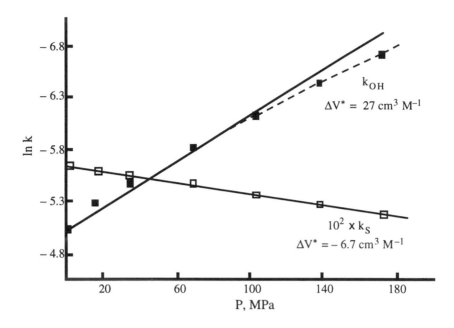

Figure 1.4. The pressure dependence of the linkage isomerization of $(H_3N)_5Co—ONO^{2+}$.

1.8 IONIC STRENGTH DEPENDENCE OF RATE CONSTANTS

For reactions of ions in solution, the variation of the activity coefficients with reagent concentrations is sometimes ignored or, more commonly, assumed to be held constant by carrying out the reaction in the presence of some "inert electrolyte", which is at a much higher concentration than that of the reactants.

The kinetic effect of ionic strength can be illustrated for a bimolecular reaction of the following type:

$$A + B \; \underset{}{\overset{K^*}{\rightleftharpoons}} \; \{ A\text{----}B \}^* \longrightarrow \text{Products} \qquad (1.73)$$

for which transition-state theory predicts that the rate is given by

$$\text{Rate} = \frac{kT}{h} K^* \left(\frac{\gamma_A \, \gamma_B}{\gamma^*} \right) [A]\,[B] = k\,[A]\,[B] \qquad (1.74)$$

where γ_A, γ_B, and γ^* are the activity coefficients of the reactants and transition state, respectively. At infinite dilution (zero ionic strength), the activity coefficients are 1 and the rate constant is defined as k_0; therefore,

$$k = \left(\frac{\gamma_A \gamma_B}{\gamma^*} \right) k_0 \qquad (1.75)$$

The simplest relationship between the activity coefficients, γ_i, and the ionic strength, μ, is given by the Debye–Hückel limiting law, which applies for $\mu \le 0.01$ M:

$$\log \gamma_i = - A z_i^2 \sqrt{\mu} \qquad (1.76)$$

where A is a constant for a given solvent (A = 0.509 $M^{-1/2}$ for water at 25°C) and z_i is the charge of the ion. Using this limiting law, and realizing that $z^* = z_A + z_B$, it follows from Eqs. (1.75) and (1.76) that

$$\log k = \log k_0 + 2 A z_A z_B \sqrt{\mu} \qquad (1.77)$$

and a plot of log k versus $\sqrt{\mu}$ should be linear with a slope of $2Az_Az_B$.

Many kinetic studies are done at ionic strengths beyond the range of applicability of the Debye–Hückel limiting law. The law was extended by Debye and Hückel to take into account the finite size of the ions to give the following relationship, which is applicable for $\mu < 0.1$ M:

$$\log k = \log k_0 + \frac{2 A z_A z_B \sqrt{\mu}}{1 + \alpha B \sqrt{\mu}} \qquad (1.78)$$

where α is the average effective diameter of the ions and B is a constant depending on the solvent properties (B = 0.328 Å^{-1} $M^{-1/2}$ for water at 25°C). Values of α for various ions in water have been tabulated by Klotz.[13] Empirical equations have been developed for higher ionic strengths; an example is the Davies equation,[14] for $\mu \le 0.5$ M.

The ionic strength dependence of k is essentially a property of the rate law. Therefore, the ionic strength dependence seldom affords new mechanistic information unless the complete rate law cannot be determined. These equations more often are used to "correct" rate constants from one ionic strength to another for the purpose of rate constant comparison. Ionic strength effects have been used to estimate the charge at the active site in large biomolecules, but the theory is substantially changed[15] because the size of the biomolecule violates basic assumptions of Debye–Hückel theory.

1.9 DIFFUSION-CONTROLLED RATE CONSTANTS

The upper limit on a rate constant for a reaction is imposed by the rate at which the reactants can diffuse together. This limit can be of significance when a particular mechanism would require a rate constant beyond the diffusion-controlled limit; then, the mechanism can be eliminated as a reasonable possibility. In addition, certain classes of reactions are known to proceed at or near diffusion-controlled rates, and this information can be useful in constructing and analyzing mechanistic models.

If two reactants A and B with radii r_A and r_B diffuse together and react at an interaction distance $(r_A + r_B)$, then theories developed from Brownian motion predict that the diffusion controlled second-order rate constant is given by

$$k_{diff} = \frac{4 \pi N (D_A + D_B)(r_A + r_B)}{1000} \left(\frac{U}{\exp(U) - 1} \right) \qquad (1.79)$$

where

$$U = \frac{z_A z_B e^2}{4 \pi \varepsilon_0 \varepsilon (r_A + r_B) k T}$$

D_A and D_B are the diffusion coefficients for A and B, respectively, z_A and z_B are their charges, N is Avogadro's number, k is Boltzmann's constant (1.38×10^{-23} J K^{-1}), e is the electron charge (1.6×10^{-19} C), ε is the dielectric constant of the solvent, and $4\pi\varepsilon_0 = 1.11 \times 10^{-10}$. If one or both of the species are neutral, then $U = 0$ and the right-hand term in parentheses in Eq. (1.79) is equal to 1.

The diffusion coefficients can be approximated from the Stokes–Einstein equation, $D = kT/6\pi\eta r$, where $k = 1.38 \times 10^{-16}$ erg K^{-1} and η is the solvent viscosity, so that

$$k_{diff} = \frac{2 R T (r_A + r_B)^2}{3000 \eta r_A r_B} \left(\frac{U}{\exp(U) - 1} \right) \qquad (1.80)$$

with r in centimeters, η in poise, and $R = 8.31 \times 10^7$ erg mol^{-1} K^{-1}. This equation shows that k_{diff} will be relatively independent of the size of the reactants as long as $r_A \approx r_B$ and its magnitude will depend inversely on the solvent viscosity. The temperature dependence of k_{diff} will be governed largely by that of the solvent viscosity, so that apparent activation energies for diffusion-controlled processes are found to be in the 1 to 3 kcal mol^{-1} range for common solvents. It should be noted that the Stokes–Einstein equation greatly underestimates the diffusion coefficients of the proton and hydroxide ion in water.

One can estimate k_{diff} from Eq. (1.80) without knowing the diffusion coefficients. Some values for various reactant sizes and charge products in water ($\eta = 0.00894$ poise, $\varepsilon = 78.3$), are given in Table 1.1. It is apparent from these data that diffusion-controlled rate constants in water can be expected to be in the range of 10^9 to 10^{10} M^{-1} s^{-1}.

Table 1.1. Estimated Diffusion-Controlled Rate Constants (25°C) in Water

$r_A = 5.0$ (Å) $z_A \times z_B$	$r_B = 2.0$ (Å)	$r_B = 5.0$ (Å) k_{diff} (M^{-1} s^{-1})	$r_B = 8.0$ (Å)
−2	9.15 x 10^9	7.44 x 10^9	7.85 x 10^9
−1	9.10 x 10^9	7.42 x 10^9	7.83 x 10^9
0	9.05 x 10^9	7.39 x 10^9	7.81 x 10^9
+1	9.01 x 10^9	7.36 x 10^9	7.78 x 10^9
+2	8.96 x 10^9	7.34 x 10^9	7.76 x 10^9

For a unimolecular dissociation, such as A—B forming A + B, the rate is controlled by the diffusion of the products out of the solvent cage. Theory predicts that the limiting dissociation rate constant is given by Eq. (1.81), which can be further simplified using the Stokes–Einstein equation.

$$k_{diff} = \frac{3(D_A + D_B)}{(r_A + r_B)^2}\left(\frac{U}{1 - \exp(-U)}\right) \qquad (1.81)$$

The predicted unimolecular dissociation rate constants (s^{-1}) are of the same magnitude as the bimolecular constants in Table 1.1.

The most important general class of reactions that have diffusion-controlled rates in water are protonation of a base by H_3O^+ and deprotonation of an acid by OH^-, as shown in the following reactions, with rate constants at 25°C in M^{-1} s^{-1}:

$$B: + H_3O^+ \xrightleftharpoons{k \approx 4 \times 10^{10}} B:H^+ + H_2O \qquad (1.82)$$

$$A:H + OH^- \xrightleftharpoons{k \approx 1 \times 10^{10}} A:^- + H_2O \qquad (1.83)$$

These results stem from the pioneering work of Eigen and co-workers.[16] The reverse rate constants for these reactions can be calculated from the equilibrium constants that are known for a wide range of such acids and bases. It is important to note that the reverse rate constants may not be extremely large. For example, trimethylamine is a strong base with a protonation rate constant of 6×10^{10} M^{-1} s^{-1}, but the K_a of the trimethylammonium ion is 1.6×10^{-10} M, so that the deprotonation by water has k = $(6 \times 10^{10})(1.6 \times 10^{-10}) \approx 10$ s^{-1}.

The main exceptions to the preceding generalizations are so-called carbon acids, such as nitromethane or acetylacetone, for which the rate constants are usually much smaller and dependent on the nature of the acid. Such reactions are thought to be slower because of the bonding rearrangements required at carbon as the conjugate base is formed. The same constraint may apply to the deprotonation of organometallic hydrides.

References

1. Levine, I. N. *Physical Chemistry*, 3rd ed.; McGraw-Hill: New York, 1988.
2. Atkins, P. W. *Physical Chemistry*, 5th ed.; Freeman: New York, 1994.
3. Moore, J. W.; Pearson, R. G. *Kinetics and Mechanism*, 3rd ed.; Wiley-Interscience: New York, 1980.
4. Laidler, K. J. *Chemical Kinetics*, 3rd ed.; Harper & Row: New York, 1987.
5. Pilling, M. J.; Seakins, P. W. *Reaction Kinetics*; Oxford University Press: Oxford, 1995.
6. *Comprehensive Chemical Kinetics*; Bamford, C. H.; Tipper, C. F. H., Eds.; Elsevier: Amsterdam, 1969; Vol. 1, 2.
7. Mangelsdorf, P. C. *J. Appl. Phys.* **1959**, *30*, 443.
8. Espenson, J. E. *Chemical Kinetics and Reaction Mechanisms*; McGraw-Hill: New York, 1981.
9. Rodiguin, N. M.; Rodiguina, E. N. *Consecutive Chemical Reactions*, English Ed., translated by R. F. Schneider; Van Nostrand: Princeton, N.J., 1969.
10. Capellos, C.; Bielski, B. H. *Kinetic Systems*; McGraw-Hill: New York, 1972.
11. Jackson, W. G.; Lawrance, G. A.; Lay, P. A.; Sargeson, A. M. *Inorg. Chem.* **1980**, *19*, 904.
12. Asano, T.; le Noble, W. J. *Chem. Rev.* **1978**, *78*, 407.
13. Klotz, I. *Chemical Thermodynamics*; Prentice-Hall: New York, 1950; pp. 330-331.
14. Davies, C. W. *Prog. React. Kin.* **1961**, *1*, 129.
15. Rosenberg, R. C.; Wherland, S.; Holwerda, R. A.; Gray, H. B. *J. Am. Chem. Soc.* **1976**, *98*, 6364.
16. Eigen, M. *Angew. Chem., Int. Ed.* **1964**, *3*, 1.

2

Rate Law and Mechanism

Once the experimental rate law has been established, the next step is to formulate a mechanism that is consistent with the rate law. The rate law will not uniquely define the mechanism but will limit the possibilities. The proposed mechanism will lead to predictions of trends in reactivity and other types of experiments that can be done to test the proposal.

Except for the simplest cases, the development of the rate law from the mechanism can be a messy exercise. The following sections describe some of the assumptions and tricks that can be used. Further discussions can be found in standard textbooks on kinetics.[1-3]

2.1 QUALITATIVE GUIDELINES

Once the experimental rate law has been determined, the problem is to determine the most reasonable mechanism(s) that will predict a rate law that is consistent with the observations. Very often this is done by analogy to previous studies on related systems, but there are some general guidelines that can be useful for writing a mechanism that will produce the desired form of the rate law.

The mechanism is composed of elementary reactions whose rate laws are implied from the stoichiometry of each reaction. The elementary mechanistic steps are usually unimolecular or bimolecular reactions; termolecular reactions are very rare because of the improbability of bringing three species together.

The form of the experimental rate law provides some guidelines for the construction of a mechanism. The following generalizations assume that the reaction is monophasic, but they may apply to individual steps in a multiphasic reaction. It also should be remembered that the experimental rate law may be incomplete because of experimental constraints. Then, the predicted rate law may contain terms not observed experimentally, but it should be possible to show that the extra terms are minor contributors under the conditions of the experiment.

For the simplest cases, in which rate = $k_{exp}[A][B]$ or rate = $k_{exp}[A]$, the kinetics only requires a one-step mechanism involving the species in the rate law. In the second case, the solvent also may be involved, since its concentration will be a constant that could be included in k_{exp}.

If the rate law has a half-order term, such as $[A]^{1/2}$, then the mechanism probably involves a step in which A is split into two reactive species before the rate-determining step.

If rate $= k_{exp}[A][B][C]^{-1}$, then a mechanism in which C is produced from A and B prior to the rate-determining step will generate such a rate law. If there is a denominator in the rate law that consists of the sum of several terms, then the mechanism may involve consecutive steps that produce reactive intermediates.

If the rate is the sum of several terms, such as $k_{exp}[A] + k_{exp}'[A][B]$, then a number of parallel reaction pathways equal to the number of terms in the sum will predict the experimental rate law.

Once the general outline of the mechanism is established, it is necessary to show that the proposal does give the required rate law. The following sections describe common methods for deriving the rate law from the mechanism.

2.2 STEADY-STATE APPROXIMATION

A mechanism often invokes an unstable intermediate of some defined structure, and a general mechanism might take the form of

$$A \underset{k_2}{\overset{k_1}{\rightleftharpoons}} \{B\} \underset{k_4}{\overset{k_3}{\rightleftharpoons}} C \qquad (2.1)$$

where B is an unstable and therefore reactive intermediate. The steady-state approximation assumes that this intermediate will disappear as quickly as it is formed:

$$\text{Rate of Appearance of B} = \text{Rate of Disappearance of B} \qquad (2.2)$$

so that

$$k_1[A] + k_4[C] = k_2[B] + k_3[B] \qquad (2.3)$$

With Eq. (2.3), one can solve for [B] in terms of the reactant and product concentration, [A] and [C], to give

$$[B] = \frac{k_1[A] + k_4[C]}{k_2 + k_3} \qquad (2.4)$$

The total concentration of reagents can be defined as [T] and will remain constant. Since B is a reactive intermediate, its concentration will always be small relative to $[A] + [C]$, so that

$$[T] = [A] + [B] + [C] = [A] + [C] \tag{2.5}$$

Substitution for [C] from Eq. (2.5) into Eq. (2.4) gives

$$[B] = \frac{k_1 [A] + k_4 [T] - k_4 [A]}{k_2 + k_3} \tag{2.6}$$

The rate of disappearance of A is given as follows (note that the mechanism has specified that all steps have first-order or pseudo-first-order rate constants):

$$-\frac{d[A]}{dt} = k_1 [A] - k_2 [B] = \frac{(k_2 k_4 + k_1 k_3)}{k_2 + k_3} [A] - \frac{k_2 k_4}{k_2 + k_3} [T] \tag{2.7}$$

where the steady-state expression for [B] has been used to eliminate [B] from the differential equation; this gives a form that is integratable because [A] is the only concentration variable. However, instead of integrating at this point, it is useful to introduce the equilibrium (final) concentrations, $[A]_e$ and $[C]_e$, through

$$[T] = [A]_e + [C]_e \tag{2.8}$$

and

$$K_e = \frac{[C]_e}{[A]_e} = \frac{k_1 k_3}{k_2 k_4} \tag{2.9}$$

Substitution for $[C]_e$ from Eq. (2.9) into Eq. (2.8) and rearranging gives

$$k_2 k_4 [T] = (k_1 k_3 + k_2 k_4)[A]_e \tag{2.10}$$

Substitution for [T] from Eq. (2.10) into Eq. (2.7) yields

$$-\frac{d[A]}{dt} = \frac{(k_2 k_4 + k_1 k_3)}{(k_2 + k_3)}([A] - [A]_e) \tag{2.11}$$

which is the mathematical equivalent of the first-order rate law and can be integrated directly to obtain

$$-\ln([A]_e - [A]) + \ln([A]_e - [A]_0) = k_{exp} t \tag{2.12}$$

where

$$k_{exp} = \frac{(k_2 k_4 + k_1 k_3)}{(k_2 + k_3)} \qquad (2.13)$$

Note that the right-hand side of Eq. (2.13) is the same as the coefficient for [A] on the right-hand side of Eq. (2.7). Therefore, it really was not necessary to go through the equilibrium conditions in order to find the expression for k_{exp}. *It is always true that once one has an integratable equation with only one concentration variable in first-order form, then the coefficient of the concentration variable in that equation will be the expression for* k_{exp}.

A limiting form of Eq. (2.13) that is often encountered assumes that $k_2 \gg k_3$. Then, the k_{exp} is given by

$$k_{exp} = \left(\frac{k_1}{k_2}\right) k_3 + k_4 = K_{12} k_3 + k_4 \qquad (2.14)$$

The steady-state approximation can be applied to systems with any number of reactive intermediates. King and Altman[4] have presented a general development for k_{exp} in steady-state systems that is very useful for complex reaction networks.

2.3 RAPID EQUILIBRIUM ASSUMPTION

The rapid equilibrium treatment assumes that the reactants are part of a rapidly attained equilibrium that is always maintained during the course of the reaction, as shown by

$$A \underset{}{\overset{K_{12}}{\rightleftharpoons}} B \overset{k_3}{\longrightarrow} C \qquad (2.15)$$

where B is not a reactive intermediate but a species with a finite concentration. For example, B might be the conjugate base of A, an isomer of A, or an ion pair. Since B may be present at significant concentrations, the total concentration of the species at equilibrium can be defined as [R] and is given by

$$[R] = [A] + [B] \qquad (2.16)$$

Since [T] = [A] + [B] + [C], it follows that

$$[R] = [T] - [C] \qquad (2.17)$$

Normally, one will know [R] but not [A] or [B], unless K_{12} is known.

The rate of formation of C is easily written down as

$$\frac{d[C]}{dt} = k_3[B] \tag{2.18}$$

and the problem is to express [B] in terms of [C], in order to obtain an equation that can be integrated. A useful trick can be used to get an expression for the concentration of one of the partners in the equilibrium, [B], in terms of the total concentration of the species involved in the equilibrium, [R]. Since $K_{12} = [B]/[A]$, then

$$\frac{1}{K_{12}} + 1 = \frac{[A]+[B]}{[B]} = \frac{[R]}{[B]} \tag{2.19}$$

Rearrangement and substitution for [R] from Eq. (2.17) into Eq. (2.19) gives

$$[B] = \frac{[R]K_{12}}{(K_{12}+1)} = \frac{([T]-[C])K_{12}}{(K_{12}+1)} \tag{2.20}$$

Substitution for [B] into Eq. (2.18) yields an equation with only [C] as the concentration variable:

$$\frac{d[C]}{dt} = \frac{k_3 K_{12}}{(K_{12}+1)}([T]-[C]) \tag{2.21}$$

so that

$$k_{exp} = \frac{k_3 K_{12}}{(K_{12}+1)} \tag{2.22}$$

The expression for k_{exp} may be compared to that derived from the steady-state assumption under the condition that $k_2 \gg k_3$. The k_4 is missing in the present example because we have assumed an irreversible model, but otherwise the steady-state and equilibrium models are the same if $K_{12} \ll 1$ (in which case the concentration of B is small).

The preceding discussion can leave the incorrect impression that B is like a particularly stable intermediate on the reaction pathway from A to C. A somewhat different perspective is gained if one views B as the starting material and A as some unreactive form of B. This situation produces the same rate law as Eq. (2.21). *The important general lesson is that all rapid equilibria involving the reactant(s) will enter into the rate law, even if the species involved are not on the net reaction pathway.*

2.4 RAPID EQUILIBRIUM OR STEADY STATE?

In many cases, the decision as to whether to use the rapid equilibrium or steady-state conditions will be obvious. If the mechanism proposes some intermediate that is thought to be very reactive, then a steady-state assumption for its concentration is probably appropriate, as long as there is no detectable concentration of the intermediate. Proton transfer reactions between acids and bases are generally treated as equilibria.

For less obvious situations, it is helpful to have some approximate idea of the rate constants involved in the formation and destruction of the intermediate in order to choose the most appropriate approach. The criteria to use have been the subject of much discussion that is summarized and further delineated in recent work by Viossat and Ben-Aim[5] and by Gellene.[6] These authors discuss the following system:

$$A \; \underset{k_2}{\overset{k_1}{\rightleftharpoons}} \; \{B\} \; \overset{k_3}{\longrightarrow} \; C \qquad\qquad (2.23)$$

For the steady-state approximation to apply, $k_1 \ll k_2 + k_3$ and Gellene notes that the reaction time scale must be such that $t \gg (k_2 + k_3)^{-1}$. For the rapid equilibrium, $k_3 \ll k_1 + k_2$ and $t \gg (k_1 + k_2)^{-1}$, according to Gellene. It is noteworthy that the condition $k_3 \ll k_1 + k_2$ only requires that either k_1 or k_2 be much larger than k_3. This results because the rate of attainment of equilibrium is determined by $k_1 + k_2$, as shown in Section 1.2.c. In the application of these criteria to real systems, it should be remembered that k_1, k_2, and k_3 may be pseudo-first-order rate constants that are the product of some species concentration times a specific rate constant.

2.5 NUMERICAL INTEGRATION METHODS

The availability of desktop computers has made numerical integration of differential equations an increasingly popular tool for kinetic analysis. One simply needs to decide on a mechanistic scheme, write the appropriate differential equations for the time dependence of the species, establish initial conditions, and then let the computer calculate the species concentrations over a chosen time range. The calculated results can be compared with the experimental ones, visually or by least-squares fitting. The main advantage of such methods is that complex kinetic schemes are easily modeled and that second-order conditions, which might otherwise be impossible to integrate, can be included.

This appears to be an ideal method, since there is no need to do integrations or worry about steady-state or rapid equilibrium assumptions. However, problems can arise in the numerical analysis. Most of these procedures use the fourth-order Runge–Kutta method in which the integration is done in a series of small time steps. The step size must be small relative to the time dependence of all the concentration variables; this can lead to problems in systems with mechanistic steps of widely different rates, because there is a tendency to shorten the time for calculation by lengthening the step size. Since the rapid equilibrium and steady-state conditions can cause rapid initial concentration changes, it can be advantageous to apply such assumptions to the differential equations before doing the numerical integration. A problem of numerical significance can also arise for species of very small concentrations, such as steady-state intermediates, unless these are removed from the model by appropriate assumptions.

Some examples of numerical integrations for the system in (2.23) are shown in Figure 2.1 for different relative rates in the rapid equilibrium model. The solid lines show the calculated time dependence of [B] and [C] for relative rate constants that satisfy the rapid equilibrium conditions. The circles are calculated from Eq. (2.21). Note that [B] initially increases rapidly to the equilibrium value; this type of fast initial change can be a problem for numerical integration. The dotted curves present the same functions for relative values of k_1 and k_3 that do not satisfy the rapid equilibrium conditions and show a slower increase of [B] and an initial induction period for [C].

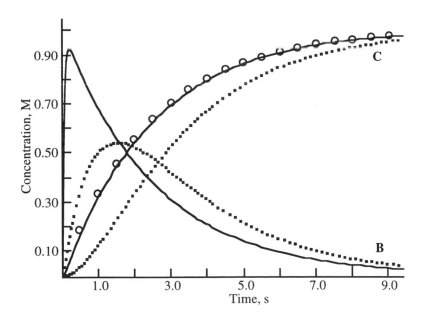

Figure 2.1. The time dependence of the concentrations of intermediate B and product C in (2.23) with $[A]_0 = 1.0$ M: for $k_1 = 40$, $k_2 = 0.001$, $k_3 = 0.4$ s^{-1} (—) by numerical integration and (O) calculated from Eq. (2.21); for $k_1 = 1.0$, $k_2 = 0.001$, $k_3 = 0.4$ s^{-1} (•••••) by numerical integration.

2.6 PRINCIPLE OF DETAILED BALANCING

The principle of detailed balancing states that when a system is at equilibrium, the rate in the forward direction equals the rate in the reverse direction for each individual step in the process as well as for the overall reaction.

This can be of use in simple systems because it makes it possible to express one of the rate constants in terms of the others and the overall equilibrium constant. For the reaction

$$A + B \underset{k_r}{\overset{k_f}{\rightleftarrows}} C + D \tag{2.24}$$

$K_e = k_f/k_r$, so that $k_r = k_f/K_e$. For cyclic systems, such as those in (2.25), a less obvious consequence is that the product of the rate constants going in one direction around the cycle must equal the product of rate constants in the other direction.

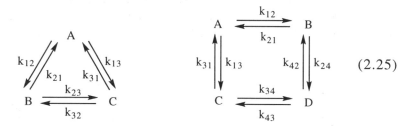

$$\tag{2.25}$$

For the three-species system, $k_{12}k_{23}k_{31} = k_{13}k_{32}k_{21}$, so that one needs to know only five of the six rate constants in order to define the system. Similarly, for the four-species system, one obtains $k_{13}k_{34}k_{42}k_{21} = k_{12}k_{24}k_{43}k_{31}$.

2.7 PRINCIPLE OF MICROSCOPIC REVERSIBILITY

The mechanism of a reverse reaction must be the same as the mechanism of the forward reaction under the same conditions. This results because *the least energetic pathway in one direction must be the least energetic pathway in the other direction*. The intermediates and transition state must be the same in either direction. One consequence of this is that a catalyst for a forward reaction will be a catalyst for the reverse reaction.

The proper application of the principles of microscopic reversibility and detailed balancing can be helpful in mechanistic assessments, as illustrated by the CO exchange in $Mn(CO)_5X$ systems. Johnson and co-workers[7] initially claimed that all the CO ligands were being exchanged at a similar rate and proposed the mechanism in Scheme 2.1.

Scheme 2.1

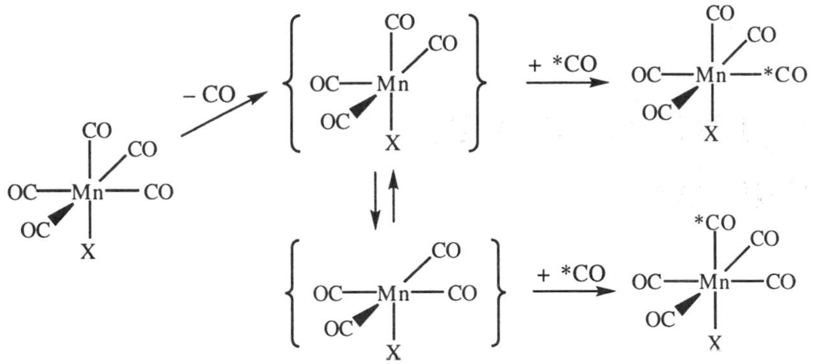

Brown[8] pointed out that this mechanism violates the principle of microscopic reversibility because, if dissociation of a cis CO is more favorable kinetically, then addition of a CO to the cis position also must be more favorable.

Subsequent work[9] using IR detection indicates that the exchange of the cis CO is faster. Jackson[10] has suggested that the more recent analysis transgresses the principle of detailed balancing, but this criticism arises from an incorrect extension[11] of Brown's arguments by Espenson.[12] The detailed analysis by Jackson, allowing for initial dissociation of both cis- and trans-CO ligands, shows that the ratio of cis to trans products is independent of time if the intermediates are in rapid equilibrium, but the ratio varies with time otherwise, unless the two dissociation rates happen to be equal.

References

1. Moore, J. W.; Pearson, R. G. *Kinetics and Mechanism,* 3rd ed.; Wiley-Interscience: New York, 1980.
2. Laidler, K. J. *Chemical Kinetics,* 3rd ed.; Harper & Row: New York, 1987.
3. Espenson, J. H. *Chemical Kinetics and Reaction Mechanisms*; McGraw-Hill: New York, 1981.
4. King, E. L.; Altman, C. *J. Phys. Chem.* **1956,** *60,* 1375.
5. Viossat, V.; Ben-Aim, R. I. *J. Chem. Educ.* **1993,** *70,* 732.
6. Gellene, G. I. *J. Chem. Educ.* **1995,** *72,* 196.
7. Johnson, B. F. G.; Lewis, J.; Miller, J. R.; Robinson, B. H.; Robinson, P. W.; Wojcicki, A. *J. Chem. Soc. A* **1968,** 522.
8. Brown, T. L. *Inorg. Chem.* **1968,** *7,* 2873.
9. Atwood, J. T.; Brown, T. L. *J. Am. Chem. Soc.* **1975,** *97,* 3380.
10. Jackson, W. G. *Inorg. Chem.* **1987,** *26,* 3004.
11. Brown, T. L. *Inorg. Chem.* **1989,** *28,* 3229.
12. Espenson, J. H. *Chemical Kinetics and Reaction Mechanisms*; McGraw-Hill: New York, 1981; pp. 128–131.

3

Ligand Substitution Reactions

In ligand substitution reactions one or more ligands around a metal ion are replaced by other ligands. In many ways, all inorganic reactions can be classified as either substitution or oxidation–reduction reactions, so that substitution reactions represent a major type of inorganic process. Some examples of substitution reactions follow:

$$Fe(OH_2)_6{}^{3+} + SCN^- \rightleftharpoons (H_2O)_5Fe-SCN^{2+} + H_2O$$

$$Ni(OH_2)_6{}^{2+} + en \rightleftharpoons \left[\begin{array}{c} H_2O \quad H_2 \\ | \quad N-CH_2 \\ H_2O-Ni-N-CH_2 \\ H_2O \quad | \quad H_2 \\ OH_2 \end{array} \right]^{2+} + 2\,H_2O \qquad (3.1)$$

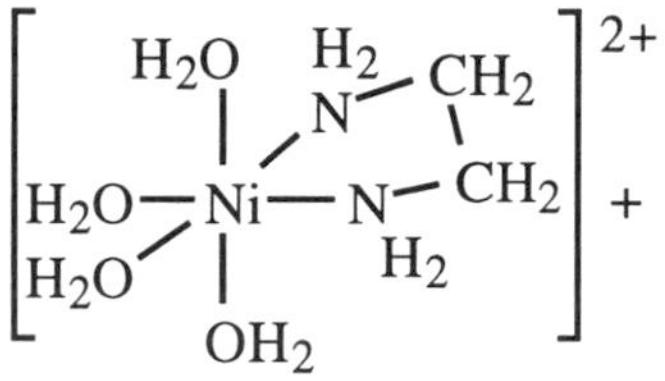

$$Cr(CO)_6 + P(C_6H_5)_3 \longrightarrow \begin{array}{c} CO \quad CO \\ | \; / \\ OC-Cr-P(C_6H_5)_3 \\ OC \quad | \\ CO \end{array} + CO$$

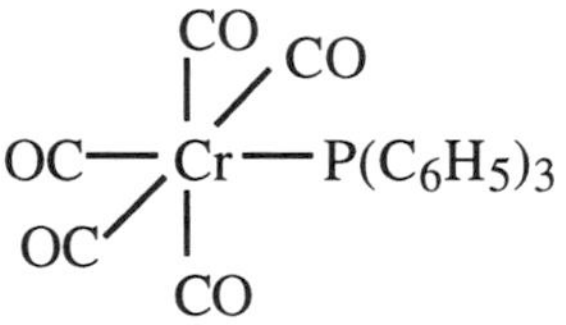

3.1 OPERATIONAL APPROACH TO CLASSIFICATION OF SUBSTITUTION MECHANISMS

The operational approach was first expounded in 1965 in a monograph by Langford and Gray.[1] It is an attempt to classify reaction mechanisms in relation to the type of information that kinetic studies of various types can provide. It delineates what can be said about the mechanism on the basis of the observations from certain types of experiments. The mechanism is classified by two properties, its stoichiometric character and its intimate character.

Stoichiometric Mechanism

The stoichiometric mechanism can be determined from the kinetic behavior of one system. The classifications are as follows:

1. *Dissociative* (**D**): an intermediate of lower coordination number than the reactant can be identified.
2. *Associative* (**A**): an intermediate of larger coordination number than the reactant can be identified.
3. *Interchange* (**I**): no detectable intermediate can be found.

Intimate Mechanism

The intimate mechanism can be determined from a series of experiments in which the nature of the reactants is changed in a systematic way. The classifications are as follows:

1. *Dissociative activation* (**d**): the reaction rate is more sensitive to changes in the leaving group.
2. *Associative activation* (**a**): the reaction rate is more sensitive to changes in the entering group.

This terminology has largely replaced the S_N1, S_N2, and so on type of nomenclature that is still used in physical organic chemistry. These terminologies are compared and further explained as follows:

Dissociative [**D** $\equiv$ S_N1 (limiting)]: there is definite evidence of an intermediate of reduced coordination number. The bond between the metal and the leaving group has been completely broken in the transition state without any bond making to the entering group.

Dissociative interchange ($\mathbf{I_d}$ $\equiv$ S_N1): there is no definite evidence of an intermediate. In the transition state, there is a large degree of bond breaking to the leaving group and a small amount of bond making to the entering group. The rate is more sensitive to the nature of the leaving group.

Associative interchange ($\mathbf{I_a}$ $\equiv$ S_N2): there is no definite evidence of an intermediate. In the transition state, there is some bond breaking to the leaving group but much more bond making to the entering group.

Associative [**A** $\equiv$ S_N2 (limiting)]: there is definite evidence of an intermediate of increased coordination number. In the transition state, the bond to the entering group is largely made while the bond to the leaving group is essentially unbroken.

The general goal of a kinetic and mechanistic study of a substitution reaction is to classify the reaction as **D**, $\mathbf{I_d}$, $\mathbf{I_a}$, or **A**.

3.2 OPERATIONAL TESTS FOR THE STOICHIOMETRIC MECHANISM

According to the original definitions, it should be possible to establish the stoichiometric mechanism on the basis of a study of one system. In practice, this has been expanded to include studies with one metal ion complex system.

3.2.a Dissociative Mechanism Rate Law

The **D** mechanism can be described by the following sequence of reactions:

$$R\!-\!X \underset{k_2}{\overset{k_1}{\rightleftharpoons}} \{R\} + X \xrightarrow[Y]{k_3} R\!-\!Y \qquad (3.2)$$

where X and Y are the leaving and entering groups, respectively, and {R} is the intermediate of reduced coordination number.

The k_{exp} for this mechanism can be derived from the previous solution of the system A $\rightleftharpoons$ B $\rightleftharpoons$ C with a steady state for B, by replacing k_2 and k_3 in Eq. (2.13) with $k_2[X]$ and $k_3[Y]$, respectively, and setting $k_4 = 0$. Then, if [X] and [Y] >> [RX], k_{exp} is given by

$$k_{exp} = \frac{k_1 k_3 [Y]}{k_2 [X] + k_3 [Y]} \qquad (3.3)$$

If this rate law is to provide a successful test of the **D** mechanism, it is necessary for the conditions to be such that $k_2[X] \approx k_3[Y]$. Then, for example, if [X] is held constant and [Y] is varied in a series of experiments, k_{exp} should change with [Y], as shown in Figure 3.1. This type of variation is often referred to as "saturation" behavior, and k_{exp} approaches a limiting value of k_1 when $k_3[Y] >> k_2[X]$. It is possible to rearrange Eq. (3.3) to give

$$\left(k_{exp}\right)^{-1} = \frac{k_2}{k_1 k_3} \left(\frac{[X]}{[Y]} \right) + \frac{1}{k_1} \qquad (3.4)$$

Therefore, a plot of $(k_{exp})^{-1}$ versus [X]/[Y] should be linear, and k_1 and k_2/k_3 can be determined.

If several different entering groups (Y) are studied, they should all yield the same value of k_1 as a further condition of a **D** mechanism.

3.2.b Ion Pair or Preassociation Problem

The success of the preceding test of a **D** mechanism depends on the assumption that there are no other reaction sequences that produce the same rate law. Unfortunately, this is *not* true.

Many metal ion complexes are positively charged, and many common entering groups are anions. Such oppositely charged species can form association complexes, commonly called ion pairs. The phenomenon of preassociation is not limited to ions and may be appreciable for polar species in nonpolar solvents due to dipole–dipole interactions and hydrogen bonding.

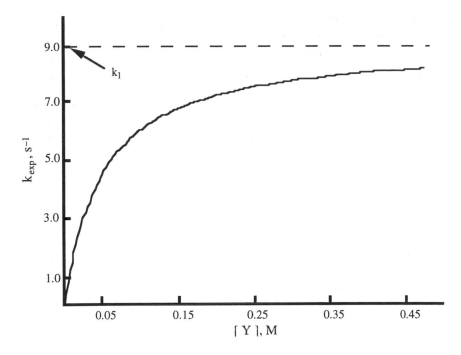

Figure 3.1. Predicted variation of k_{exp} with [Y] for a **D** mechanism with $k_1 = 9$ s^{-1} and $k_2[X]/k_3 = 0.1$ M.

The general process can be described by the following sequence of reactions:

$$R\text{—}X + Y \underset{}{\overset{K_i}{\rightleftharpoons}} (R\text{—}X{\cdot}Y) \xrightarrow{k_3} R\text{—}Y + X \quad (3.5)$$

where $(R\text{—}X{\cdot}Y)$ is the ion pair or preassociation complex formed in a fast pre-equilibrium with an ion pair formation constant, K_i.

The rate law for this type of system was developed in Eq. (2.22) and, if [Y] >> [RX], the pseudo-first-order rate constant is given by

$$k_{exp} = \frac{k_3 K_i [Y]}{K_i [Y] + 1} \quad (3.6)$$

This equation predicts the same type of variation of k_{exp} with [Y] as that from the **D** mechanism if [X] is constant in Eq. (3.3). The latter is often the case because X is the solvent. A plot of $(k_{exp})^{-1}$ versus [X]/[Y] will give the values of k_1 and K_i. It may be possible to distinguish this from the preceding model by comparing K_i to known or estimated values from analogous systems.

Table 3.1. Calculated Ion Pair Formation Constants ($a = 5 \times 10^{-8}$ cm, $T = 298$ K)

Ionic Strength $z_1 z_2$	H_2O ($\varepsilon = 78.5$)			CH_3OH ($\varepsilon = 32$)		CH_2Cl_2 ($\varepsilon = 9.1$)
	0.01	0.10	1.0	0.01	0.10	0.01
-1	1.08	0.81	0.54	5.1	2.2	1.3×10^3
-2	3.67	2.07	0.93	83	15	5.2×10^6
-3	12.5	5.3	1.6	1400	104	2.1×10^{10}
-4	42.7	13.6	2.7			
-5	146	34.7	4.7			
-6	497	88.9	8.1			

For ion pairs, Eigen[2] and Fuoss[3] developed an equation to estimate K_i based on extended Debye–Hückel theory and a hard-sphere model for the ions. It is given by

$$K_i = \frac{4 \pi N a^3}{3000} \exp\left(-\frac{U}{kT}\right) \tag{3.7}$$

where

$$U = \frac{z_1 z_2 e^2}{\varepsilon}\left(\frac{1}{a(1 + \kappa a)}\right) \quad \text{and} \quad \kappa = \sqrt{\frac{8 \pi N e^2 \mu}{1000 \, \varepsilon \, k T}}$$

and N is Avogadro's number (6.022×10^{23}), a is the contact distance of the ions (cm), k is Boltzmann's constant (1.38×10^{-16} erg K^{-1}), z_1 and z_2 are the ionic charges, e is the electron charge (4.803×10^{-10} esu), ε is the solvent dielectric constant, and μ is the ionic strength. Some calculated values of K_i are given in Table 3.1 for various charge products, solvents, and ionic strengths. Experience indicates that these calculated values are reasonable approximations when compared to the few experimental values. The main point to note is that the value of $K_i[Y]$ can easily be of the same magnitude as 1 for typical charge types and for reasonable concentrations of Y.

3.2.c Competition Studies for the Intermediate

These studies attempt to test the prediction of a **D** mechanism that a particular metal center should produce the same intermediate, independent of the leaving group. For example, one might study the solvolysis in water of $Co(NH_3)_5Cl^{2+}$ and $Co(NH_3)_5(NO_3)^{2+}$, where Cl^- and NO_3^- are the leaving groups, in the presence of some added nucleophile Y. The object is to get the same product distribution if a common intermediate $\{Co(NH_3)_5{}^{3+}\}$ is formed. The confidence in the conclusions depends on studying a significant range of leaving groups.

The principles of the method are described by the following sequence:

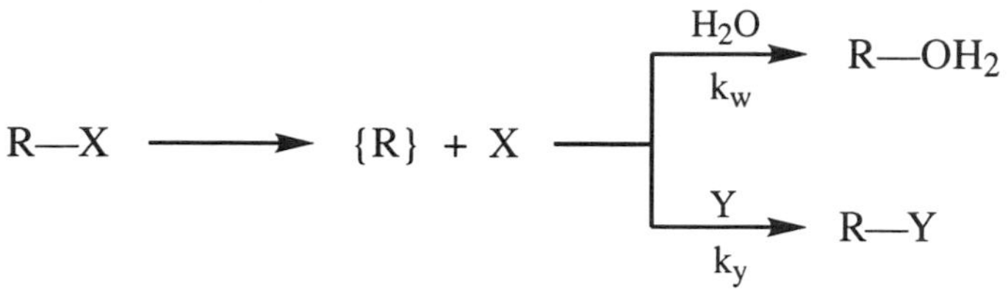

$$R\!-\!X \longrightarrow \{R\} + X \quad \begin{cases} \xrightarrow[k_w]{H_2O} R\!-\!OH_2 \\[2em] \xrightarrow[k_y]{Y} R\!-\!Y \end{cases} \tag{3.8}$$

The product ratio $[RY]/[ROH_2]$ can be calculated as follows:

$$\frac{d\,[RY]}{d\,t} = k_y\,[R]\,[Y] \quad \text{and} \quad \frac{d\,[ROH_2]}{d\,t} = k_w\,[R]\,[H_2O]$$

$$\frac{d\,[RY]}{d\,[ROH_2]} = \frac{k_y}{k_w}\,\frac{[Y]}{[H_2O]} = \frac{k_y\,[Y]}{k_w'} \tag{3.9}$$

Integration and rearrangement yields

$$\frac{[RY]}{[ROH_2]\,[Y]} = \frac{k_y}{k_w'} \tag{3.10}$$

If a **D** mechanism is operative, the ratio on the left should be a constant for different concentrations of Y and for different leaving groups.

A similar analysis can be applied if the product complex has different structural isomers or stereoisomers. Then, the isomers should be produced in a proportion independent of the leaving group.

3.2.d Constant Thermodynamic Properties for the Intermediate

In this approach, it is hoped to show that the thermodynamic parameters of the intermediate are constant and independent of the leaving group, and thereby establish the independent nature of the intermediate.

The following development is in terms of enthalpy, but the same can be done for free energy, entropy, partial molar volume, and so on. The reaction energetics are defined by Figure 3.2. From this diagram it should be apparent that

$$\Delta H^* - \Delta H_{stab} = \Delta H_f^o(R) + \Delta H_f^o(X) - \Delta H_f^o(RX) \tag{3.11}$$

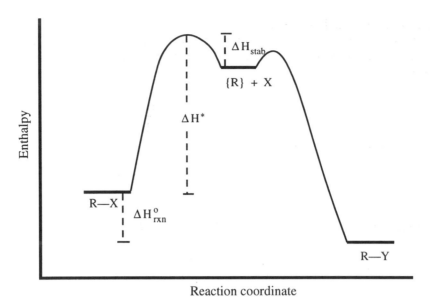

Figure 3.2. Reaction coordinate diagram for a **D** mechanism.

For the overall reaction RX + Y $\rightarrow$ RY + X, the enthalpy change is

$$\Delta H^{o}_{rxn} = \Delta H^{o}_{f}(RY) + \Delta H^{o}_{f}(X) - \Delta H^{o}_{f}(RX) - \Delta H^{o}_{f}(Y) \qquad (3.12)$$

Then Eqs. (3.11) and (3.12) can be combined to eliminate $\Delta H^{o}_{f}(RX)$ and $\Delta H^{o}_{f}(X)$ and give

$$\Delta H^{o}_{f}(R) = \Delta H^{*} - \Delta H_{stab} - \Delta H^{o}_{rxn} + \Delta H^{o}_{f}(RY) - \Delta H^{o}_{f}(Y) \qquad (3.13)$$

If a series of leaving groups is examined using the same Y (e.g., the solvent), then

$$\Delta H^{o}_{f}(R) = \Delta H^{*} - \Delta H_{stab} - \Delta H^{o}_{rxn} + \text{Constant} \qquad (3.14)$$

This equation is not truly independent of X because of ΔH_{stab}, but this term is assumed to be small, so that

$$\Delta H^{*} - \Delta H^{o}_{rxn} \approx \Delta H^{o}_{f}(R) - \text{Constant} \qquad (3.15)$$

For a **D** mechanism, $\Delta H^{*} - \Delta H^{o}_{rxn}$ is expected to be constant for a particular entering group.

3.3 EXAMPLES OF TESTS FOR A DISSOCIATIVE MECHANISM

3.3.a Dissociative Rate Law

The first example of the full **D** rate law was published by Wilmarth and co-workers.[4,5] They studied the anation of $Co(CN)_5(OH_2)^{2-}$ in water with the idea that the negative charge on the metal complex would suppress the ion pair formation and might favor a **D** mechanism. Their observations were consistent with the following mechanism:

$$Co(CN)_5(OH_2)^{2-} \underset{k_2}{\overset{k_1}{\rightleftharpoons}} \{Co(CN)_5^{2-}\} + H_2O$$

$$\{Co(CN)_5^{2-}\} + Y^- \underset{k_4}{\overset{k_3}{\rightleftharpoons}} Co(CN)_5Y^{3-}$$

(3.16)

The predicted pseudo-first-order rate constant for this mechanism can be obtained by analogy to Eq. (2.13), to give

$$k_{exp} = \frac{k_1 k_3 [Y] + k_2 [H_2O] k_4}{k_2 [H_2O] + k_3 [Y]} = \frac{k_1 k_3 [Y] + k_2' k_4}{k_2' + k_3 [Y]}$$

(3.17)

The value of k_4 was determined independently by studying the rate of aquation of $Co(CN)_5Y^{3-}$ with $[Y] \approx 0$, in which case $k_{exp} = k_4$. If k_4 is subtracted from both sides of Eq. (3.17) and the reciprocal is taken, then one obtains Eq. (3.18), which predicts that a plot of $(k_{exp} - k_4)^{-1}$ versus $[Y]^{-1}$ should be linear, allowing one to calculate k_2'/k_3 and k_1.

$$\left(k_{exp} - k_4 \right)^{-1} = \left(k_1 - k_4 \right)^{-1} + \frac{k_2'}{k_3} \left(k_1 - k_4 \right)^{-1} [Y]^{-1}$$

(3.18)

Wilmarth and co-workers found that their data satisfied this rate law and yielded reasonably constant values of k_1 for a range of Y such as Br^-, NH_3, I^-, SCN^-, and N_3^-. Note that, if Y is H_2O, then $k_{exp} = k_1$, so that a study of the water exchange rate would provide a further test.

Unfortunately, all of the preceding results have been thrown into serious doubt by recent work. Burnett and Gilfillian[6] and then Haim[7] found that the rate law with $Y = N_3^-$ has a simple first-order dependence on $[N_3^-]$. Haim's observations indicate that the early work may be in error because of the presence of $(NC)_5CoO_2Co(CN)_5^{6-}$, which has been avoided in the recent studies through modified preparative procedures. The original observations with regard to SCN^- have been confirmed by later work.[8]

Table 3.2. Values of k_3/k_2 for Systems with a **D** Mechanism Rate Law

Entering Group	$Rh(Cl)_5(OH_2)^{2-}$ in Water[a]	Entering Group	$Cr(TPP)(Cl)(py)$ in Toluene
H_2O	1.0	Pyridine	1.0^b
I^-	0.018	PPh_3	0.0017^b
Br^-	0.016	$P(C_2H_4CN)_3$	0.0085^b
Cl^-	0.021	$P(OPr)_3$	0.075^b
SCN^-	0.079	N-Methylimidazole	1.7^b, 1.1^c
NO_2^-	0.10	H_2O	1.44^c
N_3^-	0.14	3-Methylpyridine	0.93^c
		Quinoline	0.0089^c

[a] Reference 11. [b] Reference 13. [c] Reference 15.

The water exchange rate on $Co(CN)_5(OH_2)^{2-}$ has been measured recently by Swaddle and co-workers,[9] who find $k_{exch} = 5.8 \times 10^{-4}$ s^{-1} at 25°C, with $\Delta H^* = 90.2$ kJ mol^{-1} and $\Delta S^* = -4$ J mol^{-1} K^{-1}. This predicts that at 40°C, $k_{exch} = 3.5 \times 10^{-3}$ s^{-1}, whereas the results of the substitution studies give $k_1 = 2 \times 10^{-3}$ s^{-1} (Haim) or 6×10^{-4} s^{-1} (Burnett).[10] The current status of this system is that it is probably using a dissociative interchange mechanism, $\mathbf{I_d}$, and that there is some preassociation of the metal complex and the entering group despite their unfavorable charge product.

Still, there are systems for which the rate law indicates a **D** mechanism. Some examples are $Rh(Cl)_5(OH_2)^{2-}$,[11] $Co(en)_2(SO_3)(OH_2)^+$,[12] tetraphenylporphine chromium(III) chloride, $Cr(TPP)(Cl)L$,[13] and bis(dimethyglyoxime) cobalt(III) complexes.[14] Some representative values of k_3/k_2 are given in Table 3.2. Recent work[15] on $Cr(TPP)(Cl)L$ systems has given k_3/k_2 values for a number of substituted pyridines. The {Cr(TPP)Cl} intermediate has been generated by photolysis[16] and found to react at nearly diffusion-controlled rates with the pyridine entering groups, so that $k_3/k_2 \approx 1$. The small k_3/k_2 values for the phosphines and quinoline may be due to steric hindrance.

3.3.b Competition Studies for a Dissociative Intermediate

The main limitation for these studies is that the products must be stable enough that their amounts can be accurately determined. The favorite systems for these studies have been cobalt(III) amine complexes, because of their stability and the extensive documentation of their properties.

The early work was done on the hydroxide ion catalyzed hydrolysis of cobalt(III) amines, for which there was evidence that the reaction proceeds by a dissociative conjugate base mechanism (S_N1CB in earlier terminology), as shown in Scheme 3.1.

Scheme 3.1

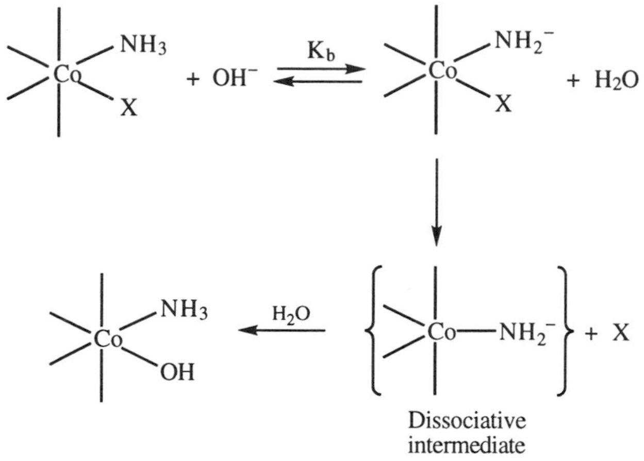

Dissociative
intermediate

An analysis of early results on the hydrolysis of cis and trans isomers of $Co(en)_2(Y)X$ complexes by Sargeson and Jordan[17] indicated that if Y is the same, then the percentage of cis and trans isomers in the product $Co(en)_2(Y)(OH)$ is fairly constant. Further work by Buckingham et al.[18] on stereoisomers of *cis*-$Co(en)_2(NH_3)X$ is summarized in Table 3.3. The percentage of cis and trans products is quite constant, but the percentage retention is significantly greater with neutral leaving groups. This can be rationalized if the anionic leaving groups are retained longer within the immediate solvation sheath of the "intermediate" and tend to inhibit entry from the position they have vacated, thereby giving less retention. In the same study, using azide ion as a competing ligand, it was found that neutral leaving groups give about 5 percent more $Co(en)_2(NH_3)(N_3)^{2+}$ than anionic leaving groups. The preceding rationale can also be used to explain this observation.

Table 3.3. Product Distribution for *cis*-$Co(en)_2(NH_3)X$ + OH^-

Leaving Group	% trans	% cis	% Retention	% Racemate
Cl^-	22	78	48	30
NO_3^-	23	77	47	30
Br^-	22	78	44	34
$(H_3C)_2SO$	23	77	52	25
$(H_3CO)_3PO$	23	77	54	23

The status of this and related work has been summarized by Jackson et al.,[19] and the earlier observations have been revised and expanded by Buckingham and co-workers.[20] It now appears that an intermediate does form, but that it is very reactive and scavenges its immediate coordination sphere rather than sensing the stoichiometric amounts of various species in the bulk solution. The intermediate may not be truly independent of its source nor of the "inert" ionic medium because it reacts while the leaving group is still in the vicinity, and the products essentially reflect the ionic atmosphere around the reactant. However, the leaving group does not dramatically change the reactivity pattern of the intermediate.

Basolo and Pearson[21] suggested that the $\{(H_3N)_4Co(NH_2)^{2+}\}$ intermediate is stabilized by back π bonding from the NH_2^- ligand, as shown in the following structure:

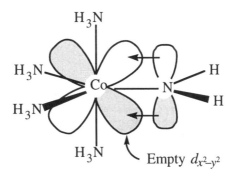

The competition results for the hydrolysis in acidic aqueous solution (aquation) show a greater sensitivity to the leaving group than those for base-catalyzed hydrolysis. The studies have also involved reactions designed to rapidly produce an intermediate of reduced coordination number, so-called induced aquations. The following are examples of typical induced reactions:

$$(H_3N)_5Co\!-\!Cl^{2+} + Hg^{2+} \longrightarrow \{(H_3N)_5Co^{3+}\} \xrightarrow[Z]{H_2O}$$
$$+ HgCl^+$$

$$(H_3N)_5Co\!-\!N_3{}^{2+} + NO^+ \longrightarrow \{(H_3N)_5Co^{3+}\} \xrightarrow[Z]{H_2O}$$
$$+ N_2 + N_2O$$

(3.19)

The intermediate here is different than that for the conjugate base mechanism because the NH_2^- ligand has been replaced by NH_3. It appears that the $\{(H_3N)_5Co^{3+}\}$ intermediate is more reactive than its conjugate base so that $\{(H_3N)_5Co^{3+}\}$ shows a greater dependence of the competition ratios on the leaving group. The induced aquation of

$Co(NH_3)_5(N_3)^{2+} + NO^+$ in the presence of NCS^- yields 12 percent $Co(NH_3)_5(NCS)^{2+}$, whereas spontaneous aquation with trimethyl phosphate as the leaving group yields 4.6 percent $Co(NH_3)_5(NCS)^{2+}$. However, the distribution of linkage isomers in the product is nearly the same, with about 60 percent $(H_3N)_5Co-NCS^{2+}$ and 40 percent $(H_3N)_5Co-SCN^{2+}$. The status of $\{(H_3N)_5Co^{3+}\}$ is still a subject of controversy and is discussed in detail by Jackson and Dutton[22] for the reaction of the azido complex with NO^+. For spontaneous aquations, differing views have been expressed by Jackson et al.[23] and Buckingham and co-workers[24] on the effect of ion pairing on the NCS^- competiton for the intermediate.

3.3.c Constant Thermodynamic Parameters

As shown previously, for a **D** mechanism one can expect that

$$\Delta H^* - \Delta H^o_{rxn} = \Delta H^o_f(R) - \Delta H^o_f(RY) + \Delta H^o_f(Y) \qquad (3.20)$$

House and Powell[25] analyzed enthalpy data for the aquation reaction

$$(H_3N)_5Co-X + H_2O \longrightarrow (H_3N)_5Co-OH_2^{3+} + X \qquad (3.21)$$

and found that $\Delta H^* - \Delta H^o_{rxn}$ (kcal mol^{-1}) varied with X, from 22.2 for SO_4^{2-} to ~25 for Cl^-, Br^- and NO_3^-, to 27 for H_2O. This variation was taken as evidence against a simple **D** mechanism. On the other hand, for the reaction

$$(H_3N)_5Co-X + OH \longrightarrow (H_3N)_5Co-OH^{2+} + X \qquad (3.22)$$

the same enthalpy difference is 32.4 ± 0.5 kcal mol^{-1} for the same range of ligands. This is quite constant and independent of the leaving group, therefore consistent with a **D** mechanism.

The activation volumes, ΔV^*, have been analyzed for the same reactions[26,27] with similar conclusions. However, the analysis is more complex than was originally anticipated because of solvent electrostriction effects and the effect of the volume of the leaving group on the volume of the reactant.[28]

3.4 OPERATIONAL TEST FOR AN ASSOCIATIVE MECHANISM

3.4.a Associative Mechanism Rate Law

The **A** mechanism proceeds by formation of an intermediate with the entering group followed by elimination of the leaving group, as shown by the following sequence:

$$R\!-\!X + Y \; \underset{k_2}{\overset{k_1}{\rightleftharpoons}} \; \left\{ R\!\begin{smallmatrix} \diagup Y \\ \diagdown X \end{smallmatrix} \right\} \; \xrightarrow{k_3} \; R\!-\!Y + X \quad (3.23)$$

If a steady state is assumed for the intermediate and pseudo-first-order conditions are maintained with [Y] >> [RX], then

$$k_{exp} = \frac{k_1 k_3 [Y]}{k_2 + k_3} = \text{Constant [Y]} \quad (3.24)$$

The rate should always be first order in [Y], and the rate law contains no information that uniquely defines an **A** mechanism.

A rather unlikely but feasible possibility is that the intermediate is formed in a rapid pre-equilibrium, as shown in the following:

$$R\!-\!X + Y \; \overset{K_{12}}{\rightleftharpoons} \; \left\{ R\!\begin{smallmatrix} \diagup Y \\ \diagdown X \end{smallmatrix} \right\} \; \xrightarrow{k_3} \; R\!-\!Y + X \quad (3.25)$$

where, if [Y] >> [RX], then

$$k_{exp} = \frac{k_3 K_{12} [Y]}{K_{12} [Y] + 1} \quad (3.26)$$

This expression has the same [Y] dependence as that for the **D** mechanism and the ion pair pathway. It might be distinguished from the latter by comparing the values of K_{12} to those expected for K_i. In addition, the spectral properties of the intermediate are likely to be much different from those of the reactant, whereas an ion pair is not much different because no bonds have been made or broken in the ion pair. The value of k_3 should depend on the nature of the entering group and thus could be distinguished from its mathematical equivalent k_1 in the **D** rate law.

3.4.b Examples of Associative Rate Laws

Examples of the full **A** rate law in (3.26) are rare because the "intermediate" must be quite stable if $K_{12}[Y] \geq 1$. Coordinatively unsaturated systems are most likely to satisfy this condition.

One apparent example was reported by Cattalini et al.[29] for reaction (3.27), where the diene is cyclooctadiene. The rate is first order in the Rh complex concentration and independent of the amine concentration.

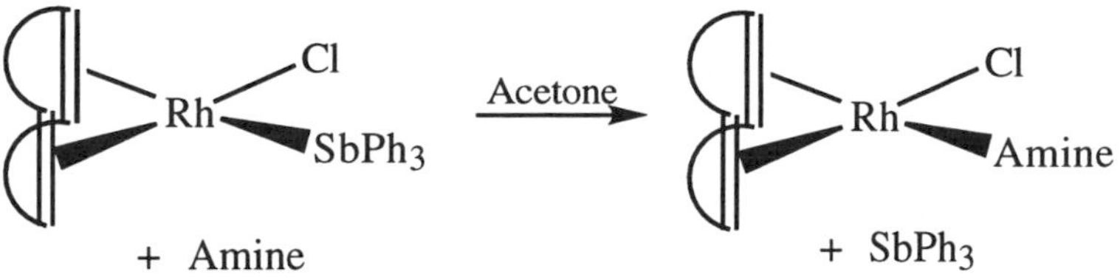

$$(3.27)$$

The rate constant varies with the nature of the amine, and there is a rapid spectral change on mixing the reactants. If K_{12}[amine] $\gg$ 1, then k_{exp} = k_3 and should depend on the nature of amine. The values (in acetone at 25°C) of k_{exp} range from 1.58×10^{-2} s^{-1} for 3-cyanopyridine to 4.57×10^{-2} s^{-1} for *n*-butylamine and do not show a large variation.

Another apparent example appears in the work of Toma and Malin[30] on the reaction

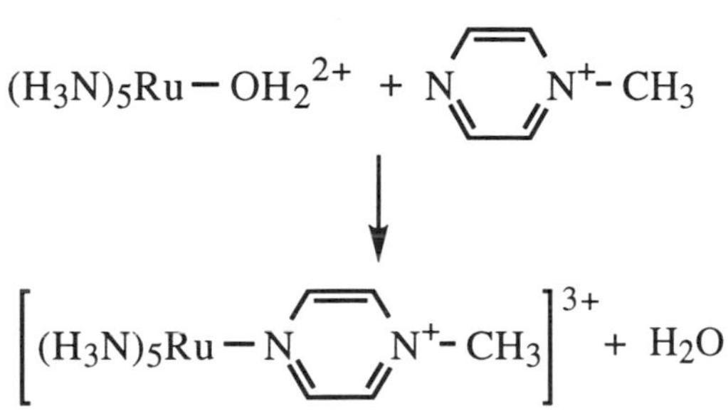

$$(3.28)$$

The dependence of k_{exp} on the concentration of methylpyrazinium ion is consistent with the *A* rate law. The authors argue that this is not due to ion pairing because of the like charges of the reactants. They suggest that the intermediate is a charge-transfer complex due to donation of t_{2g} electrons on Ru(II) to the empty π^* orbitals on the entering group. However, it is debatable whether this should be considered as an intermediate of expanded coordination number.

Species of expanded coordination number have been isolated and structurally characterized by Maresca et al.[31] as products of the reaction of Zeise's salt with bis hydrazones, as shown in reaction (3.29). These five-coordinate products slowly lose ethylene in a first-order process.

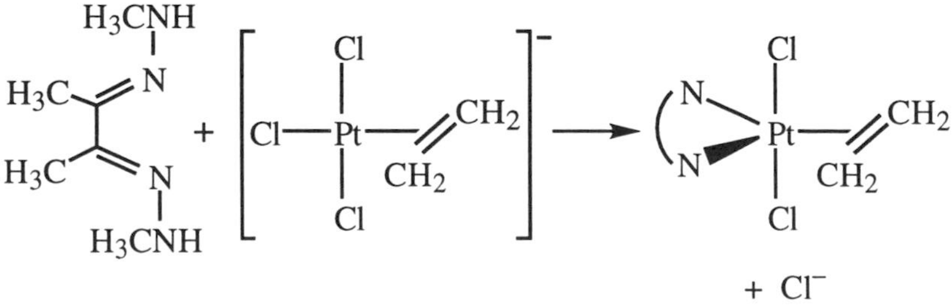

$$(3.29)$$

Tobe and co-workers[32] have provided evidence that reaction (3.30) proceeds through a stable intermediate, which may be the five-coordinate species shown or one of its structural isomers.

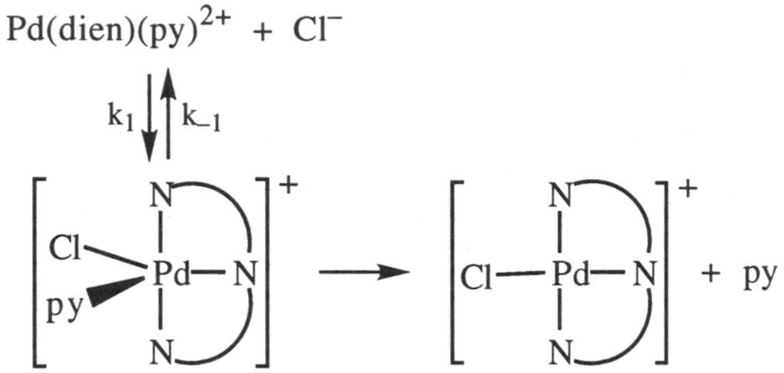

$$Pd(dien)(py)^{2+} + Cl^-$$

(3.30)

The intermediate forms reasonably quickly, with a rate that is first order in [Cl$^-$] ($k_1 = 9.3 \times 10^{-2}$ M^{-1} s^{-1}, $k_{-1} = 2.5 \times 10^{-2}$ s^{-1} in 1 M NaClO$_4$ at 25°C), but too slowly to be considered as an ion pair. The ^{1}H NMR shows that the intermediate has not released pyridine nor undergone ring opening of the dien chelate.

3.5 OPERATIONAL TESTS FOR THE INTIMATE MECHANISM

These tests are concerned with the sensitivity of the reaction rate constant to the chemical nature of the entering and leaving groups for a general reaction, such as (3.31), in which there is no definite evidence for an intermediate.

$$R{-}X + Y \longrightarrow R{-}Y + X \qquad (3.31)$$

Associative activation, **a**, requires more sensitivity to the nature of Y and dissociative activation, **d**, requires more sensitivity to the nature of X.

These effects appear to be easy to test, but there is always a somewhat subjective decision in evaluating the degree of sensitivity to variations in X and Y. For example, in the associative case, since X is still present in the transition state, the rate constant must show some variation with X, but the variation with changes in Y must be greater. A further problem is that X or Y is often the solvent, and it cannot be changed without a major perturbation on the whole system. It is also necessary to ensure that the changes in X or Y have not been so trivial that the interaction with "R" in the transition state would not vary by much. For example, changing from Cl$^-$ to Br$^-$ would probably not produce much change in the rate constant for either type of activation. In order to avoid such possibilities, scales of nucleophilicity for various ligands are very useful. It is assumed that a better nucleophile will make a stronger bond to "R", so that one should choose entering or leaving groups of significantly different nucleophilicity in testing for the type of activation.

3.5.a Inorganic Nucleophilicity Scales

Several variables are believed to generally affect nucleophilicity:
1. *Basicity towards H+*: the commonly available pK_a values of ligands measure this and it seems to parallel the nucleophilicity toward many metal centers.
2. *Polarizability*: a more polarizable ligand should be a better electron donor and therefore a better nucleophile.
3. *Oxidizability*: a more easily oxidized ligand is more willing to give up electrons and therefore is expected to be a better nucleophile. This factor is measured by standard reduction potentials or polarographic half-wave potentials.
4. *Solvation energy*: a ligand that is more strongly solvated in a given solvent will be a poorer nucleophile in that solvent because some solvation change must occur during the formation of the metal complex.
5. *Metal at reaction center*: this factor greatly limits the generality of nucleophilicity scales in inorganic chemistry when compared to those in organic chemistry.

3.5.a.i Edwards Scale

Edwards[33] proposed that a kinetic nucleophilicity should be correlated by a combination of the factors mentioned earlier for a particular metal center, using the equation

$$\log\left(\frac{k_Y}{k_{solvent}} \right) = \alpha\left(-E^0_{Y_2} \right) + \beta\left(pK_a + 1.74 \right) \qquad (3.32)$$

where k_Y and $k_{solvent}$ are rate constants with Y and solvent, respectively, $E^0_{Y_2}$ is the reduction potential for $Y_2 + 2e^- \rightarrow 2Y^-$ in water, and α and β are empirical constants that depend on the reaction. This scale is often mentioned but has limited applicability because of the lack of $E^0_{Y_2}$ values for all but the halogens and pseudohalogens.

3.5.a.ii Methyl–Mercury(II) Scale

The methyl–mercury(II) scale is based on the equilibrium constant for the reaction[34]

$$H_3C\!-\!Hg\!-\!Y^+ + H_2O \; \underset{}{\overset{K}{\rightleftharpoons}} \; H_3C\!-\!Hg\!-\!OH_2^+ + Y \quad (3.33)$$

and nucleophilicity is taken as proportional to $-\log(K) = pK$. These pK values correlate well with the n_{Pt} scale, described in the next section, and they have a similar range of applicability.

3.5.a.iii n_{Pt} Scale

The n_{Pt} scale is a kinetic scale based on the reaction

$$trans\text{-}Pt(py)_2(Cl)_2 + Y \xrightarrow{\text{Methanol}} trans\text{-}Pt(py)_2(Cl)Y + Cl^- \quad (3.34)$$

and the n_{Pt} for Y is related to the rate constant with Y, k_Y, and with the solvent methanol, $k_{methanol}$, by

$$n_{Pt} = \log\left(\frac{k_Y}{k_{methanol}}\right) \quad (3.35)$$

Some typical values[35] of n_{Pt} are Cl^- (3.04); NH_3 (3.07); N_3^- (3.58), I^- (5.46); CN^- (7.14); PPh_3 (8.93). Clearly, if one is testing for entering group effects in Pt(II) chemistry, one should not choose Cl^- and NH_3 as test nucleophiles because their nucleophilicities are almost identical and one would not see much entering or leaving group effect. This scale works well for Pt(II) reactions, but is at best a qualitative indicator for other metals and not even that for the first-row transition-metal ions.

3.5.a.iv Gutmann Donor Numbers

This scale defines the donor number, DN, of a Lewis base as equal to $-\Delta H^o_{rxn}$ (kcal mol^{-1}) for its reaction with $SbCl_5$ in a 10^{-3} M solution in dichloroethane.[36] The larger the DN the stronger the base and therefore the stronger the nucleophile. The scale has been expanded to include acceptor numbers, AN, for Lewis acids. Some donor numbers are given in Table 3.4.

Table 3.4. Donor Numbers for Some Solvents and Anions

Solvent	DN	Anion	DN
Nitromethane	2.7	ClO_4^-	8.4
Benzonitrile	11.9	NO_3^-	21.1
Acetonitrile	14.1	CN^-	27.1
Dioxane	14.8	I^-	28.9
Propylene carbonate	15.1	$CH_3O_2^-$	29.5
Acetone	17.0	SCN^-	31.9
Water	19.5	Br^-	33.7
Ether	19.2	N_3^-	34.3
Methanol	19.1	OH^-	34.9
Dimethylformamide	26.6	Cl^-	36.2
Dimethylsulfoxide	29.8		

This donor number scale is widely referenced in relation to thermodynamic arguments. It has been criticized because of the neglect of solvent effects and side reactions that contribute to ΔH^o_{rxn} and because a one-parameter scale can never be entirely adequate. Recent measurements[37] with BF_3 as the acid have provided some points of comparison and criticism for the original donor numbers. Recently, Linert et al.[38] have used the solvatochromic shifts of a Cu(II) complex to define donor numbers for anions in dichloromethane. They also have suggested how these values can be converted for use in other solvents through a correlation with the acceptor number of the solvent. Some anion donor numbers in dichloromethane are included in Table 3.4, and the values in water are ~21 kcal mol^{-1} smaller than these.

3.5.a.v Drago E and C Scale

The Drago E and C scale[39] is based on the enthalpy change for the interaction of a Lewis acid (A) and base (B). Each acid and base is characterized by two parameters, E_A and C_A for acids and E_B and C_B for bases, and the enthalpy change for the reaction $A + B \rightarrow (A:B)$ is given by

$$- \Delta H^o_{rxn} = E_A E_B + C_A C_B \qquad (3.36)$$

The parameters E and C are determined by weighted least-squares fitting of values of ΔH^o_{rxn} for appropriate series of acids and bases. In early versions, the parameters were based on the reference values for I_2 of $E_A = C_A = 1.0$. More recently,[40] these were changed to 0.5 and 2.0, respectively, and other values were scaled accordingly. Furthermore, a constant, W, that is characteristic of the acid, has been added to Eq. (3.36). These continuing refinements and changes in parameter values may be one reason that this scale has not been more widely used. The justification for this scale is that it is able to correlate a large number and range of reaction enthalpies. Some recent values of E and C are given in Table 3.5.

Drago has suggested that the C parameter is related to the covalent part of the interaction and that E is related to the ionic or electrostatic part. Therefore, a strong base that complexes with an acid through a largely ionic interaction will have a large E_B and probably a small C_B. These ideas are potentially useful in selecting appropriate nucleophiles for mechanistic tests, but have not been widely used. There has been a recent application to heterogeneous adsorption and catalysis.[41]

Recently, Hancock and Martell[42] have developed E and C values for metal ions and ligands in aqueous solution. These are based on the assumption that aqueous F^- has $E_B = 1.0$ and $C_B = 0$. Values of the logarithm of the first complex formation constant are fitted to a model analogous to Eq. (3.36) with an additional steric term of $-D_A D_B$.

Table 3.5. E and C Values (kcal mol^{-1})$^{1/2}$ for Representative Acids and Bases

Acid	E_A	C_A	Base	E_B	C_B
I_2	0.50	2.00	NH_3	2.31	2.04
C_6H_5OH	2.27	1.07	$N(CH_3)_3$	1.21	5.61
C_6H_5SH	0.58	0.37	C_5H_5N	1.78	3.54
H_2O	1.31	0.78	$NCCH_3$	1.64	0.71
$B(CH_3)_3$	2.90	3.60	$O=C(C_6H_5)_2$	2.01	0.55
$Al(CH_3)_3$	8.66	3.68	$O=S(CH_3)_2$	2.40	1.47
$Ga(CH_3)_3$	6.95	1.48	$O=P(C_6H_5)_3$	2.59	1.67
$Cu(hfac)_2$	1.82	2.86	$P(CH_3)_3$	1.46	3.44

3.5.a.vi Solvent Property Scales

There have been a number of attempts to develop solvent parameter scales that could be used to correlate thermodynamic and kinetic results in terms of these parameters. Some examples are collected in Table 3.6. Kamlet and Taft and co-workers[43] developed the solvatochromic parameters, α_1, β_1, and π^*, which are related to the hydrogen bonding acidity, basicity, and polarity, respectively, of the solvent. Correlations with these parameters also use the square of the Hildebrand solubility parameter, $(\delta_H)^2$, which gives the solvent cohesive energy density.

A thermochemical scale of hydrogen bond basicity has been proposed based on the differences in the heats of solution, $\delta(\Delta H^\circ)$, of pyrrole, *N*-methylpyrrole, benzene, and toluene.[44] The thermochemical scale correlates well with the β_1 parameter of Kamlet and Taft, with the exception of dioxane and especially of triethylamine. The correlation gives a value for water of $\beta_1 \approx 0.2$, while interpolation of hydrogen abstraction rates[45] has given a value of 0.31.

The Reichardt E_T^N scale provides another set of parameters that are related to solvent polarity and basicity.[46] This parameter has been used to correlate the properties and reactivity of $Co(CO)_3(L)_2$ systems.[47]

Drago[48] has proposed a "unified scale of solvent polarities" based on an extension of the E and C acid and base parameters discussed above. Each solvent is characterized by a parameter S', and the change in some property, $\Delta \chi$, of a probe system is given by $\Delta \chi = P S' + W$, where P and W are constants for the probe. If the probe is an acceptor (Lewis acid) in a donor solvent, then

$$\Delta \chi = E_A^* E_B + C_A^* C_B + P S' + W \tag{3.37}$$

where E_A^* and C_A^* are constants for the acceptor probe and E_B and C_B are the solvent values, such as those in Table 3.5. If the probe is a donor (Lewis base), then

Table 3.6. Measures of Hydrogen Bond Basicity

Solvent	$(\delta_H)^2$	α_1	β_1	π^*	E'_A	C'_A	S'
Benzene	0.838	0	0.1	0.59			1.73
Acetone	1.378	0.08	0.48	0.71			2.58
CCl$_4$	0.738	0	0	0.28			1.49
CHCl$_3$	0.887	0.44	0	0.58	1.56	0.44	1.74
CH$_2$Cl$_2$	0.977	0.30	0	0.82	0.86	0.11	2.08
EtOH	1.621	0.83	0.77	0.54	1.33	1.23	2.80
McOH	2.052	0.93	0.62	0.60	1.55	1.59	2.87
THF	0.864	0	0.55	0.58			2.08
Et$_2$O	0.562	0	0.47	0.27			1.73
Dioxane	1.00	0	0.37	0.55			1.93
H$_3$CNO$_2$	1.585	0.22	0.25	0.85			3.07
DMSO	1.688	0	0.76	1.00			3.0
DMF	1.389	0	0.69	0.88			2.8
DMA	1.166	0	0.76	0.88			2.70
CH$_3$CN	1.378	0.19	0.37	0.75			3.0
C$_6$H$_5$CN	1.229	0	0.37	0.90			2.63
Prop Carb				0.83			3.1
H$_2$O	5.49	1.17	0.47	1.09	1.91	1.78	3.53

$$\Delta\chi = E'_A E^*_B + C'_A C^*_B + P\, S' + W \qquad (3.38)$$

where E'_A and C'_A are solvent acid parameters determined by Drago and co-workers, with some examples given in Table 3.6. The E^*_B and C^*_B are determined by fitting $\Delta\chi$ values for the probe system to Eq. (3.38).

3.5.a.vii Hard and Soft Acid–Base Theory

This terminology was first proposed by Pearson,[49] and the basic idea is related to the earlier separation of metal ions into (a) and (b) classes as suggested by Arhland et al.[50] Acids and bases were qualitatively classified by Pearson as "soft", "hard", or "borderline". Soft acids and bases were suspected of using covalent bonding in their interactions and hard species of using predominantly ionic forces. The rule of thumb was that *"hard acids interact most strongly with hard bases and soft acids interact most strongly with soft bases"*. The main criticism of this scale is that it is purely qualitative. In selecting nucleophiles for mechanistic tests, it would be inappropriate to use a soft base with a hard acid because the interaction would be inherently weak, but there is no basis for selecting an appropriate range of nucleophiles to test for an **I** or **I$_d$** mechanism.

The following selection gives a general idea of the types of hard and soft acids and bases:

Hard Acids: H^+, Li^+, Mg^{2+}, Cr^{3+}, Co^{3+}, Fe^{3+}
Soft Acids: Cu^+, Ag^+, Pd^{2+}, Pt^{2+}, Hg^{2+}, Tl^{3+}
Borderline: Mn^{2+}, Fe^{2+}, Zn^{2+}, Pb^{2+}

Hard Bases: F^-, Cl^-, H_2O, NH_3, OH^-, $H_3CCO_2^-$
Soft Bases: I^-, CO, $P(C_6H_5)_3$, C_2H_4, H_5C_2SH
Borderline: N_3^-, C_5H_5N, NO_2^-, Br^-, SO_3^{2-}

More recently, Pearson[51] has attempted to establish a quantitative scale of hardness and softness based on ionization potentials, I, and electron affinities, A. The absolute hardness is defined as $\eta = (I - A)/2$, softness as $\sigma = 1/\eta$, and the absolute electronegativity uses Mulliken's definition of $\chi = (I + A)/2$. For an interaction between a Lewis acid (1) and base (2), the strength of the interaction is assumed to be related to the fractional electron transfer, given by $\Delta N = (\chi_1 - \chi_2)/2(\eta_1 + \eta_2)$. These results are too recent to have been tested, except to note that the quantitative scale generally conforms to the ideas of practicing chemists. Pearson has compiled an extensive list of hardness values, and a few examples are given in Table 3.7.

3.5.a.viii Summary

None of these scales has received universal acceptance by inorganic chemists, and it may be that the heterogeneity of the field will defy anyone to establish a truly general scale. As yet, there seems to be nothing as widely applicable as the Hammet and Taft parameters in organic chemistry. Within certain areas and types of applications one finds one of these scales more often used than others, presumably because it has proven more successful in correlating information.

It has been recognized that a two-parameter scale is necessary, in general, to correlate solvent basicity.[52,53] The conditions under which a one-parameter correlation may appear to work have been discussed by Drago,[54] and the correlations between various basicity scales have been analyzed by Maria et al.[55]

In mechanistic studies, it is common to fall back on qualitative information relating to the particular metal or nonmetal center. After working with particular types of compounds, a lore develops about what are good, not-so-good, and poor nucleophiles. Sometimes, it is recognized that this correlates with one of the basicity scales. It can be of special interest when a particular system or class of ligands fails to follow a reasonably established correlation. This may point to some factor that was overlooked and may provide some mechanistic insight.

Table 3.7. Absolute Electronegativity, χ, and Hardness, η, Values (eV)

Acid	χ	η	Base	χ	η
Ca^{2+}	31.39	19.52	F^-	10.41	7.01
Cr^{2+}	23.73	7.23	Cl^-	8.31	4.70
Mn^{2+}	24.66	9.02	Br^-	7.60	4.24
Fe^{2+}	23.42	7.24	I^-	6.76	3.70
Co^{2+}	25.28	8.22	OH^-	7.50	5.67
Zn^{2+}	28.84	10.88	H_2O	3.1	9.5
Fe^{3+}	42.73	12.08	NH_3	2.6	8.2
Ru^{3+}	39.2	10.7	$N(CH_3)_3$	1.5	6.3
Os^{3+}	35.2	7.5	C_5H_5N	4.4	5.0
Co^{3+}	42.4	8.9	CH_3CN	4.7	7.5
Cr^{3+}	40.0	9.1	$(CH_3)_2O$	2.0	8.0
Pd^{2+}	26.18	6.75	$(CH_3)_3P$	2.8	5.9
Pt^{2+}	27.2	8.0	$(CH_3)_2NCHO$	3.4	5.8

3.5.b Linear Free-Energy Relationships

This general area, known as LFER, has been reviewed recently by Linert.[56] For a simple reaction, the equilibrium constant and the forward and reverse rate constants are related by $K_e = k_f/k_r$, therefore

$$\log(k_f) = \log(K_e) + \log(k_r) \tag{3.39}$$

If a reaction has dissociative activation and one varies the nature of the leaving group X while keeping the entering group Y constant, then k_r should be constant, and a plot of $\log(k_f)$ versus $\log(K_e)$ should be linear with a slope of +1. For associative activation, the same type of experiments should have a constant k_f and a plot of $\log(k_r)$ versus $\log(K_e)$ should be linear with a slope of −1.

This type of analysis has been applied to the aquation of $(H_3N)_5Co^{III}$—X complexes[57,58] with the conclusion that the mechanism is I_d. The LFER for $(H_3N)_5Cr^{III}$—X and $(H_2O)_5Cr^{III}$—X indicates an I_a mechanism.[59]

Proton transfer reactions are an important class that generally satisfy a LFER. Although they are not directly related to ligand substitution processes, proton transfer steps are often involved and usually treated as rapidly maintained equilibria:

$$HB + H_2O \underset{k_r}{\overset{k_f}{\rightleftharpoons}} B^- + H_3O^+ \tag{3.40}$$

$$HB + OH^- \underset{k_r}{\overset{k_f}{\rightleftharpoons}} B^- + H_2O \qquad (3.41)$$

For reactions such as (3.40), the reverse rate constant is essentially diffusion limited so that k_r is fairly constant at ~5 x 10^{10} M^{-1} s^{-1} (see p. 23). However, if there are significant structural and bonding differences between HB and B^-, as with carbon acids, then k_r may be smaller. For a normal acid, with an acid dissociation constant K_a, one can calculate $k_f \approx 5$ x 10^{10} x K_a. Note that, if K_a is small (e.g., 10^{-10} M), then k_f can be rather modest and could become rate limiting for a subsequent process that consumes B^-. The same analysis applies to reactions such as (3.41), where $k_f \approx 2$ x 10^{10} M^{-1} s^{-1}.

3.5.c Reagent Charge Effects

It is often possible to change the net charge on a metal complex by changing the "nonreacting" ligands (e.g., $Pt(Cl)_4^{2-}$, $Pt(NH_3)(Cl)_3^-$, $Pt(NH_3)_2(Cl)_2$, etc.). The variation of substitution rate constant with charge appears to give a clear distinction between associative and dissociative activation. Increasing positive charge on the metal complex should favor bonding to the entering nucleophile and therefore increase the rate of an I_a process, whereas the opposite would be expected for an I_d process.

Unfortunately, when the charge is changed, the ligands must change, and these types of studies can be difficult to interpret. For the Pt(II) examples noted above, other evidence indicates an I_a mechanism, but there are only minor differences[60] in the aquation rates. It can be argued that the increasing positive charge increases the bonding to the entering group but has a similar effect on the leaving group, and the effects tend to cancel.

3.5.d Solvent Dielectric Constant Effects

In organic systems, an increase of rate with increasing dielectric constant of the solvent is associated with the formation of an electrically polar transition state. The situation is more complex with inorganic systems where the metal complex and the leaving and entering groups are often charged, and one must consider both desolvation of the reactants and solvation of the transition state. The dielectric constant of the solvent also has a substantial influence on the formation of ion pairs that may affect the apparent reactivity. A further complication is that the solvent may be a potential ligand and therefore part of the reacting system, rather than just a reaction medium. As a result, the effect of solvent variation on the rate has not been a generally useful criterion for the mechanism. More commonly, such observations are used to assess the solvent effect, when the mechanism is thought to be known, and to test this variation for various theoretical models.

In a system discussed later, Wax and Bergman[61] used a series of methyl-substituted tetrahydrofurans, of presumed constant dielectric constant, to test for the involvement of coordinated solvent in the "intermediate" formed after ligand dissociation. In a study more related to an organic chemistry type of application, Rerek and Basolo[62] attempted to use solvent variations to differentiate between the following rhodium intermediates:

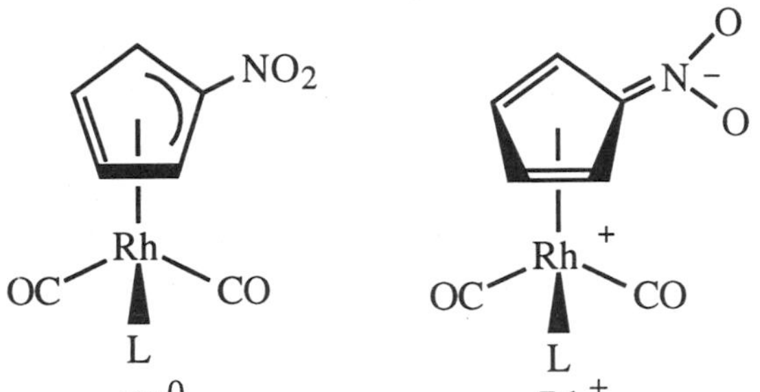

Clearly, the $Rh^{\pm}$ intermediate is more polar and should be favored by solvents with higher dielectric constants. The observations are given in Table 3.8. The authors favor the $Rh^{\pm}$ intermediate on the basis of a comparison of the rate constants in THF and methanol. But when the data are presented as in Table 3.8, it could be argued that the solvents in the left-hand column show an inverse dependence of k on ε and favor Rh^0, whereas the solvents in the right-hand column might be going through a solvent-coordinated intermediate with methanol being the most strongly coordinating ligand.

As shown in Table 3.1, the solvent dielectric constant can be expected to have a significant effect on reactions involving ion pairing. Tobe and co-workers[63] found that $[Pt(Me_4en)(DMSO)Cl]Cl$ is ~100 times more reactive in $CHCl_3$ or CH_2Cl_2 than in water or methanol because of complete ion pairing in the former solvents. Moreover, the reactivity is enhanced further if $[Ph_4As]Cl$ is added due to ion triplet formation.

Table 3.8. Variation of the Rate Constant (25°C) with the Solvent Dielectric Constant for the Reaction of $Rh(\eta^x\text{-}C_5H_4NO_2)(CO)_2$ with PPh_3

Solvent	ε	k (M^{-1} s^{-1})	Solvent	ε	k (M^{-1} s^{-1})
Hexane	1.88	4.44	Dichloromethane	8.93	2.81
Cyclohexane	2.02	3.92	Methanol	32.7	10.4
Toluene	2.38	1.26	Acetonitrile	38.8	9.58
Tetrahydrofuran	7.58	0.963			

3.5.e Steric Effects

Changes in the steric bulk of the nonreacting ligands appear to provide a clear distinction between I_d and I_a mechanisms. Increased steric bulk of the ligands should enhance an I_d mechanism by pushing the dissociating ligand away and relieving steric strain. On the other hand, it should make bonding with the entering group more difficult and inhibit an I_a mechanism. This is perhaps the most successful method of distinguishing these mechanisms, but there are some difficulties.

Parris and Wallace[64] studied the aquation of $M(NH_3)_5Cl^{2+}$ and $M(NH_2CH_3)_5Cl^{2+}$ with M = Co and Cr; the kinetic results of this and subsequent work are summarized in Table 3.9. The original interpretation of the rate constants was that the introduction of the -CH$_3$ group increases k for Co(III), consistent with an I_d mechanism, whereas the opposite effect is seen with Cr(III), as expected for an I_a mechanism for this metal center. This view has persisted until very recently and has caused many other types of information to be interpreted in a way consistent with this kinetic pattern for these two metal ions. Over the years, structural information has been accumulating, and this has been analyzed recently by Lay.[65] The structures of the metal complexes are shown in Figure 3.3.

There are two structural features of note. First, the bond lengths to Co(III) are all shorter than those to Cr(III); this is a general feature and has been part of the rationalization that the mechanisms are different for these two metal ions. Second, the M—Cl bond is actually 0.03 Å shorter in $Cr(NH_2CH_3)_5Cl^{2+}$ than in $Cr(NH_3)_5Cl^{2+}$; this is opposite to what would have been predicted on steric grounds. Lay's interpretation of the structural and kinetic results is that all these systems are reacting by a common I_d mechanism. The $Cr(NH_2CH_3)_5Cl^{2+}$ reacts more slowly because of the shorter, and presumably stronger, Cr—Cl bond compared to the NH$_3$ complex.

Table 3.9. Rate Constants (25°C) and Activation Parameters for the Aquation of Ammonia and Methylamine Complexes of Cobalt(III) and Chromium(III)

	$10^6 \times k$ (s^{-1})	ΔH^* (kJ mol^{-1})	ΔS^* (J mol^{-1} K^{-1})	ΔV^* (cm^3 mol^{-1})
$Co(NH_3)_5Cl^{2+}$	1.72	93	−44	−2.1
$Co(NH_2CH_3)_5Cl^{2+}$	39.6	95	−10	~0
$Cr(NH_3)_5Cl^{2+}$	8.70	93	−29	−1.0
$Cr(NH_2CH_3)_5Cl^{2+}$	0.26	110	−2	~0

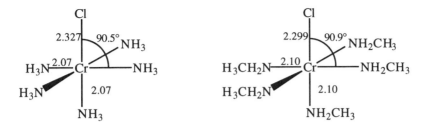

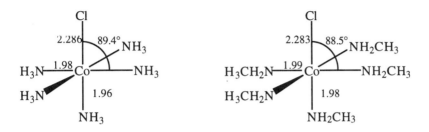

Figure 3.3. Structural parameters for chromium(III) and cobalt(III) amine complexes.

Overall, there is little evidence for steric strain in either of the NH_2CH_3 systems, and the nearly 90° bond angles confirm this impression. The $Co(NH_2CH_3)_5Cl^{2+}$ is more reactive than $Co(NH_3)_5Cl^{2+}$ because the ΔS^* is larger by 34 J mol^{-1} K^{-1} and not because the ΔH^* is smaller, as would have been expected if steric effects made the Co—Cl bond weaker in $Co(NH_2CH_3)_5Cl^{2+}$. Lay has suggested that the entropic difference is due to less effective solvation of the methylamine complex. The general conclusion is that if one is probing steric effects, one must be sure that the expected effects are present in the ground-state reactants if that state is assigned as the source of the reactivity differences.

In a somewhat different use of steric effects, Basolo et al.[66] studied reaction (3.42) with various R groups, and the results are shown in Table 3.10.

$$Pt(PEt_3)_2(R)Cl \ + \ py \ \xrightarrow{\text{EtOH}} \ Pt(PEt_3)_2(R)py^+ \ + \ Cl^- \quad (3.42)$$

The trend of decreasing rate constant with increasing steric bulk of R with both isomers is consistent with associative activation on these square planar Pt(II) complexes. However the cis isomer is much more sensitive to the steric effect (note the temperature difference for the last cis entry in Table 3.10).

Table 3.10. Rate Constants for Replacement of the Chloro Ligand by Pyridine in $Pt(PEt_3)_2(R)Cl$

R—Pt	k (M^{-1} s^{-1})	
	trans (25°C)	cis (0°C)
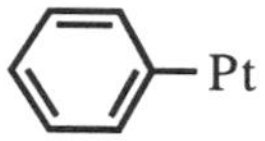 Pt (phenyl)	1.2×10^{-4}	8×10^{-2}
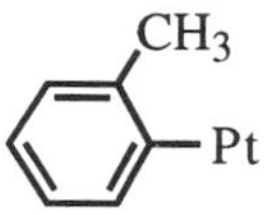 Pt (o-tolyl, CH₃)	1.7×10^{-5}	2×10^{-4}
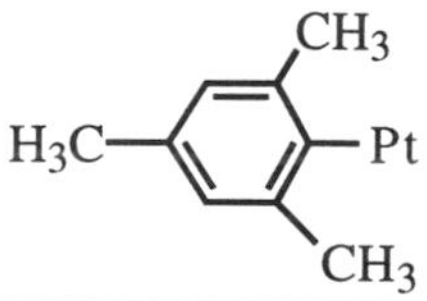 Pt (mesityl, CH₃)	3.4×10^{-6}	1×10^{-6} (25°C)

The larger effect on the cis isomer can be rationalized by a trigonal bipyramidal intermediate, assuming that the entering group (Y) and leaving group (Cl) occupy equivalent positions in the trigonal plane in order to satisfy microscopic reversibility, as shown in Scheme 3.2.

Scheme 3.2

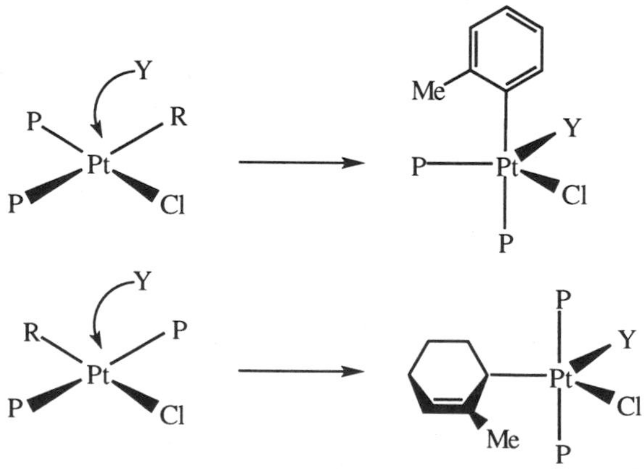

The R group in the axial position in the transition state for the cis isomer will cause more steric crowding than when it is in the equatorial position in the transition state for the trans isomer.

3.5.f Measures of Ligand Size

The problem in interpreting the "steric" effects in the ammonia and methylamine complexes of cobalt(III) and chromium(III) might have been avoided if there were some method of estimating ligand size and therefore anticipating whether steric effects were really significant in a particular system.

The systematic efforts in this regard originate with the work of Tolman[67] on phosphine and phosphite ligands. Tolman defined a *cone angle*, θ, for a number of phosphines, PR₃, based on the following diagram, with the size of the R substituents based on van der Waals radii from CPK models. Seligson and Trogler[68] have used the same methodology to determine cone angles for amines.

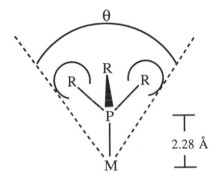

For unsymmetrical ligands P(R)(R')(R"), the angle is calculated by

$$\theta = \frac{2}{3} \sum_{i=1}^{3} \frac{\theta_i}{2} \tag{3.43}$$

where θ_i are the cone angles for the symmetrical phosphines.

The original definition chose an M—P distance based on Ni(0) chemistry and applied the steric size of the ligands to correlate equilibrium constant data for the following reaction:

$$Ni(L)_4 \underset{}{\overset{K}{\rightleftharpoons}} Ni(L)_3 + L \tag{3.44}$$

More recently, Brown and co-workers have used molecular mechanics calculations to estimate a steric repulsion energy, E_R (kcal mol⁻¹), for phosphines and other ligands on $Cr(CO)_5(L)$[69] and $Rh(Cp)(CO)(L)$.[70] Some cone angle and E_R values are given in Table 3.11. This general area has been the subject of two recent reviews,[71,72] where the strengths and weaknesses of various approaches to steric parameters are discussed in detail.

Table 3.11. Representative Steric and Electronic Parameters

Ligand	$pK_a{}^a$	δ, ppm^b	Cone Angle^c	$E_R{}^c$
P(OMe)$_3$	2.6	3.18	107 (128)^d	52
P(OEt)$_3$		3.61	109 (134)^d	59
PMe$_3$	8.65	5.05	118	39
PPhMe$_2$	6.5	4.76	122	44
P(OPh)$_3$	−2.0	1.69	128	65
PEt$_3$	8.69	5.54	132 (137)^d	61
PPh$_2$Me	4.57	4.53	136	57
PPh$_3$	2.73	4.30	145	75
P(CH$_2$Ph)$_3$	(6.0)	3.98	165	82
P(C$_6$H$_{11}$)$_3$	9.7	6.32	171	116
P(C$_6$H$_4$o-Me)$_3$	3.08	3.67	194	113
AsMe$_3$		4.46	114	27
AsPh$_3$		4.16	141	44

^{a} Reference 78. ^{b} Reference 77. ^{c} Reference 70. ^{d} Reference 74.

The original definitions have been criticized and examined by de Santo et al.,[73] and some values have been modified by Stahl and Ernst.[74] In addition, Maitlis[75] and Coville et al.[76] have given cone angles for cyclopentadienyl and arene ligands.

Since the original application, cone angles have been the basis for many attempts to correlate reactivity and steric effects. A complication in such applications is the separation of the steric and bonding or electronic effects when a ligand is changed. The electronic effect often is assumed to parallel the ligand basicity, as measured by the pK_a or by the effect on the ^{13}C NMR chemical shift in Ni(CO)$_3$L.[77] Some of these values also are given in Table 3.11. Steric effect studies should involve systems with reasonably constant electronic factors.

A typical analysis, including electronic effects, is given by Giering and co-workers,[78] who emphasize the fact that steric effects can be minimal below a certain threshold of cone angle. The effects become significant when the sum of the cone angles for two adjacent ligands causes the ligands to come into contact.

3.5.g Volumes of Activation

This type of information has been discussed with regard to the **D** and **A** mechanisms, and the expectations for **I$_d$** and **I$_a$** are qualitatively similar. For the **I$_d$** mechanism, the prediction is that the activation volume, ΔV^*, will be positive because the leaving group is being liberated into solution while there has been relatively little bonding to the entering group. On the other hand, for an $\mathbf{I_a}$ mechanism, the ΔV^* will be negative because the entering group has been captured from solution while the leaving group is still bonded to the metal center. Much of the literature bases the mechanistic differentiation on the sign (+ or −) of ΔV^*. It seems more correct to say that for a family of related reactions, those with a more positive ΔV^* are more dissociative and those with a more negative ΔV^* are more associative.

Values of ΔV^* for the following exchange reaction with M = Cr, Co, Rh,[79,80] Ru,[81,82] Ir,[83,84] and Pt[85] are given in Table 3.12.

$$(L)_nM(OH_2) + H_2O^* \rightleftharpoons (L)_nM(*OH_2) + H_2O \quad (3.45)$$

The simplest explanation of these data is that the intimate mechanism involves more bond making to the entering H_2O as the ΔV^* becomes more negative, but it is *not* necessary that the mechanism changes if the ΔV^* changes sign. For $Ru(OH_2)_6^{2+}$, $\Delta V^* = -0.4$ cm^3 mol^{-1}, but ligand substitution rates are consistent with an $\mathbf{I_d}$ mechanism.[86] There is considerable other evidence that substitution on Pt(II) has an $\mathbf{I_a}$ mechanism, but the water exchange reaction may show less bond making than usual because Pt(II) is a soft acid and water is a hard base. It would be of great value to be able to predict the ΔV^* for a particular mechanism, but this has proved to be complicated because of solvent electrostriction effects. The solvent will constrict around the reactants as the charge density increases or will expand as the charge density decreases, and this factor is very difficult to anticipate in a quantitative way. Further aspects of ΔV^* are discussed in Section 3.7.c later in this chapter and in a comprehensive review.[87]

Table 3.12. Volumes and Entropies of Activation for Some Water Exchange Reactions

$(L)_nM$	ΔV^* (cm^3 mol^{-1})	ΔS^* (J mol^{-1} K^{-1})
$(H_3N)_5Co^{3+}$	+1.2	28
$(H_3N)_5Rh^{3+}$	−4.1	3
$(H_2O)_5Rh^{3+}$	−4.1	29.3
$(H_3N)_5Ir^{3+}$	−3.2	11
$(H_2O)_5Ir^{3+}$	−5.7	2.1
$(H_3N)_5Cr^{3+}$	−5.8	0
$(H_2O)_5Cr^{3+}$	−9.6	11.6
$(H_3N)_5Ru^{3+}$	−4.0	−7.7
$(H_2O)_5Ru^{3+}$	−8.3	−48.2
$(H_2O)_5Ru^{2+}$	−0.4	16.1
$(H_2O)_3Pt^{2+}$	−4.6	−9

3.5.h Entropies of Activation

It is generally expected that an I_d mechanism will have a more positive entropy of activation, ΔS^*, than an I_a mechanism because of the increase in randomness in an I_d transition state, compared to a decrease in randomness in an I_a transition state. This criterion has been used cautiously and only for comparisons of similar reaction types. One reason for caution is the experimental error in ΔS^*, which is typically ± 8 to 12 J mol^{-1} K^{-1}. It was thought that ΔV^* would have an advantage over ΔS^* as a mechanistic criterion because of the better accuracy and our better intuitive understanding of volume changes. However, ΔS^* has a larger range of values than ΔV^*, and this tends to offset the problem of experimental uncertainty. There is some evidence of a correlation between ΔS^* and ΔV^*,[88] which could make accurate ΔS^* values of greater mechanistic importance. There is some indication of such a correlation in Table 3.12 for the systems with $(L)_n = (NH_3)_5$, but the $(H_2O)_n$ systems are different, possibly because of the usual scapegoat of solvation differences.

3.6 SOME SPECIAL EFFECTS

3.6.a Dissociative Conjugate Base Mechanism

The dissociative conjugate base, DCB, mechanism has been described previously in connection with competition studies for a **D** mechanism (Scheme 3.1). In its original form, it involved the reaction of hydroxide ion with amine complexes of cobalt(III). It occupies a special place in inorganic mechanistic work because it was the subject of a long controversy during the 1960s between Ingold, Nyholm, and Tobe on one side and Basolo and Pearson on the other. This controversy stimulated a lot of kinetic work and also generated many useful spin-offs in the areas of synthesis, spectroscopy, and theory. Ingold and colleagues believed that the reaction was a simple bimolecular displacement (S_N2) of the leaving group by hydroxide ion, whereas Basolo and Pearson were proponents of the S_N1CB (DCB) mechanism originally proposed by Garrick.[89]

In the final analysis, the DCB mechanism has passed every test so far applied and is now the accepted mechanism for these reactions. The results leading to these conclusions are summarized in several articles.[90-92]

The DCB mechanism has received an amount of attention quite out of proportion to its general importance, since it is of less kinetic significance even for amine complexes of other 3+ metal ions. Its importance for cobalt(III) may be due to the small size of low-spin cobalt(III) and its ability to stabilize the dissociative intermediate through π bonding. However, it is generally observed that the conjugate bases of hydrated metal ions, shown in the following reaction, are much more reactive toward substitution:

$$M(OH_2)_n^{z+} \rightleftharpoons M(OH_2)_{n-1}(OH)^{(z-1)+} + H^+ \quad (3.46)$$
$$\text{More active form}$$

3.6.b Trans Effect

It is well documented that the ligand in the position trans to the leaving group has a significant influence on the rate of substitution reactions. This effect has been most thoroughly studied in Pt(II) chemistry, where the ordering of trans labilization is

$$CN^-, CO > R_3P, H^-, (H_2N)_2CS > CH_3^- > NCS^- > I^- > Cl^- > OH^-$$

The sources of the trans effect can be classified into thermodynamic and kinetic factors. The thermodynamic factor refers to weakening of the bond to the leaving group in the ground state of the reactant. The kinetic factor refers to the stabilization of the transition state by the trans ligand.

3.6.b.i Thermodynamic Trans Effect

Structural evidence for this effect was summarized by Appleton et al.[93] The bond length changes generally are not dramatic. For example, the Pt—Cl bond in *cis*-Pt(PR$_3$)$_2$(Cl)$_2$ is 0.08 Å longer than the Pt—Cl bond in *trans*-Pt(PR$_3$)$_2$(Cl)$_2$ although PR$_3$ is one of the stronger trans labilizing ligands. However H$^-$ causes trans bond lengthening of 0.15 to 0.20 Å.[94] Evidence for the thermodynamic trans effect was obtained by Chatt et al.[95] from the N—H stretching frequencies in T–Pt–NHR$_2$ systems with various trans ligands, T. The C—O stretching frequencies and NMR chemical shifts also have been used to probe for this effect.

It was recognized at an early stage that, in order to rationalize the order of the ligands in the trans effect series, it would be necessary to invoke both σ- and π-bonding effects. These effects work in the ground state through the metal orbitals shared by the leaving group (X) and the trans ligand (T), as shown in the following diagrams:

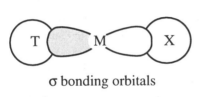

σ bonding orbitals

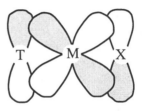

π bonding orbitals

The metal σ orbital is empty and both T and X donate electrons into it. If the T ligand is a stronger σ donor, then the σ bond to X will be weaker. The large trans effects of CH_3^- and H^- are directly attributed to this situation.

The metal π orbital is a filled d_{xy}, d_{xz}, or d_{yz} orbital and electrons are being donated from this orbital into empty orbitals on the ligands T and X. These empty orbitals might be π antibonding on CO and CN^- or d orbitals on phosphorus-donor ligands, for example. This "back donation" strengthens the metal–ligand bond, and, if T is a better π acceptor (π acid), then its bond to M will be strengthened at the expense of the M—X bond. This effect could explain the positions of CO and PR_3 in the trans effect series, since they are good π acids. However, such systems probably derive most of their reactivity from the kinetic trans effect described in the next section.

In classic studies to separate the σ and π effects, Parshall[96] studied the ^{19}F NMR chemical shifts for a series of T ligands in the following types of complexes:

meta   para

The ^{19}F shift in the meta isomer should only depend on the σ effect, but the shift in the para isomer will be sensitive to both the σ and π effects. Essentially similar NMR methods have been applied to a number of different systems. The results of these studies generally confirm the presence of both types of effects, with magnitudes consistent with expectations for different T ligands. The difficulty is that these observations do not translate easily into a direct measure of the extent of bond weakening in the ground state caused by a trans group; therefore the kinetic consequences are uncertain.

3.6.b.ii Kinetic Trans Effect

The kinetic trans effect is assumed to operate because of better bonding between M and T in the transition state. The effect can be explained pictorially, but there have been attempts at more quantitative quantum mechanical calculations.[97–99] The contributions have been divided into σ and π effects.

Most explanations have taken square planar Pt(II) complexes as models because the effect is best documented for these systems and the mechanism is taken to be I_a. The transition state is assumed to be a trigonal bipyramid, formed as follows:

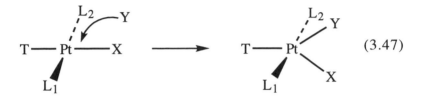

$$(3.47)$$

In the transition state, the T ligand is no longer sharing σ or π orbitals with the X ligand; therefore, the T ligand can improve its bonding to Pt(II) if it is a good σ donor or π acceptor. This can help to offset the loss in bonding to X. In addition, the transition state can be further stabilized because the empty $d_{x^2-y^2}$ orbital on Pt(II) becomes of π symmetry and can accept π electrons from T if the latter is capable of π donation, as shown in the following diagram:

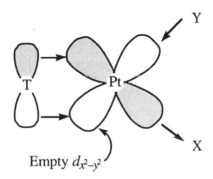

Empty $d_{x^2-y^2}$

3.6.c Cis Effect

The effect of the cis ligand(s) on substitution rates has been noted especially in the chemistry of octahedral organometallic complexes. In general, the effect is less than the trans effect. The order of cis labilization for various ligands is

$$NO_3^- > CH_3CO_2^- > Cl^-, Br^- > Py > I^- > PR_3 > CO, H^-$$

This ordering is almost the exact opposite of that for the trans effect. The reaction mechanism also is different in that the systems used as examples show I_d characteristics. Evidence for a ground-state effect is very limited, as expected, because the cis ligands share only one orbital $(d_{x^2-y^2})$ of minor significance in their σ bonding. The effect is normally attributed to stabilization of the I_d transition state by better bonding with the cis ligand. Davy and Hall[100] have given a theoretical treatment of the cis effect for $Mn(CO)_5(L^-)$ and $Cr(CO)_5L$ systems. This approach indicates that the π-donor ability of the cis ligand is the most important feature, which is consistent with the ordering of ligands. Ligands such as nitrate and carboxylates may stabilize the intermediate by coordination of the distal oxygen.[101]

It should be remembered that for systems such as

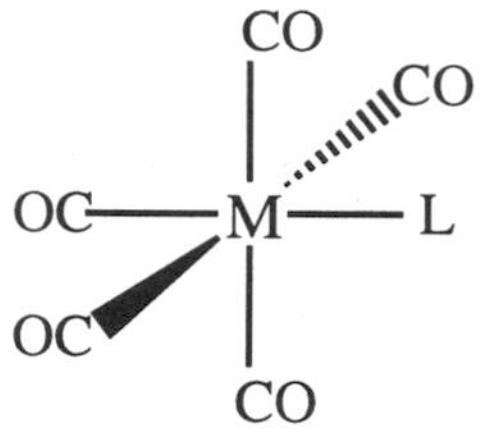

the observation is that the cis-CO ligands exchange, or are replaced more easily, than the trans one. This relative rate difference could be assigned, at least in part, to a *trans-stabilizing* influence of the L group.

3.6.d Reactions Without Metal–Ligand Bond Breaking

Some reactions of metal complexes that appear to involve substitution may actually occur without breaking a bond to the metal. The ^{18}O-labeling studies[102] in reaction (3.48) show that the Co—O bond is retained. This seems to be a general characteristic of CO_2 addition and release from such inert metal complexes. For example, it appears that $W(CO)_5(OH)$ reacts similarly with CO_2.[103]

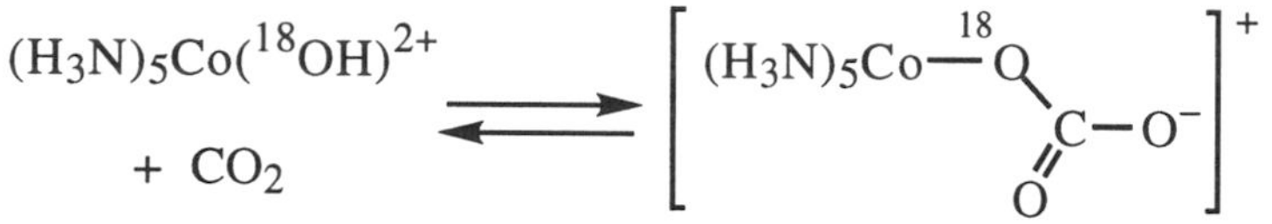

For chelated carbonate complexes, Posey and Taube[104] found that initial ring opening proceeds with Co—O bond breaking, and then CO_2 is lost by O—C bond breaking, as follows:

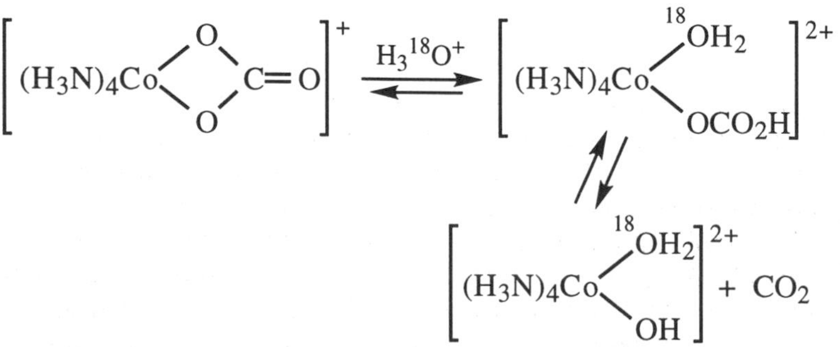

There is considerable interest in the reactions of such carbonato complexes because of their possible relevance to the action of carbonic anhydrase, and the area has been reviewed[105] and reanalyzed.[106]

Buckingham and Clark[107] have used multiwavelength stopped flow and the [H$^+$] dependence of the rate to show that these reactions are biphasic and proceed through a protonated intermediate, as shown in Scheme 3.3.

Scheme 3.3

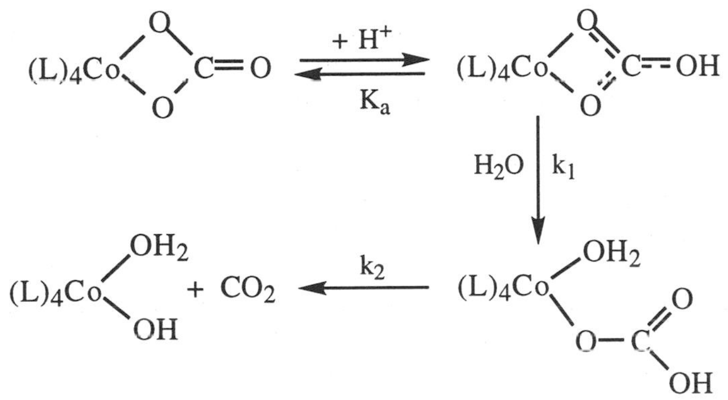

For the various systems that were studied, such as $Co(NTA)(O_2CO)^{2-}$, $Co(gly)_2(O_2CO)^-$, $Co(tren)(O_2CO)^+$, and $Co(NH_3)_4(O_2CO)^+$, it was determined that K_a ~1 M and k_2 ~1 s^{-1} (25°C, 1.0 M NaClO$_4$). This insensitivity to the ancilliary ligands is consistent with protonation at a site remote from the Co(III) and for the loss of CO_2, k_2, proceeding by breaking of the O—C rather than the Co—O bond. However, k_1 changes by ~10^4 s^{-1} for the different (L)$_4$ systems and reflects the effect of the (L)$_4$ ligands on breaking a Co—O bond. A chelated bicarbonate complex has been characterized by crystallography.[108]

Studies by Harris and co-workers[109] indicate that sulfur dioxide reacts in a fashion similar to CO_2, as shown by the example in reaction (3.50). The immediate product is the O-bonded sulfito complex, which may undergo linkage isomerism to the S-bonded form or reduce cobalt(III) to cobalt(II).

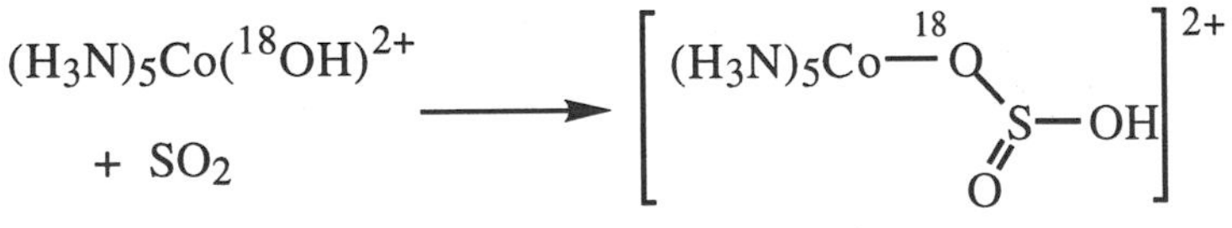

The aliphatic amine complexes react in this way, but the bis(phenanthroline) and bis(bipyridyl) complexes of cobalt(III) also show reactivity toward HSO$_3^-$ and SO$_3^{2-}$.[110]

The reactions with nitrous acid follow a more complicated route, described by

$$2 \, HNO_2 \rightleftharpoons N_2O_3 + H_2O$$

$$N_2O_3 \rightleftharpoons NO^+ + NO_2^-$$

(3.51)

$$(H_3N)_5Co\text{–}OH^{2+} + NO^+ \longrightarrow (H_3N)_5Co\text{–}O\text{–}NO^{2+} + H^+ \quad (3.52)$$

Murmann and Taube[111] found that the oxygen bonded to cobalt(III) in the reactant remains in the product, and it was long assumed that this oxygen remained bound to cobalt(III). Subsequent ^{17}O NMR studies of Jackson et al.[112] confirmed the original observations and showed that the cobalt-bound oxygen is scrambled between the two possible sites in the subsequent linkage isomerism process.

The examples in reactions (3.48), (3.50), and (3.52) can be viewed as intermolecular nucleophilic attack of coordinated OH^- on an electrophile. The same process can occur intramolecularly, as in the case of oxalate chelate ring opening studied by Andrade and Taube,[113] whose isotope tracer results are summarized by

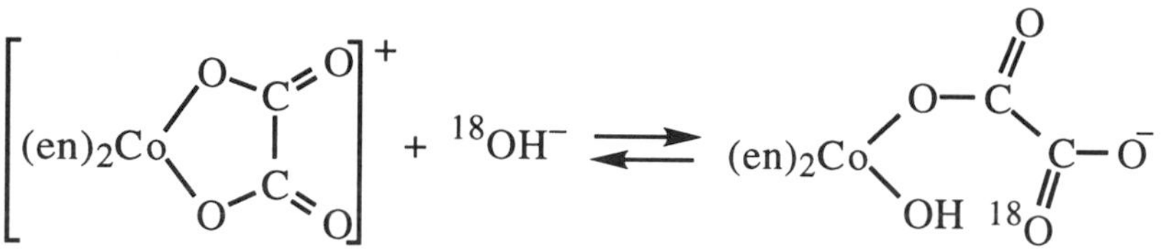

(3.53)

Microscopic reversibility requires that the ring-closing process must occur by O—C bond cleavage after attack of coordinated OH^- on the uncoordinated end of the oxalate ligand.

Similar intramolecular attack by OH^- and ring closing, with retention of the Co—O bond, is observed with esters,[114] as shown in the following reaction:

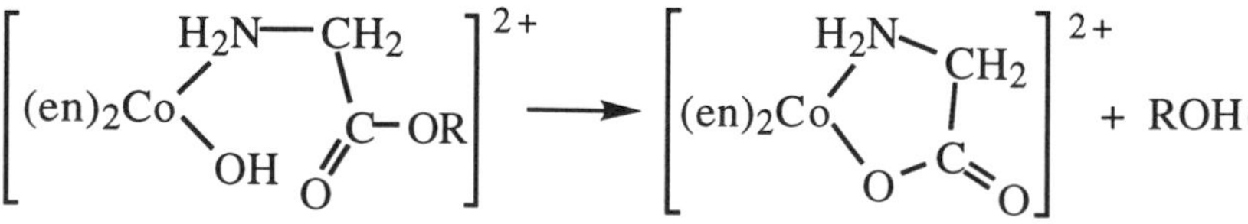

(3.54)

The amides[115] are more complex because they can produce stable chelated amide or hydrolyzed amide products, as shown in reaction (3.55), with the product distribution depending on the amide substituent.[116] In acidic solution, there is no general acid or base catalysis, and the rate-controlling step is attack of coordinated water on the carbonyl carbon followed by fast elimination of NH_2R. In alkaline solution, there is general base catalysis, which is interpreted as attack by coordinated OH^- followed by rate-controlling deprotonation of C—OH in the cyclic intermediate.

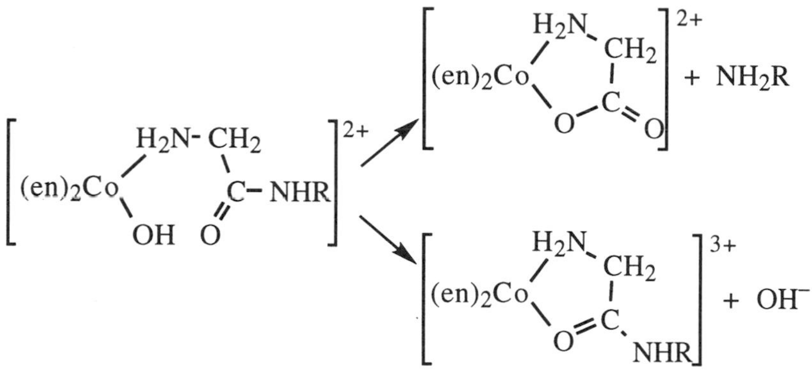

$$
\left[(en)_2Co \underset{OH \quad O}{\overset{H_2N-CH_2}{\diagdown}} C-NHR \right]^{2+}
\begin{array}{c}
\nearrow \\
\\
\searrow
\end{array}
\begin{array}{l}
\left[(en)_2Co \underset{O-C=O}{\overset{H_2N \diagup CH_2}{\diagdown}} \right]^{2+} + NH_2R \\
\\
\left[(en)_2Co \underset{O=C \diagdown NHR}{\overset{H_2N \diagup CH_2}{\diagdown}} \right]^{3+} + OH^-
\end{array}
\qquad (3.55)
$$

Bifunctional bases, such as $HPO_4{}^{2-}$, $HCO_3{}^-$, and $HAsO_4{}^{2-}$, are particularly good catalysts for amide hydrolysis. This is attributed to the ability of such bases to simultaneously remove the C—OH proton and protonate the NHR leaving group, as shown in Scheme 3.4.

Scheme 3.4

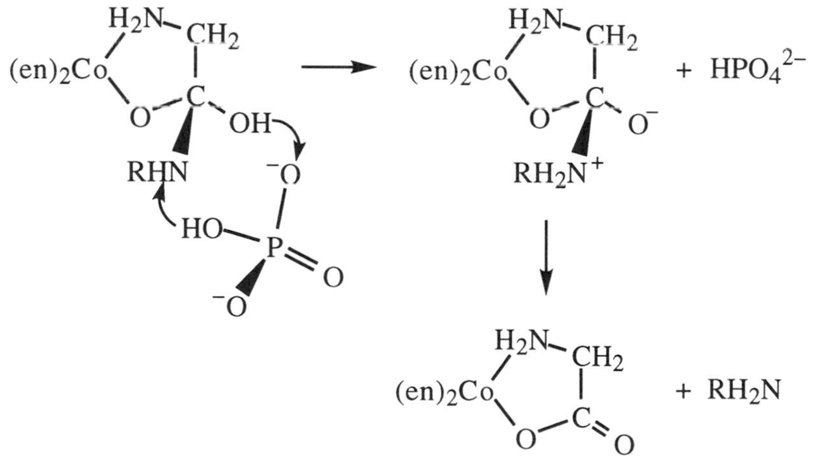

The chelate ring closing of glycine has been the subject of extensive isotope and kinetic studies,[117] and the results are summarized in Scheme 3.5 (k, s^{-1} at 25°C). The reactions proceed with retention of the Co—O bond, and it is noteworthy that coordinated water is an active nucleophile in this intramolecular process. The acidic pathways, given by k_1 and k_2 in Scheme 3.5, show general acid catalysis, which the authors interpret as due to rate-controlling elimination of water from the tetrahedral carbon in a cyclic intermediate.

Scheme 3.5

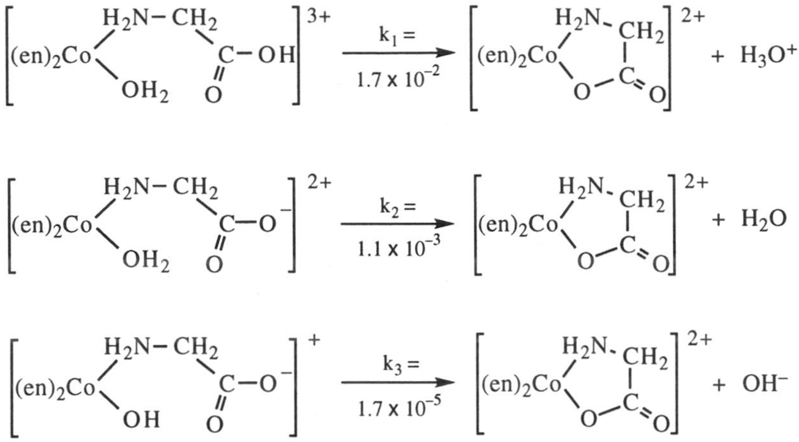

$$\left[(en)_2Co\begin{array}{c} H_2N-CH_2 \\ \diagup \diagdown \\ \diagup \quad \diagdown C-OH \\ OH_2 \quad \| \\ \qquad O \end{array}\right]^{3+} \xrightarrow[1.7 \times 10^{-2}]{k_1 =} \left[(en)_2Co\begin{array}{c} H_2N_\diagdown CH_2 \\ \diagup \diagup \\ \diagdown \diagup C \\ O-C\!\!\stackrel{\diagdown}{\diagup}O \end{array}\right]^{2+} + H_3O^+$$

$$\left[(en)_2Co\begin{array}{c} H_2N-CH_2 \\ \diagup \diagdown \\ \diagup \quad \diagdown C-O^- \\ OH_2 \quad \| \\ \qquad O \end{array}\right]^{2+} \xrightarrow[1.1 \times 10^{-3}]{k_2 =} \left[(en)_2Co\begin{array}{c} H_2N_\diagdown CH_2 \\ \diagup \diagup \\ \diagdown \diagup C \\ O-C\!\!\stackrel{\diagdown}{\diagup}O \end{array}\right]^{2+} + H_2O$$

$$\left[(en)_2Co\begin{array}{c} H_2N-CH_2 \\ \diagup \diagdown \\ \diagup \quad \diagdown C-O^- \\ OH \quad \| \\ \qquad O \end{array}\right]^{+} \xrightarrow[1.7 \times 10^{-5}]{k_3 =} \left[(en)_2Co\begin{array}{c} H_2N_\diagdown CH_2 \\ \diagup \diagup \\ \diagdown \diagup C \\ O-C\!\!\stackrel{\diagdown}{\diagup}O \end{array}\right]^{2+} + OH^-$$

In other cases, coordination to a metal may activate the ligand to nucleophilic attack. The following equations give examples[118–120] of such processes that are generally much slower or not even observed with the free ligand:

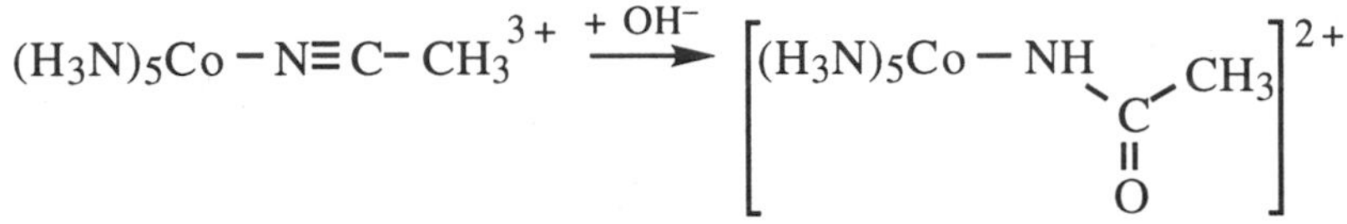

$$(H_3N)_5Co-N\!\!\equiv\!\!C-CH_3{}^{3+} \xrightarrow{+ OH^-} \left[(H_3N)_5Co-NH_{\diagdown C\diagup CH_3} \atop \qquad \qquad \|\atop \qquad \qquad O\right]^{2+} \quad (3.56)$$

$$(H_3N)_5Co-N\!\!\equiv\!\!C-CH_3{}^{3+} \xrightarrow{+ N_3^-} \left[(H_3N)_5Co-N{\diagup N\!\!=\!\!N \atop \diagdown C\!\!=\!\!N} \atop \qquad H_3C\right]^{2+} \quad (3.57)$$

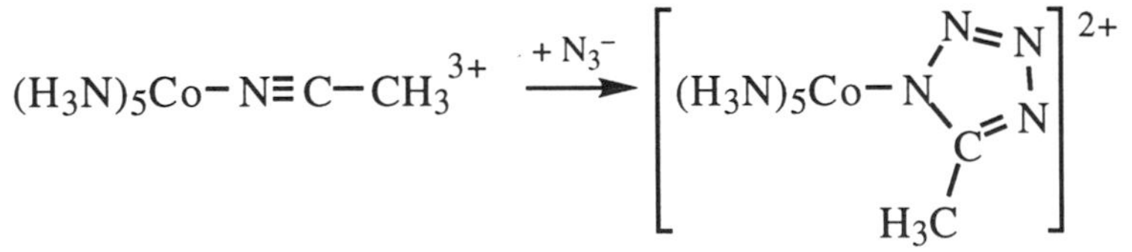

Phosphate ester hydrolysis[121] is an important biological process that is catalyzed by metal-containing enzymes and model systems. An early example from Co(III) chemistry is shown in the following reaction, where tn is trimethylenediamine and RO^- is *p*-nitrophenolate.[122] The chelated phosphate ester hydrolyzes about 10^9 times faster than the free ester, and the acceleration is attributed to relief of ring strain in the five-coordinate phosphorus intermediate as well as to the inductive effect of the $Co(tn)_2{}^{3+}$. Either or both of these effects may be operating in the natural and model systems.

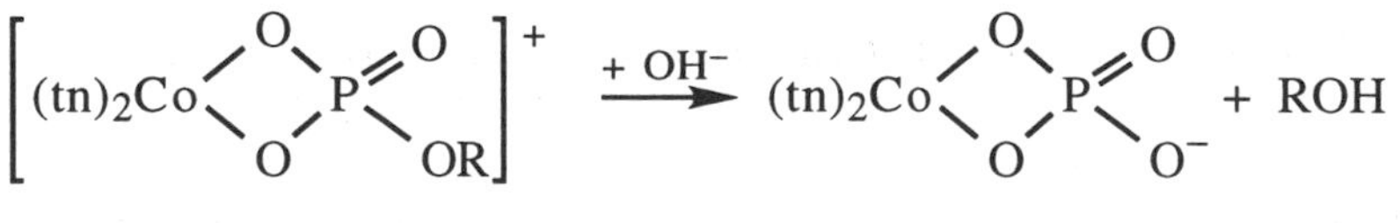

$$\left[(tn)_2Co \underset{O}{\overset{O}{<}} \underset{OR}{\overset{O}{>}} P \overset{O}{\underset{}{=}} O \right]^+ \xrightarrow{\text{+ OH}^-} (tn)_2Co \underset{O}{\overset{O}{<}} \underset{O^-}{\overset{O}{>}} P \overset{O}{\underset{}{=}} O + ROH \quad (3.58)$$

It has been found[123] that chelated amino acid esters are activated so that they are useful for peptide synthesis, as shown in reaction (3.59). This work has been reviewed recently.[124]

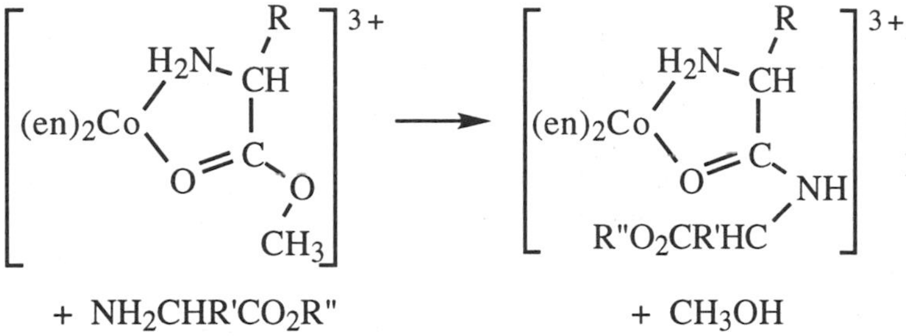

$$\left[(en)_2Co \begin{array}{c} H_2N \\ \\ O \end{array} \begin{array}{c} R \\ | \\ CH \\ | \\ C \\ O \\ | \\ CH_3 \end{array} \right]^{3+} \longrightarrow \left[(en)_2Co \begin{array}{c} H_2N \\ \\ O \end{array} \begin{array}{c} R \\ | \\ CH \\ | \\ C \\ NH \\ R''O_2CR'HC \end{array} \right]^{3+} \quad (3.59)$$

$$+ \; NH_2CHR'CO_2R'' \qquad\qquad\qquad + \; CH_3OH$$

3.7 VARIATION OF SUBSTITUTION RATES WITH METAL ION

The variation in rates of substitution with different metal ions has been an area of interest for many years. This type of information on the qualitative level is very useful for synthetic purposes and for studies on equilibrium properties. It provides an indication of the conditions required for a particular preparative procedure or of how long one must wait for a system to reach equilibrium. The wide range of rates for the transition-metal ions has been of considerable interest, both to rationalize the relative rates and to apply these rationalizations to the probable reaction mechanism. This topic has been reviewed recently.[125]

3.7.a Water Exchange Rates

Since the pioneering work of Plane and Hunt,[126] Taube and co-workers,[127] and Swift and Connick,[128] one of the most widely studied reactions for transition-metal complexes has been that of water solvent exchange, using either $^{18}OH_2$ or $^{17}OH_2$:

$$(L)_nM(OH_2) + {}^*OH_2 \; \rightleftharpoons \; (L)_nM({}^*OH_2) + H_2O \quad (3.60)$$

The results of such studies are summarized in Table 3.13, and the field has been reviewed and discussed by Merbach.[129] Results of NMR studies prior to about 1970 should be viewed with caution, and most of the early work has been repeated with modern instrumentation and methods of analysis.

Table 3.13. Water Exchange Rate Constants (25°C) and Activation Parameters

$M(L)_n(OH_2)$	k (s^{-1})	ΔH^* (kJ mol^{-1})	ΔS^* (J mol^{-1} K^{-1})	ΔV^* (cm^3 mol^{-1})	Ref.
$Ti(OH_2)_6^{3+}$	1.8×10^5	43.4	1.2	-12.1	a
$V(OH_2)_6^{2+}$	8.7×10^1	61.8	-0.4	-4.1	b
$V(OH_2)_6^{3+}$	5.0×10^2	49.4	-27.8	-8.9	c
$Cr(OH_2)_6^{2+}$	$>10^8$				
$Cr(OH_2)_6^{3+}$	2.4×10^{-6}	108.6	11.6	-9.6	e
$Cr(OH_2)_5(OH)^{2+}$	1.8×10^{-4}	111	55.6	2.7	e
$Mn(OH_2)_6^{2+}$	2.1×10^7	32.9	5.7	-5.4	f
$Fe(OH_2)_6^{2+}$	4.4×10^6	41.4	21.2	3.8	f
$Fe(OH_2)_6^{3+}$	1.6×10^2	63.9	12.1	-5.4	g
$Fe(OH_2)_5(OH)^{2+}$	1.2×10^5	42.4	5.3	7.0	g
$Co(OH_2)_6^{2+}$	3.2×10^6	46.9	37.2	6.1	f
$Ni(OH_2)_6^{2+}$	3.2×10^4	56.9	32.0	7.2	f
$Cu(OH_2)_6^{2+}$	4.4×10^9	11.5	-21.8		d
$Zn(OH_2)_6^{2+}$	$>10^7$				
$Ru(OH_2)_6^{2+}$	1.8×10^{-2}	87.4	16.1	-0.4	h
$Ru(OH_2)_6^{3+}$	3.5×10^{-6}	89.8	-48.3	-8.3	h
$Ru(OH_2)_5(OH)^{2+}$	5.9×10^{-4}	95.8	14.9	0.9	h
$Rh(OH_2)_6^{3+}$	2.2×10^{-9}	131	29	-4.2	i
$Rh(OH_2)_5(OH)^{2+}$	4.2×10^{-5}	103		1.5	i
$Ir(OH_2)_6^{3+}$	1.1×10^{-10}	131	2.7	-5.7	j
$Ir(OH_2)_5(OH)^{2+}$	1.4×10^{-11}	139	11.5	-0.2	j
$Pt(OH_2)_4^{2+}$	3.9×10^{-4}	89.7	-9	-4.6	k
$Pd(OH_2)_4^{2+}$	5.6×10^2	49.5	-26	-2.2	k
$Cr(NH_3)_5(OH_2)^{3+}$	5.2×10^{-5}	97	0	-5.8	l
$Co(NH_3)_5(OH_2)^{3+}$	5.7×10^{-6}	111	28	1.2	m
$Rh(NH_3)_5(OH_2)^{3+}$	8.4×10^{-6}	103	3	-4.1	l
$Ir(NH_3)_5(OH_2)^{3+}$	6.1×10^{-8}	118	11	-3.2	n

[a] Hugi, A. D.; Helm, L.; Merbach, A. E. *Inorg. Chem.* **1987**, *26*, 1763.

[b] Ducommun, Y.; Zbinden, D.; Merbach, A. E. *Helv. Chim. Acta* **1982**, *65*, 1385.

[c] Hugi, A. D.; Helm, L.; Merbach, A. E. *Helv. Chim. Acta* **1985**, *68*, 508.

[d] Powell, D. H.; Helm, L.; Merbach, A. E. *J. Chem. Phys.* **1991**, *95*, 9258.

[e] Xu, F.-C.; Krouse, H. R.; Swaddle, T. W. *Inorg. Chem.* **1985**, *24*, 267.

[f] Ducommun, Y.; Newman, K. E.; Merbach, A. E. *Inorg. Chem.* **1980**, *19*, 3696.

[g] Grant, M.; Jordan, R. B. *Inorg. Chem.* **1981**, *20*, 55; Swaddle, T. W.; Merbach, A. E. *Inorg. Chem.* **1981**, *20*, 4212.

[h] Rapaport, I.; Helm, L.; Merbach, A. E.; Bernhard, P.; Ludi, A. *Inorg. Chem.* **1988**, *27*, 873.

[i] Laurenczy, G.; Rapaport, I.; Zbinden, D.; Merbach, A. E. *Magn. Reson. Chem.* **1991**, *29*, S45.

[j] Cusanelli, A.; Frey, U.; Richens, D.; Merbach, A. E. *J. Am. Chem. Soc.* **1996**, *118*, 5265.

[k] Helm, L.; Elding, L. I.; Merbach, A. E. *Inorg. Chem.* **1985**, *24*, 1719.

[l] Swaddle, T. W.; Stranks, D. R. *J. Am. Chem. Soc.* **1972**, *94*, 8357.

[m] Hunt, H. R.; Taube, H. *J. Am. Chem. Soc.* **1958**, *80*, 2642.

[n] Tong, S. B.; Swaddle, T. W. *Inorg. Chem.* **1974**, *13*, 1538.

A noteworthy feature of the data in Table 3.13 is the wide range of rate constants: for the 2+ ions, k varies from 4×10^{-4} s^{-1} for Pt(II) to 4.4×10^{9} s^{-1} for Cu(II); for the 3+ ions, k varies from 1.1×10^{-10} s^{-1} for Ir(III) to 1.8×10^{5} s^{-1} for Ti(III).

3.7.a.i Labile and Inert Classification of Taube

Taube[130] was the first to attempt an explanation of the substitution lability of these metal ions by classifying them qualitatively on the basis of their reactivity.

Labile metal ions react essentially on mixing of the metal ion and ligand solutions, that is, within a few seconds at most. *Inert metal ions* require at least a few minutes for their substitution reactions to be complete. This very operational classification was about all that was possible in 1952, but it provides a useful practical classification that has endured as a qualitative description of the reactivity of a metal ion.

Taube offered a theoretical explanation for the qualitative differences in reactivity. The original explanation was in terms of Pauling's valence bond theory, but the same arguments can be framed in terms of crystal field theory for octahedral complexes. The terminology is defined in Figure 3.4. *Labile metal ions* have either an empty low-energy t_{2g} orbital or at least one electron in a high-energy e_g orbital. The rationalization for this is that the empty t_{2g} orbital can be used by the entering group in an **A** or **I$_a$** transition state. Electrons in the higher-energy orbital set will favor a **D** or **I$_d$** mechanism because the ligand bonds in the ground state will be weaker. *Inert metal ions* must have at least one electron in each t_{2g} orbital and no electrons in the e_g orbitals.

These ideas are consistent with the inertness of the octahedral complexes of chromium(III) (t_{2g}^3), low-spin cobalt(III), iron(II) (t_{2g}^6), and iron(III) (t_{2g}^5), as well as the complexes of the second- and third-row transition metals with more than two d electrons, which are low spin. They also provide an explanation for the fact that vanadium(III) (t_{2g}^2)

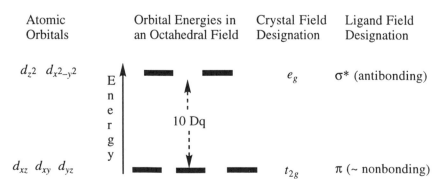

Atomic Orbitals	Orbital Energies in an Octahedral Field	Crystal Field Designation	Ligand Field Designation
$d_{z^2}\ d_{x^2-y^2}$		e_g	σ^* (antibonding)
$d_{xz}\ d_{xy}\ d_{yz}$	10 Dq	t_{2g}	π (~ nonbonding)

Figure 3.4. Energies and designations of d orbitals in an octahedral complex.

is more labile than vanadium(II) ($t_{2g}{}^3$), whereas chromium(III) ($t_{2g}{}^3$) is inert and chromium(II) ($t_{2g}{}^3 e_g{}^1$) is labile.

The predictions of this theory are qualitatively correct, but it does not explain the wide range of reactivities, especially of the labile systems. For example, why is the water exchange rate for nickel(II) 10^2 times slower than that for cobalt(II) and $>10^4$ times slower than that for copper(II)?

3.7.a.ii Crystal Field Theory Applications

This application of crystal field theory was put forward first in the textbook by Basolo and Pearson[131] in an attempt to explain the finer details of the reactivity differences between various metal ions.

The energies of the valence d orbitals were calculated for various ideal geometries of possible transition states for the substitution reactions. The calculations assume a pure crystal field model (no covalent bonding) and the same bond lengths and crystal field parameter, Dq, as in the ground state of the metal ion complex, with normal bond angles for the various geometries of the intermediates. The crystal field stabilization energy, CFSE, was calculated for the ground state and the intermediate, and the difference between these was defined as the *crystal field activation energy*, CFAE. The differences in reactivity were assigned to the difference in this electronic factor for various numbers of d electrons.

The energies of the d orbitals in units of Dq are given in Table 3.14. Based on these orbital energies, one can calculate the CFAE for each of the possible transition states and then predict the order of reactivity for the metal ions and the transition state that is most favorable (the one with the lower CFAE). Such calculations are shown for square pyramid and octahedral wedge intermediates in Table 3.15.

If one disregards differences in Dq for different metal ions, these calculations predict the order of reactivity for a square pyramid transition state as (d^4, d^9) > (d^2, d^7) > (d^1, d^6) > (d^0, d^5, d^{10}) > (d^3, d^8).

Table 3.14. Energies of d Orbitals in Units of Dq

Structure	$d_{x^2-y^2}$	d_{z^2}	d_{xy}	d_{xz}	d_{yz}
Octahedron	6.00	6.00	−4.00	−4.00	−4.00
Trigonal bipyramid (T.P.)	−0.82	7.07	−0.82	−2.72	−2.72
Square pyramid (S.P.)	9.14	0.86	−0.86	−4.57	−4.57
Pentagonal bipyramid (P.P.)	2.82	4.93	2.82	−5.28	−5.28
Octahedral wedge (O.W.)	8.79	1.39	−1.51	−2.60	−6.08
Square planar	12.28	−4.28	2.28	−5.14	−5.14

Table 3.15. Crystal Field Activation Energies for Two Transition States with High-Spin Configurations

| No. d electrons | CFSE | | | CFAE | |
	Octahedron	S. P.	O. W.	S. P.	O. W.
0, 10	0.0	0.0	0.0	0.0	0.0
1, 6	4.00	4.57	6.08	−0.57	−2.08
2, 7	8.00	9.14	8.68	−1.14	−0.68
3, 8	12.00	10.00	10.20	2.00	1.80
4, 9	6.00	9.14	8.79	−3.14	−2.79
5, 10	0.0	0.0	0.0	0.0	0.0

However, an octahedral wedge gives a more favorable transition state for the (d^1, d^6) and (d^3, d^8) configurations and a pentagonal bipyramid is predicted to be the best for the (d^2, d^7) configurations with CFAE = −2.56 Dq.

Table 3.16 summarizes the predictions and results for the 2+ metal ions of the first transition series. The theory correctly predicts that nickel(II) and vanadium(II) should have the smallest exchange rates and that chromium(II) and copper(II) should have the largest. The next most labile are predicted to be cobalt(II) and iron(II), but they are actually less labile than manganese(II) and probably zinc(II).

The predictions of absolute and relative activation energies are less successful. For example, $Ni(OH_2)_6{}^{2+}$ has Dq = 850 cm^{-1} = 10 kJ and $V(OH_2)_6{}^{2+}$ has Dq = 1240 cm^{-1} = 14.8 kJ, so that their ΔH^* values are calculated to be 18 and 26.6 kJ mol^{-1}, respectively. The differences

Table 3.16. Water Solvent Exchange Rates (25°C), Activation Parameters, and Predicted Crystal Field Activation Energies for First-Row Transition-Metal Ions

		CFAE	k (s^{-1})	ΔH^* (kJ mol^{-1})	ΔS^* (J mol^{-1} K^{-1})	ΔV^* (cm^3 mol^{-1})
$V(OH_2)_6{}^{2+}$	d^3	1.80 (O.W.)	87	61.8	−0.4	−4.1
$Cr(OH_2)_6{}^{2+}$	d^4	−3.14 (S.P.)	>10^8			
$Mn(OH_2)_6{}^{2+}$	d^5	0.0	2.1 x 10^7	32.9	5.7	−5.4
$Fe(OH_2)_6{}^{2+}$	d^6	−2.08 (O.W.)	4.4 x 10^6	41.4	21.2	3.8
$Co(OH_2)_6{}^{2+}$	d^7	−2.56 (P.P.)	3.2 x 10^6	46.9	37.2	6.1
$Ni(OH_2)_6{}^{2+}$	d^8	1.80 (O.W.)	3.2 x 10^4	56.9	32.0	7.2
$Cu(OH_2)_6{}^{2+}$	d^9	−3.14 (S.P.)	>10^7			
$Zn(OH_2)_6{}^{2+}$	d^{10}	0.0	>10^7			

from the experimental numbers could be ascribed to solvation effects. The predicted difference of 8.6 kJ mol^{-1} in the ΔH^* values is larger than the observed value of 5.1 kJ mol^{-1}, but one must allow for an error of $\pm$ 2 kJ mol^{-1} in the experimental values. The larger ΔH^* for the 3+ ions would be attributed to differences in Dq. For chromium(III), the Dq = 2000 cm^{-1} = 23.7 kJ, so that the calculated ΔH^* = 42.7 kJ mol^{-1}, and the difference between this value and that for nickel(II) is predicted to be 24.7 kJ mol^{-1} compared to the observed value of 53.1 kJ mol^{-1}. Again, solvation differences with different charge types may explain this discrepancy.

The general impression is that the crystal field approach has pointed out a significant feature for the understanding of the reactivity differences. However, this is not the whole story, and it is probably too crude an approximation in the form used by Basolo and Pearson to be capable of assigning small differences or preferred geometries for transition states.

There have been attempts to refine the theory. Breitschwerdt[132] allowed the effective charge on the metal ion to change with the number of ligands, so that Dq varied between the ground and transition states, and Dq also decreased by 40 percent for ligands in the plane of the pentagonal bipyramid compared to the axial ligands. It was found that a square pyramid was the most stable transition state for all the 2+ ions and that all the CFAEs are positive, as shown in Table 3.17.

Spees et al.[133] presented a more extensively parameterized ligand field model and included the possibility of a change of spin state in the "intermediate". However, this approach seems to be no more effective than earlier versions with regard to the water exchange reactions.

3.7.a.iii Other Theoretical Applications

Burdett[134] has applied the angular overlap bonding model to this problem. This is essentially an extended Hückel molecular orbital approach. The change in bonding energy between the octahedral ground state and a square pyramid transition state was calculated in

Table 3.17. Modified Crystal Field Activation Energies

No. d electrons	CFAE (S.P.)
0, 5	2.80 Dq
1, 6	2.87 Dq
2, 7	3.65 Dq
3, 8	4.80 Dq
4, 9	1.06 Dq

terms of the exchange integral, β, and the overlap integral, S. It was also argued that βS^2 would increase with atomic number across the transition series. The energy loss for d^0 to d^3 is $2\beta S^2$, for d^4 to d^8 is $1\beta S^2$, and for d^9 and d^{10} is zero. This predicts the abrupt change in reactivity that is observed between d^3 and d^4 and between d^8 and d^9, but quantitative predictions are not possible, nor were other transition states considered. The angular overlap model has been applied by Mønsted[135] to the aquation reactions of $Cr^{III}(OH_2)_5X$ and $Cr^{III}(NH_3)_5X$ complexes, with the conclusion that an I_a octahedral wedge "transition state" is preferred.

A more sophisticated quantum mechanical model has been applied by Rode et al.[136] to calculate the hydration energies of these metal ions, by including effects from two hydration shells beyond the first coordination sphere. The stabilization per water molecule in the first coordination sphere, $\Delta E(I)$, was calculated and was found to correlate reasonably well with the water exchange rates [i.e., $\Delta G^*(25°C)$], but the absolute energies are different by a factor of about seven. To account for this discrepancy, Rode et al. propose that there is synchronous movement of the leaving and entering groups between the first and second coordination spheres, so that there is bond making as well as bond breaking. This is taken into account in Eq. (3.61), where $\Delta E(II)$ is the stabilization energy of a water molecule in the second coordination sphere and κ is an adjustable parameter that varies with the degree of movement between the two solvent shells.

$$\Delta G^* = \kappa [\, \Delta E(I) + \Delta E(II) \,] - \Delta E(I) \tag{3.61}$$

This equation illustrates an inevitable feature of this type of calculation in that one is taking the difference between two large numbers of uncertain validity in order to obtain the parameter of relevance for the kinetic interpretation.

Connick and Alder[137] have applied molecular modeling to attempt to understand the nature of the exchange process in the $Ni(OH_2)_6^{2+}$ system. Merbach and co-workers[138] used Monte Carlo simulations for lanthanide ions to predict solvation numbers for these ions and found that the calculations predicted a dissociative mechanism for the 9-coordinate ions. Calculations such as these will benefit from the extensive structural information available from EXAFS studies that has been complied and reviewed by Ohtaki and Radnai.[139]

Åkesson et al.[140] have done SCF computations of the gas-phase energies for the ions $M(OH_2)_n^{3/2+}$ (n = 5, 6, 7) of the first transition metal series. They have combined these in a thermochemical cycle with hydration energies, estimated from the Born equation, in order to estimate solvent exchange activation energies for **D** and **I** mechanisms. The results for metal ions with experimentally measured activation energies are given in Table 3.18.

Table 3.18. Experimental and SCF Estimated ΔH^* for Water Exchange[a]

	ΔH^*_{obs}	$\Delta H^*(D)$[b]	$\Delta H^*(I)$[c]
$V(OH_2)_6^{2+}$	61.8	93 (S)	106
$Mn(OH_2)_6^{2+}$	32.9	62 (T)	54
$Fe(OH_2)_6^{2+}$	41.4	61 (T)	62
$Co(OH_2)_6^{2+}$	46.9	63 (T)	83
$Ni(OH_2)_6^{2+}$	56.9	77 (S)	132
$Cu(OH_2)_6^{2+}$	11.5	46 (T)	120
$Ti(OH_2)_6^{3+}$	43.4	170 (T)	34
$V(OH_2)_6^{3+}$	49.4	177 (T)	58
$Cr(OH_2)_6^{3+}$	108.6	204 (S)	164
$Fe(OH_2)_6^{3+}$	64	169 (T)	85
$Ga(OH_2)_6^{3+}$	67.1	150 (T)	155

[a] Activation enthalpies in kJ mol^{-1}.
[b] For a **D** mechanism with the most stable trigonal bipyramid (T) or square-based pyramid (S) intermediate.
[c] For an **I** mechanism with a distorted pentagonal bipyramid intermediate.

For each of the 2+ ions, the smaller calculated ΔH^* value is about 20 kJ mol^{-1} larger than the experimental values. For the 3+ ions, the differences between calculated and experimental values are irregular; the calculated value is smaller than the experimental one for Ti(III) but it is 55 kJ mol^{-1} larger for Cr(III). The relative values of $\Delta H^*(D)$ and $\Delta H^*(I)$ predict a **D** mechanism for V(II), Co(II), Ni(II), and Cu(II) and an **I** mechanism for Ti(III), V(III), Cr(III), and Fe(III), while Ga(III), Fe(II), and possibly Mn(II) are too close to warrant a prediction and might be I_d or I_a. These predictions are consistent with those based on ΔV^*, except for V(II), Mn(II), and Ga(III).

3.7.b Solvent Exchange in Nonaqueous Solvents

Although water is the solvent of most general interest, there has been a great deal of work in other solvents, such as acetonitrile, methanol, dimethylsulfoxide and *N,N*-dimethylformamide. In general, the purpose is to gain some understanding of the effect of ligand size, basicity, and so on, on the solvent exchange rate. The main complication is that both the bulk solvent and the exchanging ligand are changed at the same time, so that individual factors affecting the exchange rate are difficult to separate. Some representative results are given in Table 3.19.

The reactivity pattern observed in water generally is maintained in other solvents. Nickel(II) always shows substantially smaller exchange rate constants and higher ΔH^* values, and trends in ΔV^* with atomic number are also maintained.

Table 3.19. Nonaqueous Solvent Exchange Rate Constants (25°C), Enthalpies, Entropies, and Volumes of Activation[a]

Metal	Solvent	k (s^{-1})	ΔH^* (kJ mol^{-1})	ΔS^* (J mol^{-1} K^{-1})	ΔV^* (cm^3 mol^{-1})
Cr^{2+}	CH_3OH	1.2×10^8	31.6	16.6	
Mn^{2+}	CH_3OH	3.7×10^5	25.9	−50.2	−5.0
Fe^{2+}	CH_3OH	5.0×10^4	50.2	12.6	0.4
Co^{2+}	CH_3OH	1.8×10^4	57.7	30.1	8.9
Ni^{2+}	CH_3OH	1.0×10^3	66.1	33.5	11.4
Cu^{2+}	CH_3OH	3.1×10^7	17.2	−44.0	8.3
Mn^{2+}	CH_3CN	1.4×10^7	29.6	−8.9	−7.0
Fe^{2+}	CH_3CN	6.6×10^5	41.4	5.5	3.0
Co^{2+}	CH_3CN	3.4×10^5	49.5	27.1	9.9
Ni^{2+}	CH_3CN	3.1×10^3	60.8	25.8	7.3
Co^{2+}	NH_3	5.0×10^7	45.8	31.2	
Ni^{2+}	NH_3	7.0×10^4	57.3	40.2	5.9
Mn^{2+}	$(CH_3)_2NCHO$	2.2×10^6	34.6	−7.4	2.4
Fe^{2+}	$(CH_3)_2NCHO$	9.7×10^5	43.0	13.8	8.5
Co^{2+}	$(CH_3)_2NCHO$	3.9×10^5	56.9	52.7	6.7
Ni^{2+}	$(CH_3)_2NCHO$	3.8×10^3	62.8	33.5	9.1
Ti^{3+}	$(CH_3)_2NCHO$	6.6×10^4	23.6	−73.6	−5.7
Fe^{3+}	$(CH_3)_2NCHO$	6.1×10^1	42.3	12.1	−5.4
Cr^{3+}	$(CH_3)_2NCHO$	3.3×10^{-7}	97.1	−43.5	−6.3
V^{3+}	$(CH_3)_2SO$	1.3×10^1	38.5	−94.5	−10.1
Cr^{3+}	$(CH_3)_2SO$	3.1×10^{-8}	96.7	−64.5	−11.3
Fe^{3+}	$(CH_3)_2SO$	9.3	62.5	−16.7	−3.1

[a] Original references in reference 124.

There are some disturbing features of the activation parameters with regard to the conventional interpretations of the data in water. For example, the order of Dq values for the solvents is NH_3 > CH_3CN > DMF > H_2O ≈ CH_3OH. If crystal field effects are the determining factor for the ΔH^* values, then one should expect these to be in the same order for the various solvents. In fact, the order of ΔH^* values for nickel(II) is CH_3OH > DMF > CH_3CN ≈ NH_3 > H_2O and there seems to be no relationship to the Dq values.

The general interpretation of these results is that there are specific solvation effects operating in different solvents, and these are not taken into account by any of the simple models. However, it is not widely acknowledged that this greatly weakens all of the more simplistic rationalizations that are used to explain the results of these types of studies.

Table 3.20. Correlation of ΔH^* Values (kcal mol^{-1}) for Solvent Exchange[a]

Solvent	Dq	b_S	Ni(II)	Co(II)	Fe(II)	Mn(II)
NH$_3$	3.10	1.63	11.0 (14.8)[b]	11.2(11.9)	(10.4)	8.0 (7.3)
CH$_3$CN	3.03	1.52	15.0 (14.4)	11.4(11.5)	9.7 (10.1)	7.3 (7.0)
DMF	2.50	4.64	15.0 (15.3)	13.6(12.9)	11.7 (11.7)	8.9 (9.1)
H$_2$O	2.46	3.58	14.4 (14.1)	11.9(11.7)	7.7 (10.5)[b]	7.9 (8.0)
CH$_3$OH	2.43	5.44	15.8 (15.8)	13.8(13.5)	12.0 (12.3)	6.2 (9.8)[b]
DMSO	2.25	4.12	13.0 (13.7)	12.2(11.6)	11.3 (10.5)	7.4 (8.2)
c_M			4.26	3.31	2.82	1.80
a_M			4.26	3.16	2.51	1.88

[a] Values in parentheses are predicted from the correlation.
[b] These values were not used to obtain the correlation parameters.

It was found by Jordan and co-workers[141] that the ΔH^* values can be correlated by Eq. (3.62), which involves a crystal field parameter, a_M, that is fixed for the metal ion and a solvent parameter, b_S, that is a constant for a particular solvent.

$$\Delta H^* = a_M g_M (Dq_{Ni}) + b_S = c_M (Dq_{Ni}) + b_S \qquad (3.62)$$

The g_M is obtained from spectroscopic data that gives the proportionality between the widely available Dq values for nickel(II) and those of the other metal ions. The results of the 1978 correlation are shown in Table 3.20.

Since the original correlation, a number of experimental values have been redetermined and the best current values for these are Ni(II)–NH$_3$, 13.7; Co(II)–CH$_3$CN, 11.7; Fe(II)–CH$_3$CN, 9.9; Fe(II)–H$_2$O, 9.9. The newer values are equally or more consistent with the predictions of this correlation. It remains to be determined whether the Mn(II)–CH$_3$OH data are in error. Although this correlation seems to have some valuable predictive property, there is as yet no rationalization of the b_S values in terms of any solvent property.

3.7.c Volumes of Activation for Solvent Exchange

It has been noted that the ΔV^* values for the 2+ ions of the first transition series become increasingly positive with atomic number. Merbach[142] has suggested that this represents a trend from I_a for V(OH$_2$)$_6^{2+}$ to I_d for Ni(OH$_2$)$_6^{2+}$, and that the reason for this is the decreasing size of the metal ion. The values for the M(OH$_2$)$_5$(OH)$^{2+}$ ions are more positive than those for the corresponding M(OH$_2$)$_6^{3+}$ ions [M = Cr(III), Fe(III), and Ru(III)], suggesting that the former species have more dissociative character. Swaddle has noted that the partial molar volumes of $M(OH_2)_n{}^{z+}$ ions can be calculated from

$$V_{abs}^o = 2.523 \times 10^{-6} \left(r_M + \Delta r \right) - 18.07 \, n - \frac{417.5 \, z}{\left(r_M + \Delta r \right)} \qquad (3.63)$$

where V_{abs}^o is the absolute volume of the ion relative to the value of $V_{abs}^o \, (H^+) = -5.4 \, cm^3 \, mol^{-1}$, r_M is the ionic radius of the metal ion, in pm, Δr is the apparent radius of the coordinated water molecule determined empirically, and 18.07 is the partial molar volume of water; the last term accounts for solvent electrostriction around the charged ion. Swaddle[143] has used this equation, with appropriately adjusted values of r and n, to estimate a limiting value of $\Delta V^* \approx 13 \, cm^3 \, mol^{-1}$ for a **D** mechanism and $\Delta V^* \approx -13 \, cm^3 \, mol^{-1}$ for an **A** mechanism. This implies that $Ti(OH_2)_6{}^{3+}$ has a mechanism close to the **A** limit.

For the nonaqueous systems in Table 3.19, the ΔV^* values are almost invariant with changes in solvent for a given metal ion. If the mechanism is I_d for nickel(II), as is commonly assumed, then one might expect the ΔV^* values to parallel the size of the leaving group (solvent), based on a straightforward extension of Eq. (3.63) from water to other solvents. Yet the ΔV^* for DMF is only 2 $cm^3 \, mol^{-1}$ larger than that for water, although the partial molar volumes of the solvents are 115.4 and 18 $cm^3 \, mol^{-1}$, respectively.

3.8 LIGAND SUBSTITUTION ON LABILE TRANSITION-METAL IONS

The common complexes of the M(II) and M(III) transition-metal ions of the first transition series are labile, except for Cr(III) and low-spin Co(III), Fe(II), and Fe(III). These labile ions form a wide range of complexes of general chemical and biochemical importance. As a result, there have been many studies of the kinetics and equilibrium constants for reactions of the general form

$$M(Solvent)_6{}^{n+} + L^{z-} \rightleftharpoons M(Solvent)_5(L)^{(n-z)+} + Solvent \quad (3.64)$$

The majority of this work is in water, but there are an increasing number of studies in nonaqueous solvents. The results have been the subject of numerous reviews.[125,144–146]

3.8.a General Reactivity Trends

The rate constants for these reactions are normally of the same order of magnitude as those for solvent exchange. As a result, the reactions typically have half-times in the microsecond to second range at normal

Table 3.21. Rate Constants (25°C) for Some
Substitution Reactions on $Ni(OH_2)_6^{2+}$

Entering Group (L)	pK_a	k (M^{-1} s^{-1})
$H_3CPO_4^{2-}$	~2	2.9×10^5
F^-	~3	8.0×10^3
NH_3	9.2	4.5×10^3
NH_2CH_3	10	1.3×10^3
$NH_2(CH_2)N(CH_3)_3^+$	7	4.0×10^2

concentrations. Therefore, the experimental techniques are stopped-flow and various relaxation methods pioneered by Eigen (see Chapter 9). The lability of these systems means that one cannot do the type of competition studies that rely on product analysis to yield mechanistic information. Rather, these mechanistic studies are primarily concerned with the evaluation of entering group effects. Since the leaving group is the solvent, it cannot be systematically changed without introducing substantial changes to the solvation energies of the species involved.

The reactions of Ni(II) are slower than those of the other M(II) ions in this group, and this metal ion has been studied most extensively because the rates fall conveniently in the range of the stopped-flow technique. The substitution kinetic results on $Ni(OH_2)_6^{2+}$ have been reviewed by Wilkins.[145] Some representative results are given in Table 3.21. The first impression of these data is that there are significant entering group effects on the rate constant, so that the mechanism appears to be I_a. However, there is no correlation with the nucleophilicity of L as judged by the pK_a values of the entering ligands. An analysis of a more extensive set of this type of data led Wilkins and Eigen[146] to suggest that the apparent rate constant was strongly correlated with the charge on L. Within modest limits, the rate constants (M^{-1} s^{-1}) are ~3 x 10^5 for L^{2-}, ~1 x 10^4 for L^-, ~3 x 10^3 for L^0, and ~500 for L^+. These observations led Eigen and Wilkins to propose what is now the classic mechanism for this system, called the *Dissociative ion pair mechanism* or the *Eigen–Wilkins mechanism*; it is formulated in Scheme 3.6.

The rate law for this system with a rapid pre-equilibrium has been developed previously. These studies are usually carried out with total [M] >> [L], in order to prevent formation of higher complexes such as $M(OH_2)_4(L)_2$; therefore, the experimental pseudo-first-order rate constant is given by

$$k_{exp} = \frac{k_1 K_i}{1 + K_i [M(OH_2)_6^{2+}]} \qquad (3.65)$$

Scheme 3.6

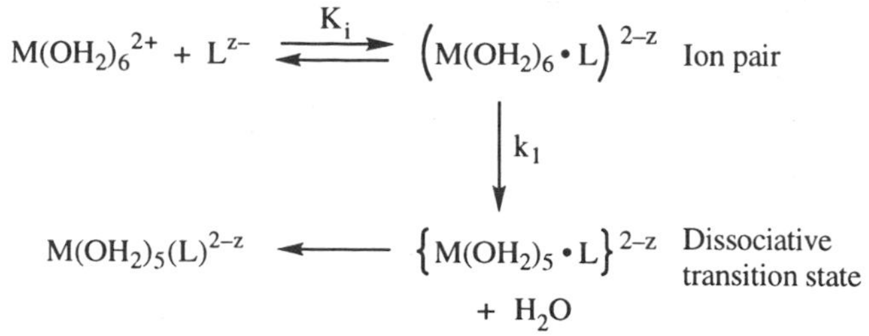

$$M(OH_2)_6{}^{2+} + L^{z-} \xrightleftharpoons{\text{K}_i} \left(M(OH_2)_6 \cdot L\right)^{2-z} \quad \text{Ion pair}$$

$$\downarrow k_1$$

$$M(OH_2)_5(L)^{2-z} \longleftarrow \left\{M(OH_2)_5 \cdot L\right\}^{2-z} \quad \begin{array}{l}\text{Dissociative} \\ \text{transition state}\end{array}$$
$$+ \; H_2O$$

The theoretical expression for K_i, discussed on p. 38, was developed in part to analyze these results. When L is neutral or uninegative, K_i is predicted to be ≤ 1, and for most studies total $[M] \approx 10^{-2}$ M, so that $K_i[M(OH_2)_6{}^{2+}] \ll 1$. Under these conditions, the expression for the experimental rate constant in Eq. (3.65) simplifies to

$$k_{exp} = k_1 K_i \tag{3.66}$$

Since K_i depends on the charge of L, this expression explains the variation of k_{exp} with the charge of the entering group. In many interpretations of experimental results, k_{exp} is divided by a calculated value of K_i in order to obtain k_1, and then k_1 is compared to the rate constant for solvent exchange, which also is assumed to be dissociatively activated. In many instances, this analysis works well. In general, since K_i is of the order of magnitude of 1, the value of k_{exp} is quite similar to that for solvent exchange.

For Mn(II), Fe(II), and Co(II), the preceding type of analysis also indicates a lack of entering group effects and the mechanism appears to be I_d. However, Merbach and co-workers have interpreted the changes in ΔV^* to indicate that there is a mechanistic trend across the first transition series from I_a for V(II) to I_d for Ni(II).

The best case for an I_a mechanism from entering group effects comes from the results of Diebler and co-workers[147] for $Ti(OH_2)_6{}^{3+}$, as given in Table 3.22. These results do not correlate simply with the charge of the entering group, but do parallel the nucleophilicity as measured by the pK_a. These reactions are still presumed to proceed through an ion pair, but the rate-controlling step has the entering group dependence expected for an I_a mechanism. Merbach has suggested that the mechanism is **A** on the basis of the ΔV^* for the water exchange.[148]

Substitution reactions on vanadium(III) have been of long-standing interest since Taube's prediction that such a d^2 system should be labile with an **A** mechanism. These studies are difficult because $V(OH_2)_6{}^{3+}$ is readily oxidized by air and perchlorate ion. The results of several studies, as summarized by Patel and Diebler,[149] are given in Table 3.23.

Table 3.22. Rate Constants (15°C) for Substitution Reactions on $Ti(OH_2)_6^{3+}$

Entering Group	pK_a	k_{exp} (M^{-1} s^{-1})
$ClCH_2CO_2H$		6.7×10^2
CH_3CO_2H		9.7×10^2
NCS^- [a]	-1.84	8.0×10^3
$HO_2CCO_2^-$	1.23	3.9×10^5
$Cl_2CHCO_2^-$	1.25	1.1×10^5
$HO_2CCH_2CO_2^-$	2.43	4.2×10^5
$ClCH_2CO_2^-$	2.46	2.1×10^5
$HO_2CCH(CH_3)CO_2^-$	2.62	3.2×10^5
$CH_3CO_2^-$	4.47	1.8×10^6

[a] At 8–9°C, Diebler, H. *Z. Phys. Chem.* **1969**, *68*, 64.

These rate constants vary widely with the entering group and do not show the correlation with charge expected for a dissociative ion pair mechanism. More recently, Merbach and co-workers have studied the reaction of V(III) with NCS^- in water[150] and DMSO.[151] For the formation reaction in water, they obtained $\Delta H^* = 49.1$ kJ mol^{-1}, $\Delta S^* = -39.8$ J mol^{-1} K^{-1}, and $\Delta V^* = -9.4$ cm^3 mol^{-1}. In DMSO, the corresponding values are 44.6, –54.1, and –1.1, respectively. The rate constant trends and the negative ΔS^* and ΔV^* for water exchange and complexation are taken as evidence for an I_a mechanism, confirming Taube's prediction.

In the case of $Mn(OH_2)_6^{2+}$, the complexation kinetics appear more consistent with dissociative activation as seen with $Ni(OH_2)_6^{2+}$. However, the ΔV^* values are negative and have led to the assignment of an I_a mechanism.[152] It should be noted that pre-association or ion pairing causes complications in the interpretation of ΔV^* for complexation.

Table 3.23. Rate Constants (25°C) for Substitution Reactions on $V(OH_2)_6^{3+}$

Entering Group	k (M^{-1} s^{-1})
$HC_2O_4^-$	1.3×10^3
SCN^-	1.1×10^2
Br^-	≤ 10
Cl^-	≤ 3
HN_3	0.4
H_2O (in s^{-1})	5.0×10^2

3.8.b Substitution on Iron(III) and the Proton Ambiguity

Despite its common occurrence and importance, $Fe(OH_2)_6^{3+}$ was the last of the air-stable first-row transition-metal ions to have its water exchange rate determined. The difficulty is the tendency of $Fe(OH_2)_6^{3+}$ to hydrolyze and polymerize in dilute aqueous acid, forming $Fe(OH_2)_5(OH)^{2+}$ and $(H_2O)_4Fe(OH)_2Fe(OH_2)_4^{4+}$, respectively, as the major species. There have been numerous kinetic studies of substitution on $Fe(OH_2)_6^{3+}$, but the results were difficult to interpret because of the lack of a water exchange rate until 1981 and because of the proton ambiguity discussed later.

For the typical system in which the entering group has an ionizable proton, and for iron(III) concentrations sufficiently low to avoid the hydrolyzed dimer, the rapid equilibria involving the reactants are represented by

$$(H_2O)_5Fe(OH_2)^{3+} \underset{}{\overset{K_m}{\rightleftharpoons}} (H_2O)_5Fe(OH)^{2+} + H^+$$

$$HL \underset{}{\overset{K_a}{\rightleftharpoons}} L^- + H^+ \tag{3.67}$$

where $K_m = 1.6 \times 10^{-3}$ M (25°C). In the following development, the charges of the iron and ligand species have been omitted. The total iron(III) concentration in the reactants is given by

$$[Fe]_{tot} = [FeOH_2] + [FeOH] \tag{3.68}$$

From reaction (3.67) and Eq. (3.68), the following equations for the concentrations of the iron(III) species in terms of $[Fe]_{tot}$ can easily be developed:

$$[FeOH_2] = \frac{[H^+]}{K_m + [H^+]} [Fe]_{tot} \quad \text{and} \quad [FeOH] = \frac{K_m}{K_m + [H^+]} [Fe]_{tot} \tag{3.69}$$

Similarly, since the total entering group concentration is given by

$$[L]_{tot} = [HL] + [L] \tag{3.70}$$

the concentrations of the ligand species in terms of $[L]_{tot}$ are

$$[HL] = \frac{[H^+]}{K_a + [H^+]} [L]_{tot} \quad \text{and} \quad [L] = \frac{K_a}{K_a + [H^+]} [L]_{tot} \tag{3.71}$$

The possible substitution reactions in this system are given by

$$Fe(OH)^{2+} + L \xrightarrow{\ k_1\ }$$

$$Fe(OH)^{2+} + HL \xrightarrow{\ k_2\ }$$

$$Fe(OH_2)^{3+} + L \xrightarrow{\ k_3\ } \tag{3.72}$$

$$Fe(OH_2)^{3+} + HL \xrightarrow{\ k_4\ }$$

The rate of formation of product, P, is given by Eq. (3.73), so that substitution from Eqs. (3.68–3.71) into Eq. (3.73) leads to Eq. (3.74):

$$\frac{d[P]}{dt} = (k_1[L] + k_2[HL])[FeOH] + (k_3[L] + k_4[HL])[FeOH_2] \tag{3.73}$$

$$= \left(\frac{k_1 K_m K_a + (k_2 K_m + k_3 K_a)[H^+] + k_4[H^+]^2}{(K_m + [H^+])(K_a + [H^+])} \right)[Fe]_{tot}[L]_{tot} \tag{3.74}$$

which can be simplified to

$$\frac{d[P]}{dt} = k_{exp}[Fe]_{tot}[L]_{tot} \tag{3.75}$$

In Eq. (3.75) it has been assumed that the experimental conditions are such that $[H^+]$ is constant for a particular kinetic run. The overall experimental study involves determining k_{exp} at different $[H^+]$ and fitting this information to the $[H^+]$ dependence of k_{exp} predicted by Eq. (3.74).

For most such studies, $[H^+] \gg K_m$ (= 1.6×10^{-3} M), so that two limiting conditions for k_{exp} can be identified depending on the strength of the acid HL.

1. *Strong acid ligands*: $K_a \gg [H^+]$ and

$$k_{exp} = \frac{k_1 K_m}{[H^+]} + \frac{k_2 K_m}{K_a} + k_3 + \frac{k_4[H^+]}{K_a} \tag{3.76}$$

2. *Weak acid ligands*: $[H^+] \gg K_a$ and

$$k_{exp} = \frac{k_1 K_m K_a}{[H^+]^2} + \frac{k_2 K_m}{[H^+]} + \frac{k_3 K_a}{[H^+]} + k_4 \tag{3.77}$$

The *proton ambiguity* refers to the fact that the $[H^+]$ dependence of k_{exp} does not allow one to separate $k_2 K_m/K_a$ from k_3 in the first case, or $k_2 K_m$ from $k_3 K_a$ in the second case. This results because both transition states, Fe(OH) • HL or Fe(OH$_2$) • L, contain one ionizable proton that is in different sites. *The kinetics can only give the composition of the transition state, not its structure.*

The experimental pseudo-first-order rate constant for almost all substitution reactions on aqueous iron(III) has the general form of

$$k_{exp} = k' + k'' \, [H^+]^{-1} \qquad (3.78)$$

The specific assignment of k' and k" can be made by comparison to the theoretical expressions for the appropriate case, except for the proton ambiguity problem.

1. *Strong acid ligands*: ($K_a \gg [H^+]$). A comparison of Eqs. (3.76) and (3.78) shows that $k'' = k_1 K_m$, and k_1 can be calculated since K_m is known. However, $k' = k_3 + k_2 K_m / K_a$ and cannot be uniquely assigned except for very strong acid ligands, such as Cl^- and Br^-. For such ions, the large K_a ($\approx 10^8$ M) makes the second term negligible even if k_2 has a diffusion-limiting value of ~10^{10} M^{-1} s^{-1}.

2. *Weak acid ligands*: ($[H^+] \gg K_a$). From Eqs. (3.77) and (3.78), $k' = k_4$ and $k'' = k_2 K_m + k_3 K_a$. If K_a is very small (e.g., $<10^{-10}$ M), it may be possible to eliminate the $k_3 K_a$ term again on the grounds that the k_3 would need to be beyond the diffusion-controlled limit.

Table 3.24. Rate Constants (25°C)[a] for Substitution Reactions on Aqueous Iron(III)

Entering Group	k (M^{-1} s^{-1})	
	$(H_2O)_5Fe(OH)^{2+}$	$Fe(OH_2)_6^{3+}$
SO_4^{2-}	1.1×10^5	2.3×10^3
Cl^-	5.5×10^3	4.8
Br^-	2.8×10^3	1.6
NCS^-	5.1×10^3	9.0×10^1
$Cl_3CCO_2^-$	7.8×10^3	6.3×10^1
$Cl_2HCCO_2^-$	1.9×10^4	1.2×10^2
$ClH_2CCO_2^-$	4.1×10^4	1.5×10^3
H_3CCO_2H	$\leq 2.8 \times 10^3$	2.7×10^1
C_6H_5OH	1.5×10^3	
$C_{10}H_{12}(=O)(OH)$[b]	6.3×10^3	2.2×10^1
$H_3CC(O)NH(OH)$	2.0×10^3	1.2
H_2O (in s^{-1})	1.2×10^5	1.6×10^2

[a] Data from Grant, M.; Jordan, R. B. *Inorg. Chem.* **1981**, *20*, 55, where original references are given.

[b] 4-isopropyltropolone, Ishihara, K.; Funahashi, S.; Tanaka, M. *Inorg. Chem.* **1983**, *22*, 194.

The usual strategy in this area has been to calculate both k_2 and k_3 in ambiguous cases, and then compare k_1 and k_2, assuming that they should be similar for a dissociative mechanism, and likewise for k_4 and k_3. Until the water exchange rates were determined, this strategy led to rather ambiguous mechanistic conclusions.[153,154] The current interpretation is that $Fe(OH_2)_6^{3+}$ has an I_a mechanism [like Ti(III) and V(III)], whereas $(H_2O)_5Fe(OH)^{2+}$ has an I_d mechanism. Some of the kinetic data leading to this conclusion are given in Table 3.24. In the case of 4-isopropyltropolone, Hipt, the assignment to k_3 was made because the alternative assignment to k_2 ($Fe(OH)^{2+} + H_2ipt^+$) gives a larger value for k_2 than k_1 ($Fe(OH)^{2+} + Hipt$) and is inconsistent with expected charge effects.

3.9 KINETICS OF CHELATE FORMATION

Complexes containing bidentate, tridentate, and so on ligands are very common with metal ions. They are especially important because of their apparently enhanced stability, often referred to as the "chelate effect". Such ligands have two or more potential donor atoms separated by two, three, or four other atoms in the molecule. Some common examples are oxalate, glycine, salicylate, ethylenediamine, acetylacetonate, 2,2'-bipyridyl, and ethylenediaminetetraacetic acid.

3.9.a Rate-Controlling Step

The formation of a bidentate chelate from the solvated metal ion involves the displacement of two solvent molecules. The process is unlikely to involve concerted replacement of two solvent ligands, and is rather considered to be a stepwise process, as shown in Scheme 3.7.

Scheme 3.7

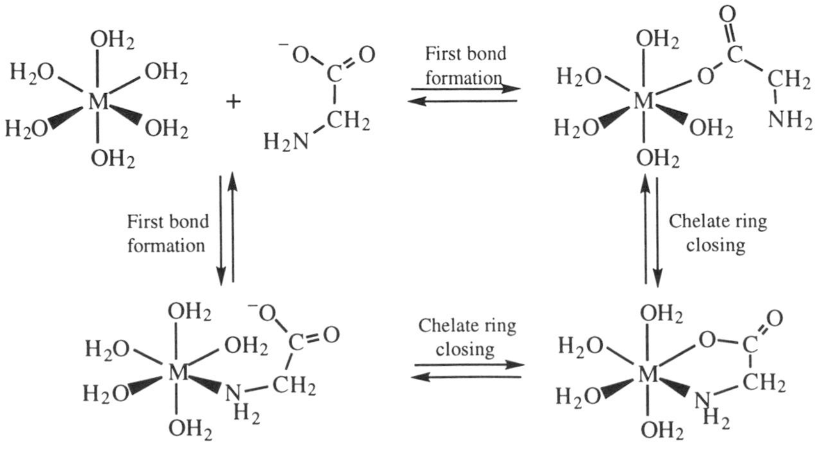

The glycine example in Scheme 3.7 illustrates two mechanistic complications that can occur in chelate formation reactions. First of all, there are two potential rate-controlling steps, either *first bond formation* or *chelate ring closing*. Second, for unsymmetrical chelates such as glycine, there is an ambiguity as to which donor group may coordinate first. The latter is not a complication for symmetrical chelates such as oxalate or ethylenediamine.

Prior to about 1970, it was commonly assumed that the chelate ring-closing step was very fast relative to the first bond formation. The rationale for this assumption was that the free end of the chelate in the monodentate intermediate would have a very high effective concentration in the neighborhood of the metal ion so that ring closing would be fast. However, there is increasing evidence that substitution reactions on the M(II) first-row transition-metal ions have considerable dissociative character, and the dissociative ion pair mechanism has achieved wider acceptance. If the mechanism is dissociative, then the rate-limiting feature is the breaking of the M—OH$_2$ bond, and the effective concentration of the entering group is no greater than it would be in the ion pair. The idea that first bond formation is always rate limiting still persists in some quarters, perhaps because the experimental rate constants for chelate formation are often quite similar to those for monodentate complex formation with a given metal ion.

A proper consideration of the kinetic situation requires an analysis of the theoretical rate law for such a system. The following analysis is somewhat simplified in that a symmetrical bidentate chelate is assumed, and the ring closing is not reversible. A complete version has been given by Letter and Jordan.[155] These reactions are usually done in dilute acid (pH 5–7), so that protonation of the entering group has been included in Scheme 3.8, but charges and coordinated solvent have been omitted.

Scheme 3.8

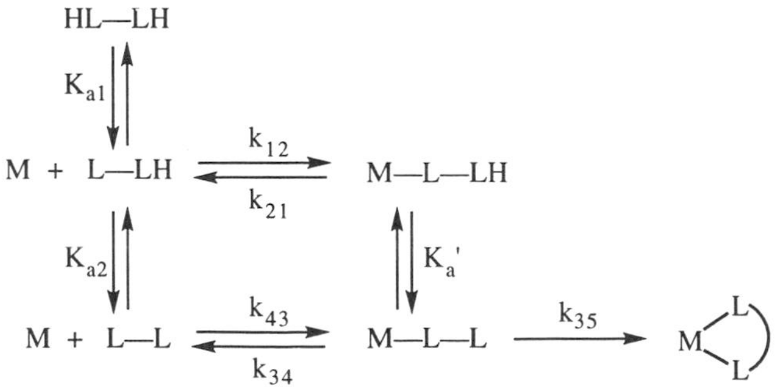

If the monodentate intermediates (M—L—LH and M—L—L) are assumed to be in a steady state, and the usual pseudo-first-order conditions of [M] $\gg$ [L]$_{tot}$ and constant [H$^+$] are assumed, then the pseudo-first-order rate constant is given by Eq. (3.79) for $K_{a1} \gg [H^+]$.

$$k_{exp} = \frac{\left(\dfrac{k_{12}[H^+] + k_{43}K_{a2}}{K_{a2} + [H^+]}\right)k_{35}K_a'}{k_{21}[H^+] + K_a'(k_{34} + k_{35})} [M] \qquad (3.79)$$

Rearrangement of this equation gives Eq. (3.80), which is of the form often used to analyze data by plotting the left-hand side versus $[H^+]^{-1}$.

$$\frac{k_{exp}}{[M]}\left(\frac{K_{a2} + [H^+]}{[H^+]}\right) = \frac{(k_{12} + k_{43}K_{a2}[H^+]^{-1})k_{35}K_a'}{k_{21}[H^+] + K_a'(k_{34} + k_{35})} \qquad (3.80)$$

In a typical study, such as that of Wilkins and co-workers,[156] this plot is linear. The traditional explanation for this is that $K_a'k_{35} \gg k_{21}[H^+]$ and $k_{35} \gg k_{34}$, so that Eq. (3.80) simplifies to Eq. (3.81), which will give k_{12} as the intercept and $k_{43}K_{a2}$ as the slope.

$$\frac{k_{exp}}{[M]}\left(\frac{K_{a2} + [H^+]}{[H^+]}\right) = k_{12} + k_{43}K_{a2}[H^+]^{-1} \qquad (3.81)$$

This rate law is consistent with the observations, but there is at least one notable problem. For amino acids such as glycine, $HO_2CCH_2NH_3^+$, it is always found that $k_{12} = 0$ (i.e., there is no significant intercept for the plot). This implies that the zwitterion, $^-O_2CCH_2NH_3^+$, is unreactive. This lack of reactivity always seemed remarkable but was attributed to intramolecular hydrogen bonding between the amino and carboxylate groups. However, monoprotonated ethylenediamine does react, although the charge is less favorable and there is a greater likelihood of hydrogen bonding.

These and other problems caused Letter and Jordan to reconsider the assumptions used to reduce the complete rate law. Is it reasonable that $K_a'k_{35} \gg k_{21}[H^+]$? If one uses the nickel(II)–glycine system for analysis, then k_{21} can be estimated from the corresponding Ni(II)–acetate system to get $k_{21} \approx 10^4$ s^{-1} and $((H_3N)_5Co-O_2CCH_2NH_3^+)^{3+}$ can be used to estimate that $K_a' \approx 3 \times 10^{-10}$ M. Then, the condition that $K_a'k_{35} \gg k_{21}[H^+]$ at pH 6 requires that $k_{35} \geq 10^8$ s^{-1}. However, water exchange on nickel(II) has $k = 3 \times 10^4$ s^{-1}, and all the available evidence indicates an $\mathbf{I_d}$ mechanism for substitution on Ni(II). Therefore, if k_{35} is $\geq 10^8$ s^{-1}, then the water ligands in the monodentate intermediate must be labilized by about a factor of 10^4, an effect that is without precedent.

On the other hand, if one assumes that $k_{21}[H^+] \gg K_a'k_{35}$, then the denominator in the complete rate law can be rearranged using the principle of detailed balancing, which requires that $K_a'(k_{12}/k_{21}) = K_{a2}(k_{43}/k_{34})$, to give

$$k_{21}[H^+] + (k_{34} + k_{35})K_a' = \frac{k_{21}[H^+]}{k_{12}}\left(k_{12} + \frac{k_{43}K_{a2}k_{35}}{k_{34}[H^+]}\right) \quad (3.82)$$

A comparison of the right- and left-hand sides of Eq. (3.82), combined with the assumed inequality, requires that $k_{12} \gg k_{43}K_{a2}k_{35}/k_{34}[H^+]$. Then, if $k_{35} \gg k_{34}$, as assumed before, $k_{12} \gg k_{43}K_{a2}/[H^+]$ and the rearranged version of Eq. (3.79) simplifies to

$$\frac{k_{exp}}{[M]}\left(\frac{K_{a2} + [H^+]}{[H^+]}\right) = \frac{k_{12}k_{35}K_a'}{k_{21}[H^+]} = \frac{k_{43}k_{35}K_{a2}}{k_{34}[H^+]} \quad (3.83)$$

This equation is consistent with the experimental observations; a zero intercept is required but does not imply that k_{12} is zero. This rate law corresponds to the condition of a rapid pre-equilibrium formation of the monodentate species, k_{43}/k_{34}, followed by *rate-controlling chelate ring closure*. The values originally associated with k_{43} are actually $k_{43}k_{35}/k_{34}$. Since these are numerically similar to rate constants observed for monodentate systems, it would appear that the ratio k_{35}/k_{34} is of the order of magnitude of 1.

Unusually large experimental rate constants ($\sim 10^7 M^{-1} s^{-1}$), assigned originally to k_{43} but actually $k_{43}k_{35}/k_{34}$, have been observed for reactions of polyamines, such as ethylenediamine. To explain this, an internal conjugate base, ICB, mechanism was proposed[157] in which the free end of the amine forms a hydrogen bond to a coordinated water, giving it some hydroxide character, as shown in reaction (3.84). This is supposed to make the remaining waters more labile to substitution according to the ICB proposal. Strangely, this rate enhancement is not observed with amino acids or with 2-methylaminopyridine.

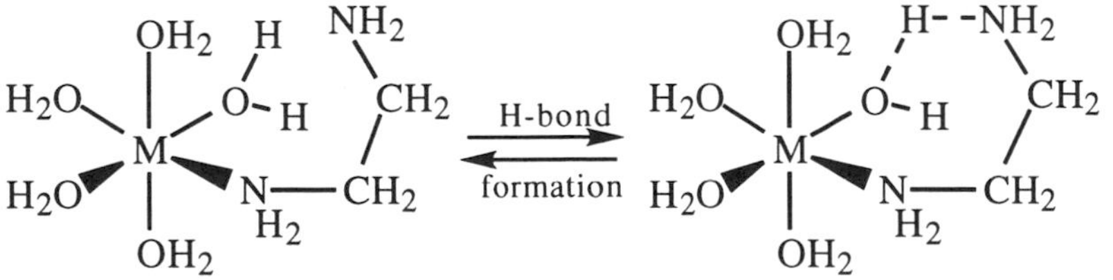

$$(3.84)$$

A reanalysis of these observations for the reaction of ethylenediamine with nickel(II)[158] has shown that $k_{43}k_{35}/k_{34} = 7.2 \times 10^6 M^{-1} s^{-1}$. One can estimate from the kinetics for the ethylamine reaction that $k_{43} \approx 900$ $M^{-1} s^{-1}$ and $k_{34} \approx 15 s^{-1}$, so that $k_{35} \approx 1.2 \times 10^5 s^{-1}$. This value is about five times larger than the rate constant for water exchange on $Ni(OH_2)_6^{2+}$, but Hunt and co-workers[159] found that the water exchange rate constant on $Ni(OH_2)_5(NH_3)^{2+}$ is 2.5×10^5 s^{-1}. It is reasonable to expect that the water exchange on the monodentate ethylenediamine complex is similar to that of the NH_3 complex. Therefore, there is no anomalous reactivity for ethylenediamine relative to monodentate models.

The conclusion is that the high reactivity of the polyamines is due to a nonreacting ligand effect in which amines greatly increase the rate of water release from Ni(II). The work of Hunt et al. also established that a coordinated pyridine or carboxylate group does not have a strong labilizing influence on Ni(II). This explains the normal reactivity of the amino acids and pyridine amines.

Moore and co-workers[160] used stopped-flow NMR to measure the rate of first bond formation and chelate ring closing in the reactions of $Al(DMSO)_6^{3+}$ with 2,2'-bipyridine and 2,2':6',2"-terpyridine systems in nitromethane. They found that the rate constant for first bond formation ($\sim 2 \times 10^3$ M^{-1} s^{-1} at 25°C) is much larger than the DMSO exchange rate (5.3×10^{-2} s^{-1} at 25°C), presumably because of strong preassociation of the reactants. The chelate ring closing rate constants ($\sim 10^{-2}$ s^{-1}) are smaller than the solvent exchange rate constant, possibly because of steric hindrance to twisting the second pyridine ring into a conformation suitable for chelation. In the same study, the reaction with 1,10-phenanthroline proceeded in a one-step process to the chelate. This may be because the fused ring of phenanthroline forces the two donor nitrogen atoms to be oriented for chelate formation.

3.9.b Kinetics and the Chelate Effect

It has been observed many times that the formation constant for a chelate such as $Ni(OH_2)_4(en)^{2+}$ ($K_f = 5 \times 10^7$ M^{-1}) is substantially greater than that for the monodentate analogue such as $Ni(OH_2)_4(NH_3)_2^{2+}$ ($K_f = \beta_2 = 1.5 \times 10^5$ M^{-2}). There have been several approaches to explaining this effect, such as differences in entropy change due to the different number of particles involved and/or a very large forward rate constant for ring closing because of local concentration effects.

It is instructive to consider the chelate formation process from a kinetic analysis of the individual steps, given by

$$M + L{-}L \underset{k_{34}}{\overset{k_{43}}{\rightleftharpoons}} M{-}L{-}L \underset{k_{53}}{\overset{k_{35}}{\rightleftharpoons}} M\overset{L}{\underset{L}{\diagdown\diagup}} \qquad (3.85)$$

For the ethylenediamine system at 25°C, the analysis in the preceding section gave $k_{43} \approx 900$ M^{-1} s^{-1}, $k_{34} \approx 15$ s^{-1}, and $k_{35} \approx 1.2 \times 10^5$ s^{-1}

and by direct measurement of the dissociation kinetics, $k_{53} \approx 0.14$ s^{-1}. The small value of k_{53} for chelate ring opening is unexpected because, from a mechanistic standpoint, one would expect k_{53} to increase relative to k_{34} roughly in proportion to k_{35}/k_{43}. But k_{53} is much smaller than k_{34}. One reaches a similar conclusion for glycine, for which the rate constants (M^{-1} s^{-1} or s^{-1} at 25°C) are given in Scheme 3.9.

Scheme 3.9

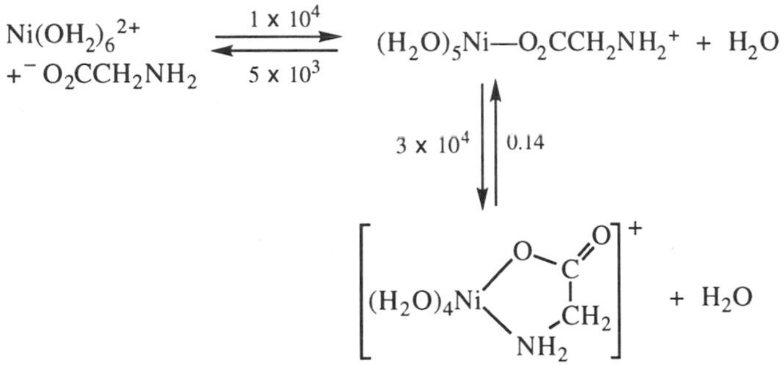

It has been noted that the magnitude of the chelate effect decreases as the chelate ring size increases. For example, the formation constants[161] for the aquanickel(II) complexes of ethylenediamine and 1,3-diaminopropane are 2.5 x 10^7 and 2 x 10^6 M^{-1}, respectively. The six-membered chelate ring complex has about a 10 times smaller formation constant, despite the fact that trimethylenediamine is a stronger base toward the proton. This effect has traditionally been assigned to a smaller chelate ring-closing rate constant, k_{35}, for the longer chelate arm in the monodentate intermediate. However, the preceding analysis implies that the effect might lie in a larger rate constant for chelate ring opening, k_{53}, with the larger chelate ring. Some evidence for this is found for amino-pyridine systems,[162,163] as shown in Figure 3.5.

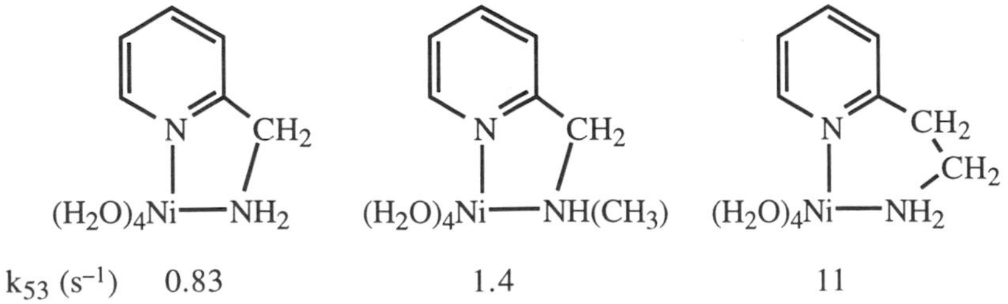

Figure 3.5. Some chelates of (H$_2$O)$_4$Ni(amino-pyridine)$^{2+}$ and their rate constants for chelate ring opening.

The rate constant for ring opening of the six-membered chelate is about 10 times larger than that for either of the five-membered rings. A methyl group on the coordinated amine has a more modest effect that could be ascribed to steric acceleration of a dissociative process.

The conclusion is that chelate ring opening, k_{53}, *is an unexpectedly slow process,* and this accounts for the high formation constants of chelate complexes, since

$$K_f = \frac{k_{43} \, k_{35}}{k_{34} \, k_{53}} \tag{3.86}$$

Previous explanations have assumed that k_{35} is unusually large. It is somewhat ironic that k_{53} has received very little attention, since kinetic studies in this area have concentrated on entering group effects. Further examples with similar conclusions are provided in a series of papers by Funahashi and co-workers on the exchange of chelating diamines.[164]

References

1.   Langford, C. H.; Gray, H. B. *Ligand Substitution Processes*; Benjamin, Inc.: New York, 1966.
2.   Eigen, M. *Z. Electrochem.* **1960**, *64*, 115.
3.   Fuoss, R. M. *J. Am. Chem. Soc.* **1958**, *80*, 5059.
4.   Haim, A.; Wilmarth, W. K. *Inorg. Chem.* **1962**, *1*, 573, 583.
5.   Haim, A.; Grassi, R. J.; Wilmarth, W. K. *Adv. Chem. Ser.* **1965**, *49*, 31.
6.   Burnett, M. G.; Gilfillian, W. M. *J. Chem. Soc., Dalton Trans.* **1981**, 1578.
7.   Haim, A. *Inorg. Chem.* **1982**, *21*, 2887.
8.   Abou-El-Wafa, M. H. M.; Burnett, M. G.; McCullagh, J. F. *J. Chem. Soc., Dalton Trans.* **1987**, 2311, and references therein.
9.   Bradley, S. M.; Doine, H.; Krouse, H. R.; Sisley, M. J.; Swaddle, T. W. *Aust. J. Chem.* **1988**, *41*, 1323.
10. Abou-El-Wafa, M. H. M.; Burnett, M. G.; McCullagh, J. F. *J. Chem. Soc., Dalton Trans.* **1986**, 2083.
11. Robb, D.; Steyn, M. M. De V.; Kruger, H. *Inorg. Chim. Acta* **1969**, *3*, 383.
12. Stranks, D. R.; Yandell, J. K. *Inorg. Chem.* **1970**, *9*, 751.
13. O'Brien, P.; Sweigart, D. A. *Inorg. Chem.* **1982**, *21*, 2094.
14. Seibles, L.; Deutsch, E. D. *Inorg. Chem.* **1977**, *16*, 2273.
15. Inamo, M.; Sumi, T.; Nakagawa, N.; Funahashi, S.; Tanaka, M. *Inorg. Chem.* **1989**, *28*, 2688; Inamo, M.; Sugiura, S.; Fukuyama, H.; Funahashi, S. *Bull. Chem. Soc. Jpn.* **1994**, *67*, 1848.
16. Inamo, M.; Hoshino, M.; Nakajima, K.; Aizawa, S.; Funahashi, S. *Bull. Chem. Soc. Jpn.* **1995**, *68*, 2293.
17. Sargeson, A. M.; Jordan, R. B. *Inorg. Chem.* **1965**, *4*, 431.
18. Buckingham, D. A.; Clark, C. R.; Lewis, T. W. *Inorg. Chem.* **1979**, *18*, 1985.

19. Jackson, W. G.; McGregor, B. C.; Jurisson, S. S. *Inorg. Chem.* **1987**, *26*, 1286; Jackson, W. G.; Hookey, C. N. *Inorg. Chem.* **1984**, *23*, 668, 2728.
20. Brasch, N. E.; Buckingham, D. A.; Clark, C. R.; Finnie, K. S. *Inorg. Chem.* **1989**, *28*, 4567.
21. Basolo, F.; Pearson, R. G. *Mechanisms of Inorganic Reactions*, 2nd ed.; Wiley & Sons: New York, 1967.
22. Jackson, W. G.; Dutton, B. H. *Inorg. Chem.* **1989**, *28*, 525.
23. Jackson, W. G.; McGregor, B. C.; Jurisson, S. S. *Inorg. Chem.* **1990**, *29*, 4677.
24. Brasch, N. E.; Buckingham, D. A.; Clark, C. R.; Simpson, J. *Inorg. Chem.* **1996**, *35*, 7728; Buckingham, D. A.; Clark, C. R.; Liddell, G. F. *Inorg. Chem.* **1992**, *31*, 2909.
25. House, D. A.; Powell, H. K. J. *Inorg. Chem.* **1971**, *10*, 1583.
26. Lawrance, G. A. *Inorg. Chem.* **1982**, *21*, 3687.
27. Curtis, N. J.; Lawrance, G. A.; van Eldik, R. *Inorg. Chem.* **1989**, *28*, 329.
28. Jordan, R. B. *Inorg. Chem.* **1996**, *35*, 3725.
29. Cattalini, L.; Ugo, R.; Orio, A. *J. Am. Chem. Soc.* **1968**, *90*, 4800.
30. Toma, H.; Malin, J. M. *J. Am. Chem. Soc.* **1972**, *94*, 4039.
31. Maresca, L.; Natile, G.; Calligaris, M.; Delise, P.; Randaccio, L. *J. Chem. Soc., Dalton Trans.* **1976**, 2386.
32. Canovese, L.; Cattalini, L.; Uguagliati, P.; Tobe, M. L. *J. Chem. Soc., Dalton Trans.* **1990**, 867.
33. Edwards, J. O. *J. Am. Chem. Soc.* **1954**, *76*, 1540.
34. Schwarzenbach, G.; Shellenberg, M. *Helv. Chim. Acta* **1965**, *48*, 28.
35. Pearson, R. G.; Sobel, H.; Songstad, J. *J. Am. Chem. Soc.* **1968**, *90*, 319.
36. Mayer, U.; Gutmann, V. *Adv. Inorg. Chem. Radiochem.* **1975**, *17*, 189; Gutmann, V. *Electrochim. Acta* **1976**, *21*, 661.
37. Maria, P.-C.; Gal, J.-F. *J. Phys. Chem.* **1985**, *89*, 1296.
38. Linert, W.; Jameson, R. F.; Taha, A. *J. Chem. Soc., Dalton Trans.* **1993**, 3181.
39. Drago, R. S. *Coord. Chem. Rev.* **1980**, *33*, 251; Drago, R. S.; Wong, N.; Bilgrien, C.; Vogel, G. C. *Inorg. Chem.* **1987**, *26*, 9.
40. Drago, R. S.; Dadmun, A. P.; Vogel, G. C. *Inorg. Chem.* **1993**, *32*, 2473.
41. Lim, Y. Y.; Drago, R. S.; Babich, M. W.; Wong, N.; Doan, P. E. *J. Am. Chem. Soc.* **1987**, *109*, 169.
42. Hancock, R. D.; Martell, A. E. *J. Chem. Educ.* **1996**, *73*, 654; Hancock, R. D.; Martell, A. E. *Adv. Inorg. Chem.* **1995**, *42*, 89.
43. Abraham, M. H.; Doherty, R. M.; Kamlet, M. J.; Taft, R. W. *Chem. Brit.* **1986**, *22*, 551; Abraham, M. H.; Grellier, P. L.; Prior, D. V.; Taft, R.W.; Morris, J. J.; Taylor, P. J.; Laurence, C.; Berthelot, M.; Doherty, R. M.; Kamlet, M. J.; Aboud, J.-L. M.; Sraidi, K.; Guihéneuf, G. *J. Am. Chem. Soc.* **1988**, *110*, 8534; Marcus, Y.; Kamlet, M. J.; Taft, R. W. *J. Phys. Chem.* **1993**, *22*, 409.
44. Catalán, J.; Gómez, J.; Couto, A.; Laynez, J. *J. Am. Chem. Soc.* **1990**, *112*, 1678.
45. Valgimigli, L.; Ingold, K.U.; Lusztyk, J. *J. Am. Chem. Soc.* **1996**, *118*, 3545.

46. Reichardt, E. C. *Solvents and Solvent Effects in Organic Chemistry*, 2nd ed.; VCH Publishers: New York, 1988; Marcus, Y. *J. Solution Chem.* **1984**, *13*, 599.
47. Schut, D. M.; Keana, K. J.; Tyler, D. R.; Rieger, P. H. *J. Am. Chem. Soc.* **1995**, *117*, 8939.
48. Drago, R. S.; Hirsch, M. S.; Ferris, D. C.; Chronister, C. W. *J. Chem. Soc., Perkin Trans. 2* **1994**, 219; George, J. E.; Drago, R. S. *Inorg. Chem.* **1996**, *35*, 239.
49. Pearson, R. G. *J. Chem. Educ.* **1968**, *45*, 643.
50. Arhland, S.; Chatt, J.; Davies, N. R. *Quart. Rev. Chem. Soc.* **1958**, *12*, 265.
51. Pearson, R. G. *Inorg. Chem.* **1988**, *27*, 734; Pearson, R. G. *Coord. Chem. Rev.* **1990**, *100*, 403; Pearson, R. G. *Inorg. Chim. Acta* **1995**, *240*, 93.
52. Doan, P. E.; Drago, R. S. *J. Am. Chem. Soc.* **1984**, *106*, 2772.
53. Kamlet, M. J.; Gal, J.-F.; Maria, P.-C.; Taft, R. W. *J. Chem Soc., Perkin Trans. 2* **1985**, 1583.
54. Drago, R. S. *Inorg. Chem.* **1990**, *29*, 1379.
55. Maria, P.-C.; Gal, J.-F.; Francheschi, J.; Fargin, E. *J. Am. Chem. Soc.* **1987**, *109*, 483.
56. Linert, W. *Chem. Soc. Rev.* **1994**, *23*, 429.
57. Langford, C. H. *Inorg. Chem.* **1965**, *4*, 265.
58. Haim, A. *Inorg. Chem.* **1970**, *9*, 426.
59. Swaddle, T. W.; Gustalla, G. *Inorg. Chem.* **1968**, *7*, 1915.
60. Tucker, M. A.; Colvin, C. B.; Martin, D. S., Jr. *Inorg. Chem.* **1964**, *3*, 1373.
61. Wax, M. J.; Bergman, R. G. *J. Am. Chem. Soc.* **1981**, *103*, 7028.
62. Rerek, M. E.; Basolo, F. *J. Am. Chem. Soc.* **1984**, *106*, 5908.
63. Alibrandi, G.; Romeo, R., Scolaro, L. M.; Tobe, M. L. *Inorg. Chem.* **1992**, *31*, 5061.
64. Parris, M.; Wallace, W. J. *Can. J. Chem.* **1969**, *7*, 2257.
65. Lay, P. A. *Inorg. Chem.* **1987**, *26*, 2144.
66. Basolo, F.; Chatt, J.; Gray, H. B.; Pearson, R. G.; Shaw, B. L. *J. Chem. Soc.* **1961**, 2207.
67. Tolman, C. A. *Chem. Rev.* **1977**, *77*, 313.
68. Seligson, A. L.; Trogler, W. C. *J. Am. Chem. Soc.* **1991**, *113*, 2520.
69. Caffery, M. L.; Brown, T. L. *Inorg. Chem.* **1991**, *30*, 3907; Brown, T. L. *Inorg. Chem.* **1992**, *31*, 1286; Choi, M. G.; White, D.; Brown, T. L. *Inorg. Chem.* **1994**, *33*, 5591.
70. Choi, M. G.; Brown, T. L. *Inorg. Chem.* **1993**, *32*, 5603.
71. Brown, T. L.; Lee, K. J. *Coord. Chem. Rev.* **1993**, *128*, 89.
72. White, D.; Coville, N. J. *Adv. Organomet. Chem.* **1994**, *36*, 95.
73. DeSanto, J. T.; Mosbo, J. A.; Storhoff, B. N.; Bock, P. L.; Bloss, R. E. *Inorg. Chem.* **1980**, *19*, 3086.
74. Stahl, L.; Ernst, R. D. *J. Am. Chem. Soc.* **1987**, *109*, 5673.
75. Maitlis, P. *Chem. Soc. Rev.* **1981**, 1.
76. Coville, N. J.; Loonat, M. S.; White, D.; Carlton, L. *Organometallics* **1992**, *11*, 1082.

77. Bodner, G. M.; May, M. P.; McKinney, L. E. *Inorg. Chem.* **1980**, *19*, 1951.
78. Rahman, Md. M.; Liu, H. Y.; Prock, A.; Giering, W. P. *Organometallics* **1987**, *6*, 650; Lorsbach, B. A.; Bennett, D. M.; Prock, A.; Giering, W. P. *Organometallics* **1995**, *14*, 869.
79. Swaddle, T. W.; Stranks, D. R. *J. Am. Chem. Soc.* **1972**, *94*, 8357.
80. Laurenczy, G.; Rapaport, I.; Zbinden, D.; Merbach, A. E. *Magn. Reson. Chem.* **1991**, *29*, S45.
81. Rapaport, I.; Helm, L.; Merbach, A. E.; Bernhard, P.; Ludi, A. *Inorg. Chem.* **1988**, *27*, 873.
82. Doine, H.; Ishihara, K.; Krouse, H. R.; Swaddle, T. W. *Inorg. Chem.* **1987**, *26*, 3240.
83. Tong, S. B.; Swaddle, T. W. *Inorg. Chem.* **1974**, *13*, 1538.
84. Cusanelli, A.; Frey, U.; Richens, D. T.; Merbach, A. E. *J. Am. Chem. Soc.* **1996**, *118*, 5265.
85. Helm, L.; Elding, L. I.; Merbach, A. E. *Inorg. Chem.* **1985**, *24*, 1719.
86. Aebischer, N.; Layrenczy, G.; Ludi, A.; Merbach, A. E. *Inorg. Chem.* **1993**, *32*, 2810.
87. van Eldik, R.; Asano, T.; Le Noble, W. J. *Chem. Rev.* **1989**, *89*, 549.
88. Twigg, M. V. *Inorg. Chim. Acta* **1977**, *24*, L84.
89. Garrick, F. J. *Nature* **1937**, *139*, 507.
90. Tobe, M. L. *Acc. Chem. Res.* **1970**, *3*, 377; Tobe, M. L. *Adv. Inorg. Bioinorg. Mech.* **1983**, *2*, 1.
91. Sargeson, A. M. *Pure Appl. Chem.* **1973**, *33*, 527.
92. Buckingham, D. A.; Clark, C. R.; Lewis, T. W. *Inorg. Chem.* **1979**, *18*, 2041.
93. Appleton, T. G.; Clark, H. C.; Manzer, L. E. *Coord. Chem. Rev.* **1973**, *10*, 335.
94. Coyle, B. A.; Ibers, J. A. *Inorg. Chem.* **1972**, *11*, 1105.
95. Chatt, J.; Duncanson, L.; Shaw, B.; Venanzi, L. *Disc. Faraday Soc.* **1958**, *26*, 131; Adams, D. M.; Chatt, J.; Shaw, B. *J. Chem. Soc.* **1960**, 2047.
96. Parshall, G. W. *J. Am. Chem. Soc.* **1966**, *88*, 704.
97. Zumdahl, S. S.; Drago, R. S. *J. Am. Chem. Soc.* **1970**, *90*, 6669.
98. Armstrong, D. R.; Fortune, R.; Perkins, P. G. *Inorg. Chim. Acta* **1974**, *9*, 9.
99. Lin, Z. Y.; Hall, M. B. *Inorg. Chem.* **1991**, *30*, 646.
100. Davy, R. D.; Hall, M. B. *Inorg. Chem.* **1989**, *28*, 3524.
101. Darensbourg, D. J.; Joyce, J. A.; Bischoff, C. J.; Reibenspies, J. H. *Inorg. Chem.* **1991**, *30*, 1137.
102. Hunt, J. P.; Rutenberg, A. C.; Taube, H. *J. Am. Chem. Soc.* **1952**, *74*, 268.
103. Darensbourg, D. J.; Jones, M. L. M.; Reibenspies, J. H. *Inorg. Chem.* **1996**, *35*, 4406.
104. Posey, F. A.; Taube, H. *J. Am. Chem. Soc.* **1953**, *75*, 4099.
105. Palmer, D. A.; van Eldik, R. *Chem. Rev.* **1983**, *83*, 561.
106. Massoud, S.; Jordan, R. B. *Inorg. Chim. Acta* **1994**, *221*, 9.
107. Buckingham, D. A.; Clark, C. R. *Inorg. Chem.* **1994**, *33*, 6171.
108. Baxter, K. E.; Hanton, L. R.; Simpson, J.; Vincent, B. R.; Blackman, A. G. *Inorg. Chem.* **1995**, *34*, 2795.

109. Van Eldik, R.; Harris, G. M. *Inorg. Chem.* **1980**, *19*, 880; Dasgupta, T. P.; Harris, G. M. *Inorg. Chem.* **1984**, *23*, 4398.
110. Joshi, V. K.; van Eldik, R.; Harris, G. M. *Inorg. Chem.* **1986**, *25*, 2229.
111. Murmann, R. K.; Taube, H. *J. Am. Chem. Soc.* **1956**, *78*, 4886.
112. Jackson, W. G.; Lawrance, G. A.; Lay, P. A.; Sargeson, A. M. *J. Chem. Soc., Chem. Commun.* **1982**, 70.
113. Andrade, C.; Taube, H. *J. Am. Chem. Soc.* **1964**, *86*, 1328.
114. Buckingham, D. A.; Foster, D. M.; Sargeson, A. M. *J. Am. Chem. Soc.* **1969**, *91*, 4102.
115. Buckingham, D. A.; Keene, F. R.; Sargeson, A. M. *J. Am. Chem. Soc.* **1974**, *96*, 4981.
116. Buckingham, D. A. In *Biological Aspects of Inorganic Chemistry*; Addison, A. W.; Cullen, W. R.; Dolphin, D.; James, B. R., Eds.; Wiley Interscience: New York, 1977; pp. 141–196.
117. Boreham, C. J.; Buckingham, D. A.; Francis, D. J.; Sargeson, A. M.; Warner, L. G. *J. Am. Chem. Soc.* **1981**, *103*, 1975.
118. Pinnel, D.; Wright, G. B.; Jordan, R. B. *J. Am. Chem. Soc.* **1972**, *94*, 6104; Buckingham, D. A.; Keene, F. R.; Sargeson, A. M. *J. Am. Chem. Soc.* **1973**, *95*, 5649.
119. Dixon, N. E.; Fairlie, D. P.; Jackson, W. G.; Sargeson, A. M. *Inorg. Chem.* **1983**, *22*, 4038.
120. Ellis, W. R., Jr.; Purcell, W. L. *Inorg. Chem.* **1982**, *21*, 834.
121. Hendry, P.; Sargeson, A. M. *Prog. Inorg. Chem.* **1990**, *38*, 201; Chin, J. *Acc. Chem. Res.* **1991**, *24*, 145; Linkletter, B.; Chin, J.; *Angew. Chem. Int. Ed. Engl.* **1995**, *34*, 472; Deal, K. A.; Burstyn, J. N. *Inorg. Chem.* **1996**, *35*, 2792; Koike, T.; Inoue, M.; Kimura, E.; Shiro, M. *J. Am. Chem. Soc.* **1996**, *118*, 3091.
122. Anderson, B.; Milburn, R. M.; MacB. Harrowfield, J.; Robertson, G. B.; Sargeson, A. M. *J. Am. Chem. Soc.* **1977**, *98*, 2652.
123. Clark, C. R.; Tasker, R. F.; Buckingham, D. A.; Knighton, D. R.; Harding, D. R. K.; Hancock, W. S. *J. Am. Chem. Soc.* **1981**, *103*, 7023.
124. Sutton, P. A.; Buckingham, D. A. *Acc. Chem. Res.* **1987**, *20*, 357.
125. Lincoln, S. F.; Merbach, A. E. *Adv. Inorg. Chem.* **1995**, *42*, 1.
126. Plane, R. A.; Hunt, J. P. *J. Am. Chem. Soc.* **1957**, *79*, 3343.
127. Jackson, J. A.; Lemons, J. F.; Taube, H. *J. Chem. Phys.* **1960**, *32*, 553; Baldwin, H. W.; Taube, H. *J. Chem. Phys.* **1960**, *33*, 206.
128. Swift, T. J.; Connick, R. E. *J. Chem. Phys.* **1962**, *37*, 307.
129. Merbach, A. E. *Pure Appl. Chem.* **1982**, *54*, 1479; Ibid. **1987**, *59*, 161.
130. Taube, H. *Chem. Rev.* **1952**, *50*, 69.
131. Basolo, F.; Pearson, R. G. *Mechanisms of Inorganic Reactions*, 2nd ed.; Wiley & Sons: New York, 1967; pp. 65–80.
132. Breitschwerdt, K. *Ber. Bunsenges. Phys. Chem.* **1968**, *72*, 1046.
133. Spees, S. T., Jr.; Perumareddi, J. R.; Adamson, A. W. *J. Am. Chem. Soc.* **1968**, *90*, 6626.
134. Burdett, J. K. *Adv. Inorg. Chem. Radiochem.* **1978**, *21*, 113.
135. Mønsted, O. *Acta Chem. Scand.* **1978**, *A32*, 297.

136. Rode, B. M.; Reihnegger, G. J.; Fujiwara, S. *J. Chem. Soc., Faraday Trans. 2* **1980**, *76*, 1268.

137. Connick, R. E.; Alder, B. J. *J. Phys. Chem.* **1983**, *87*, 2764.

138. Galera, S.; Lluch, J. M.; Oliva, A.; Bertran, J.; Foglia, F.; Helm, L. Merbach, A. E. *New J. Chem.* **1993**, *17*, 773.

139. Ohtaki, H.; Radnai, T. *Chem. Rev.* **1993**, *93*, 1157.

140. Åkesson, R.; Pettersson, L. G. M.; Sandström, M.; Wahlgren, U. *J. Am. Chem. Soc.* **1994**, *116*, 8705.

141. Rusnak, L. L.; Yang, E. S.; Jordan, R. B. *Inorg. Chem.* **1978**, *17*, 1810.

142. Meyer, F. K.; Newman, K. E.; Merbach, A. E. *J. Am. Chem. Soc.* **1979**, *101*, 5588; Ducommun, Y.; Newman, K. E.; Merbach, A. E. *Inorg. Chem.* **1980**, *19*, 3696.

143. Swaddle, T. W. *Inorg. Chem.* **1983**, *22*, 2663.

144. Eigen, M.; Wilkins, R. G. *Adv. Chem. Ser.* **1965**, *49*, 55; Kustin, K.; Swinehart, J. *Prog. Inorg. Chem.* **1970**, *13*, 107; Hewkin, D. J.; Prince, R. H. *Coord. Chem. Rev.* **1970**, *5*, 45; Swaddle, T. W. *Adv. Inorg. Bioinorg. Mech.* **1983**, *2*, 96; Hoffmann, H. *Pure Appl. Chem.* **1975**, *41*, 327.

145. Wilkins, R. G. *Acc. Chem. Res.* **1970**, *3*, 407.

146. Wilkins, R. G.; Eigen, M. *Adv. Chem. Ser.* **1965**, *49*, 55.

147. Chaudhuri, P.; Diebler, H. *J. Chem. Soc., Dalton Trans.* **1986**, 1693.

148. Hugi, A. D.; Helm, L.; Merbach, A. E. *Inorg. Chem.* **1987**, *26*, 1763.

149. Patel, R. C.; Diebler, H. *Ber. Bunsenges. Phys. Chem.* **1972**, *76*, 1035.

150. Sauvageat, P.-Y.; Ducommun, Y.; Merbach, A. E. *Helv. Chim. Acta* **1989**, *72*, 1801.

151. Dellavia, I.; Sauvageat, P.-Y.; Helm, L.; Ducommun, Y.; Merbach, A. E. *Inorg. Chem.* **1992**, *31*, 792.

152. Ducommun, Y.; Newman, K. E.; Merbach, A. E. *Inorg. Chem.* **1980**, *19*, 3696.

153. Fogg, P. G. T.; Hall, R. J. *J. Chem. Soc. A* **1971**, 1365.

154. Monzyk, B.; Crumblis, A. L. *J. Am. Chem. Soc.* **1979**, *101*, 6203.

155. Letter, J. E., Jr.; Jordan, R. B. *J. Am. Chem. Soc.* **1975**, *97*, 2381.

156. Cassatt, J. C.; Johnson, W. A.; Smith, L. M.; Wilkins, R. G. *J. Am. Chem. Soc.* **1972**, *94*, 8399.

157. Rorabacher, D. B. *Inorg. Chem.* **1966**, *5*, 1891.

158. Jordan, R. B. *Inorg. Chem.* **1976**, *15*, 748.

159. Desai, A. G.; Dodgen, H. W.; Hunt, J. P. *J. Am. Chem. Soc.* **1969**, *91*, 5001; Ibid. **1970**, *92*, 798.

160. Brown, A. J.; Howarth, O. W.; Moore, P.; Parr, W. J. E. *J. Chem. Soc., Dalton Trans.* **1978**, 1776.

161. Martell, A. E.; Smith, R. M. *Critical Stability Constants*; Plenum: New York, 1977; Vol. 2.

162. Hubbard, C. D.; Palaitis, W. *J. Coord. Chem.* **1979**, *9*, 107.

163. Jordan, R. B. *J. Coord. Chem.* **1980**, *10*, 239.

164. Aizawa, S.; Ida, S.; Matsuda, K.; Funahashi, S. *Inorg. Chem.* **1996**, *35*, 1338; Aizawa, S.; Matsuda, K.; Tajima, T.; Maeda, M.; Sugata, T.; Funahashi, S. *Inorg. Chem.* **1995**, *34*, 2042; Soyama, S.; Ishii, M.; Funahashi, S.; Tanaka, M. *Inorg. Chem.* **1992**, *31*, 536.

4

Stereochemical Change

The kinetic and mechanistic aspects of this general area tend to be strongly dependent on the particular system. This makes general treatments and explanations impossible, at least at the current stage of understanding. Various aspects of this area have been summarized in some general reviews.[1-5]

4.1 TYPES OF LIGAND REARRANGEMENTS

Ligands bonded to a metal can undergo a number of structural changes that do not involve complete breaking of the metal–ligand bond(s). Such processes are the subject of the following sections.

4.1.a Conformational Change

Many chelate ligands have conformers that can interconvert. For example, the conformers of ethylenediamine interchange by rotation about the carbon–carbon bond:

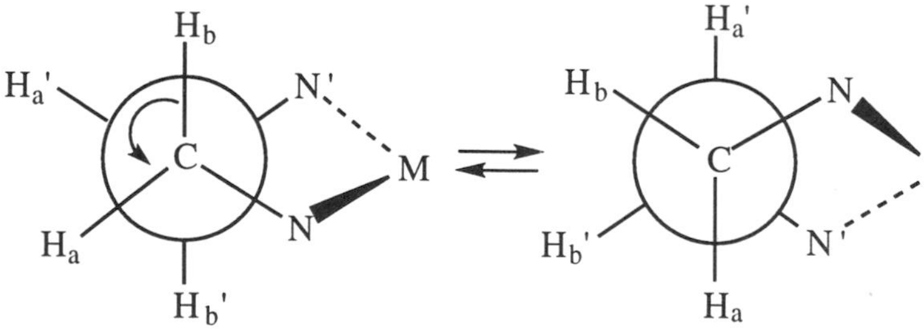

$$\text{(4.1)}$$

The H_a and H_a' protons are magnetically different from the H_b and H_b' protons, so their interconversion can, in principle, be studied by NMR. These protons may be referred to as exo and endo, respectively. In simple systems, their interconversion is too rapid ($k > 10^6$ s^{-1}) for this method. However, if there is some constraint (e.g., CH_3 groups) or if the coordinating atoms are part of a larger chelate system, then interconversion is slow enough to be detected by NMR.[6]

Aminotroponeimine   Salicylaldimine   ß-Ketoamine

Figure 4.1. Ligands that have paramagnetic tetrahedral and diamagnetic square planar isomers with nickel(II).

4.1.b Coordination Geometry Change

There are several nickel(II) complexes, such as those with the ligands shown in Figure 4.1, which exist as equilibrium mixtures of paramagnetic *tetrahedral* and diamagnetic *square planar* isomers. The planar–tetrahedral interconversions are rapid (k ~ 10^4–10^6 s^{-1}, $\Delta H^* \approx$ 10 kcal mol^{-1}, $\Delta S^* \approx 0$). The mechanism for the transformation is thought to be intramolecular rearrangement with no metal–ligand bond breaking.[7] A number of equilibria and substitution reactions on such systems have been studied by Elias and co-workers,[8] with the notable conclusion that substitution does not seem to occur on the tetrahedral form.

Octahedral to *square planar* transformations have also been observed. These must involve a substitution as well because of the change in coordination number. Again nickel(II) complexes provide the most examples, with systems having a four-coordinate ligand in the square plane and two solvent molecules occupying the other octahedral positions, as shown in the following reaction:

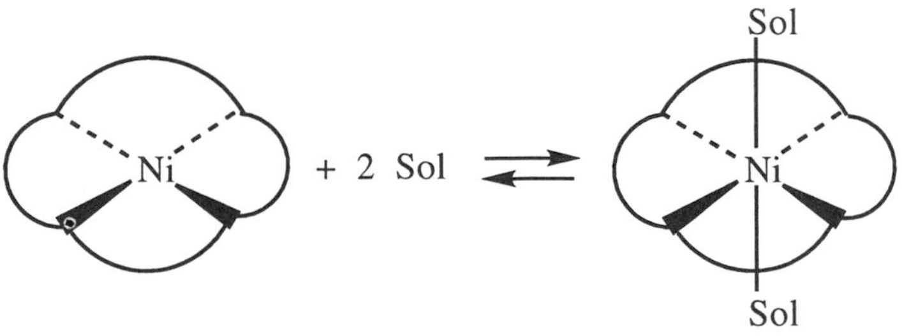

$$\text{(4.2)}$$

These interconversions are quite rapid and have been studied by laser T-jump[9] and ultrasonic relaxation.[10]

A *tetrahedral* to *octahedral* conversion has been studied for the following cobalt(II)[11] complex:

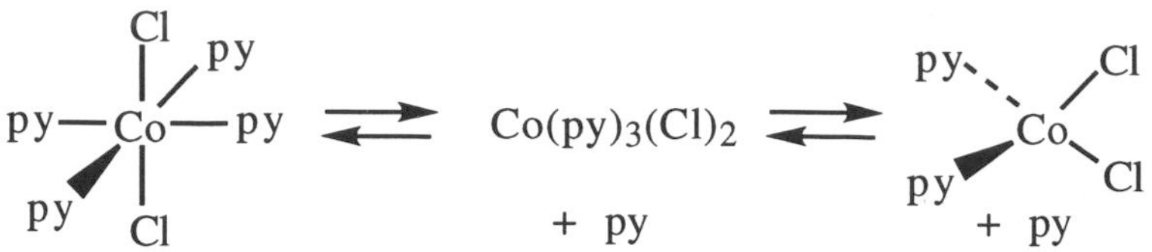

$$\text{(4.3)}$$

The results have been interpreted in terms of a five-coordinate intermediate and the reactions are rapid, as expected for the labile cobalt(II) (d^7) ion.

4.1.c Linkage Isomerism

Linkage isomerism has been studied with inert metal ions and is possible, in principle, for any metal with a ligand that has more than one type of atom with an unshared electron pair. The following reactions are some common examples:[12,13]

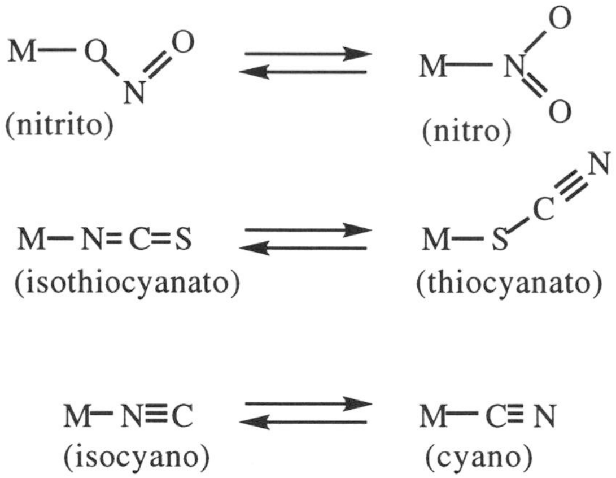

$$\text{(4.4)}$$

M—O—N=O ⇌ M—N(=O)$_2$
(nitrito)   (nitro)

M—N=C=S ⇌ M—S—C≡N
(isothiocyanato)   (thiocyanato)

M—N≡C ⇌ M—C≡N
(isocyano)   (cyano)

With inert octahedral complexes, these reactions appear to be intramolecular. This can be demonstrated by allowing the reaction to proceed in the presence of uncomplexed isotopically labeled ligand. A form of linkage isomerism that involves O-atom transfer has been observed in $M(NO)(NO_2)$[14] and $M(CO)(CO_2)$[15] systems.

The linkage isomerism of $(H_3N)_5Co(SCN)^{2+}$ has been extensively studied and recently reviewed.[16] In water in the presence of $N^{14}CS^-$, the isomerism was reported[17] to occur with ~3 percent incorporation of $N^{14}CS^-$ into the product $(H_3N)_5Co(NCS)^{2+}$. More recently,[18] these observations have been refined and it has been found that some $N^{14}CS^-$ enters the coordination sphere during the linkage isomerism; the quantitative results for 1 M NCS^- at 25°C are summarized in Scheme 4.1. It remains unclear whether the intermediate is best considered as a π complex[19] or a tight ion pair.

Scheme 4.1

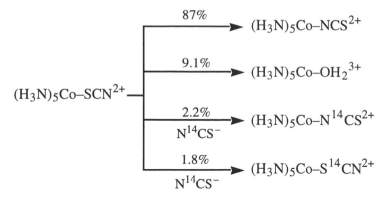

Exchange of free and bound thiocyanate has established that the linkage isomerism of square planar $(Et_4dien)Pd(SCN)^+$ occurs by an intermolecular rearrangement. The reaction probably proceeds through a solvent intermediate that reanates to give the isomerized product.[20]

Data on the linkage isomerization of $(H_3N)_5Co(ONO)^{2+}$ in water were given in Chapter 1. The reaction rate has been shown[21] to have a significant solvent dependence, which has been interpreted in terms of the Lewis basicity, acidity, and polarity of the solvents. The rearrangement of cobalt(III) complexes of N-bonded carboxamides to their O-bonded isomers have been studied.[22] The isocyano complex, $(H_3N)_5Co(NC)^{2+}$, has been prepared rather recently and found to undergo slow linkage isomerism in water and DMSO, and the reaction is catalyzed by cobalt(II) species.[23] The rearrangement of the imidazole complex is given in the following reaction:

$$
\left[\begin{array}{c} H_3C \\ \diagdown \\ C{=}CH \\ / \quad \diagdown \\ (H_3N)_5Co{-}N \quad N \\ \diagdown \diagup \\ C \\ | \\ H \end{array} \right]^{2+} \rightleftharpoons \left[\begin{array}{c} H_3C \\ \diagdown \\ C{=}CH \\ / \quad \diagdown \\ N \quad N{-}Co(NH_3)_5 \\ \diagdown \diagup \\ C \\ | \\ H \end{array} \right]^{2+} \qquad (4.5)
$$

It proceeds intramolecularly with cobalt migration across two ring atoms[24] in a 1,3-shift. The analogous methyltetrazole complexes can undergo 1,2-shifts of $(H_3N)_5Co$ between adjacent N atoms.[25]

A number of linkage isomerism reactions have been studied for complexes of $(H_3N)_5Ru^{II/III}$. These exploit the fact that Ru(II) favors the softer donor atom and π-acceptor systems while Ru(III) favors harder donor atoms and does not participate in π bonding. Thus, one isomer of Ru(II) is prepared, then oxidized to Ru(III) by chemical or electrochemical methods and the isomerism is observed. The Ru(III) isomer can be reduced and the reverse isomerization observed for the Ru(II) complex. For example, with DMSO, $(H_3N)_5Ru-S(O)(CH_3)_2^{3+}$ isomerizes to $(H_3N)_5Ru-OS(CH_3)_2^{3+}$.[26] With acetone, Ru(II) forms an η^2-$OC(CH_3)_2$ complex that rearranges to the η^1-O bonded isomer in the Ru(III) complex[27]; the acrylamide complex reacts analogously.[28] Creutz and co-workers[29] reduced a Ru(III)–nicotinamide complex to form the Ru(II) isomer and found that the latter undergoes competitive aquation and isomerization by a 1,4-shift with $k = 9.6$ s^{-1}, as shown in Scheme 4.2 for pH > 6. The length and speed of the isomerization led the observers to suggest that an η^2-arene intermediate is possible.

Scheme 4.2

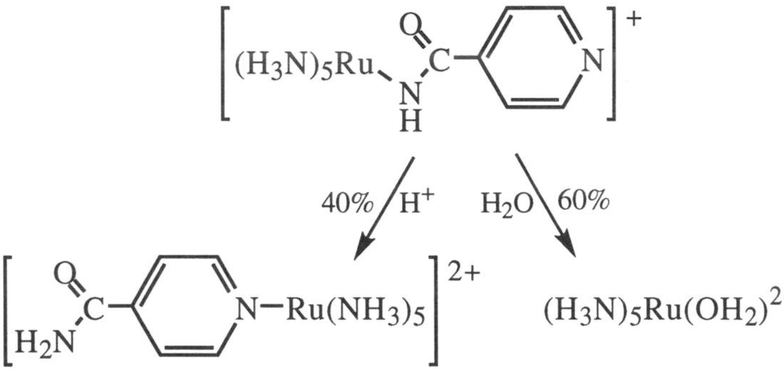

There are many analogous reactions in osmium(II/III) systems, with the major difference being that Os(II) has a greater tendency to π back bonding than Ru(II). The osmium chemistry has been reviewed recently.[30]

4.2 GEOMETRICAL AND OPTICAL ISOMERISM IN OCTAHEDRAL SYSTEMS

The interconversion of geometrical isomers (cis, trans, fac, mer) and the racemization of optical isomers ($\lambda - \Delta$) can proceed by two general mechanisms, *ligand dissociation* or *intramolecular rearrangement*.

4.2.a Dissociation Mechanisms

If the complex is an octahedral bis chelate system, $M(AA)_2(X)_2$, then two trigonal bipyramidal intermediates are possible, as shown in Scheme 4.3. The product distributions in Scheme 4.3 are based on the assumption of a statistical attack along the edges of the trigonal plane of the intermediate. Since the lower intermediate has a plane of symmetry, it must lead to racemization. Similar intermediates can be drawn for tris chelate systems if one end of a chelate dissociates. This is called *one-ended dissociation* and is actually an intramolecular process.

Scheme 4.3

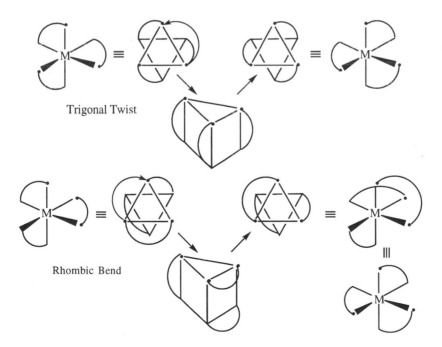

67% cis, retention
33% trans

50% cis, retention
50% cis, inversion

4.2.b Intramolecular Rearrangement Mechanisms

This mechanism will be illustrated for tris chelates of the type $M(AA)_3$ but can be extended easily to other systems. The two most commonly considered rearrangements involve rotating one trigonal face of the octahedron by 120° relative to the opposite trigonal faces. The *Bailar* or *trigonal twist* involves rotation about a C_3 axis and the *rhombic* or *Ray–Dutt bend* is rotation about an imaginary C_3 axis. Both of these are shown in Figure 4.2.

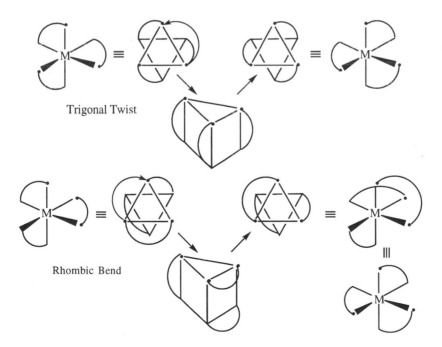

Trigonal Twist

Rhombic Bend

Figure 4.2. The trigonal twist and rhombic bend rearrangements for an $M(AA)_3$ complex.

4.2.c Differentiation of Dissociative and Intramolecular Mechanisms

The differentiation of the dissociation and intramolecular mechanisms is usually based on a comparison of the rates of ligand dissociation (exchange or solvolysis) and racemization. If the rates are quite similar, dissociation is assumed, whereas if racemization is much faster, then an intramolecular mechanism is operating. However, this involves the implict assumption that the initial chelate ring-opening step is rate limiting for both processes.

For the tris(oxalato)rhodium(III) ion in water, Damrauer and Milburn[31] have studied the kinetics of racemization, aquation, and $^{18}OH_2$ exchange of the oxalate oxygens. The racemization and aquation have acid-dependent pseudo-first-order rate constants given by

$$k_{exp} = k_2 [H^+] + k_3 [H^+]^2 \qquad (4.6)$$

The oxygen exchange is first order in $[H^+]$ and the inner, coordinated oxygens are exchanged much more slowly than the outer oxygens. The kinetic results are summarized in Table 4.1.

Damrauer and Milburn noted the similarity in the kinetic parameters of k_2 for inner oxygen exchange and racemization and suggested that both processes are proceeding through a common intermediate formed by acid-catalyzed one-ended dissociation of oxalate. The crystal structure[32] of $K_3[Rh(C_2O_4)_3]\cdot4.5H_2O$ reveals that the O—Rh—O angles are ~83°. This makes it seem more probable that racemization and inner oxygen exchange proceed by a square pyramidal intermediate rather than by a symmetrical trigonal pyramidal one (Scheme 4.3), which requires one O—Rh—O angle of 120°.

Table 4.1. Rate Constants (50°C) and Activation Parameters for the Oxygen Exchange, Racemization, and Aquation of $Rh(C_2O_4)_3{}^{3-}$ in 0.54 M NaClO$_4$/HClO$_4$

Reaction	k_2 $(M^{-1}\,s^{-1})$	k_3 $(M^{-2}\,s^{-1})$	ΔH^* (kcal mol^{-1})	ΔS^* (cal mol^{-1} K^{-1})
Outer O exchange	1.1×10^{-3}		16.9	−20.0
Inner O exchange[a]	3.6×10^{-5}		23.5	−6.4
Racemization[a]	2.0×10^{-5}		23.0	−9.1
		1.1×10^{-4}	27.9	+9.6
Aquation[a]	5.6×10^{-7}		25.6	−8.1
		5.9×10^{-6}	25.7	−3.1

[a] Values recalculated from the data in Damrauer, L.; Milburn, R. M. *J. Am. Chem. Soc.* **1971**, *93*, 6481.

It is noteworthy that the kinetics of the outer oxygen exchange are similar to those of oxalic acid,[33] for which $\Delta H^* = 15.1$ kcal mol^{-1}, $\Delta S^* = -22.7$ cal mol^{-1} K^{-1}, and k (50°C) = 4 x 10^{-3} s^{-1}. Palmer and Kelm[34] found $\Delta V^* = -6.3$ cm^3 mol^{-1} for the aquation of Rh(C$_2$O$_4$)$_3$$^{3-}$ in 1.0 M H$^+$, where k_3 dominates.

Tris(oxalato)chromium(III) is much more reactive than its rhodium(III) analogue. Odell et al.[35] indicate that all the oxygens exchange at the same rate (k = 1.26 x 10^{-3} s^{-1}, 1.0 M HClO$_4$, 25°C) and racemization is about 10 times faster (k = 1.38 x 10^{-2} s^{-1}) under the same conditions. The authors conclude that ring opening and closing must be much faster than oxygen exchange; the latter is 2.4 times faster than for free oxalic acid at 25°C. Structures of chromium(III)–oxalate complexes[36] indicate O—Cr—O angles of ~83° and a square pyramidal intermediate would again seem probable. Lawrance and Stranks[37] found a very negative value of $\Delta V^* = -16.3$ cm^3 mol^{-1} for racemization of Cr(ox)$_3$$^{3-}$ in 0.05 M HCl and rationalized this as mainly due to solvent electrostriction around the -CO$_2$$^-$ group in the one-ended dissociation transition state. The authors imply that their results are for an [H$^+$]-independent reaction, but the [H$^+$]-catalyzed path dominates the kinetics below pH 3. For Cr(phen)$_2$(ox)$^+$ and Cr(bpy)$_2$(ox)$^+$, the values of ΔV^* are -1.5 and -1.0 cm^3 mol^{-1}, respectively, and an intramolecular twist is proposed by Lawrance and Stranks.

The racemization of tris(N,N'-dimethylethylenediamine)nickel(II) appears to proceed by an intramolecular rotation because bond rupture should produce some meso isomer, which is not observed. The kinetic similarity of this and the Ni(en)$_3$$^{2+}$ racemization[38] suggests that the latter also is intramolecular. On the other hand, the tris(bipyridyl) and tris(o-phenanthroline) complexes of nickel(II)[39] and iron(II)[40] have very similar rates of racemization, ligand exchange, and aquation, and therefore are assumed to proceed by dissociation. It seems surprising that these more rigid chelates would choose dissociation while the more flexible aliphatic diamines go by intramolecular rotation. It should be remembered that the aliphatic amines are much stronger bases than the aromatic amines, and thus the former may form stronger bonds to the metal, making dissociation less favorable.

Comparisons of volumes of activation in Table 4.2 from the work of Lawrance and Stranks[41] provide some further insights. The authors suggest that the small values of ΔV^* for Ni(II) and Cr(III) imply an intramolecular twist mechanism. To rationalize the large ΔV^* for Fe(II), they suggest that there is a spin-state change accompanying the intramolecular twist (i.e., the system goes from low-spin d^6 in the ground state to high-spin d^6 in the transition state). This causes the metal-to-ligand bond lengths to increase. Note that the ΔH^* and ΔS^* are different for racemization and aquation in the Fe(II) system.

Table 4.2. Activation Parameters for the Racemization and Aquation of Some Complexes of Tris(o-phenanthroline)

Complex	Reaction	ΔH^* (kJ mol^{-1})	ΔS^* (J mol^{-1} K^{-1})	ΔV^* (cm^3 mol^{-1})
Fe(phen)$_3$$^{2+}$	Racemization	118 ($\pm$ 3)	89 ($\pm$ 8)	15.6 ($\pm$ 0.3)
	Aquation	135 ($\pm$ 2)	117 ($\pm$ 8)	15.4 ($\pm$ 0.3)
Ni(phen)$_3$$^{2+}$	Racemization	105 ($\pm$ 1)	12 ($\pm$ 3)	−1.5 ($\pm$ 0.3)
	Aquation	102 ($\pm$ 2)	3 ($\pm$ 6)	−1.2 ($\pm$ 0.2)
Cr(phen)$_3$$^{3+}$	Racemization	94 ($\pm$ 4)	−56 ($\pm$ 3)	3.3 ($\pm$ 0.3)

Such spin-state equilibria have been studied for iron(II) complexes of the ligands shown in Figure 4.3. The $\Delta V^\circ = 11$ cm^3 mol^{-1} for Fe^{II}(papth)$_2$,[42] and the kinetics for the low- to high-spin direction gave values of k = 1.7 x 10^7 s^{-1} (25°C), $\Delta H^* = 31.7$ kJ mol^{-1}, and $\Delta S^* = 0.37$ J mol^{-1} K^{-1}. Similar kinetic parameters have been found[43] for the pyimH and ppa iron(II) chelates, which have ΔV^* values in the 2 to 9 cm^3 mol^{-1} range with a significant solvent dependence.

There have been several attempts to provide a theoretical framework for describing geometrical isomerization and racemization. Vanquickenborne and Pierloot[44] used ligand field theory to calculate the electronic energies of the intermediates proposed in the dissociative and trigonal twist mechanisms for low-spin d^6 systems.

The d orbital energy diagrams from the analysis of Vanquickenborne and Pierloot for the trigonal twist are shown in Figure 4.4. The values of σ and π are measures of the σ- and π-donor strength of the ligand (e.g., for en, $\sigma = 94.3$ kJ mol^{-1}, $\pi = 0$; for H$_2$O, $\sigma = 78.5$ kJ mol^{-1}, $\pi = 12.85$ kJ mol^{-1}). The prediction is that these systems should go through a spin-state change to give a more stable transition state.

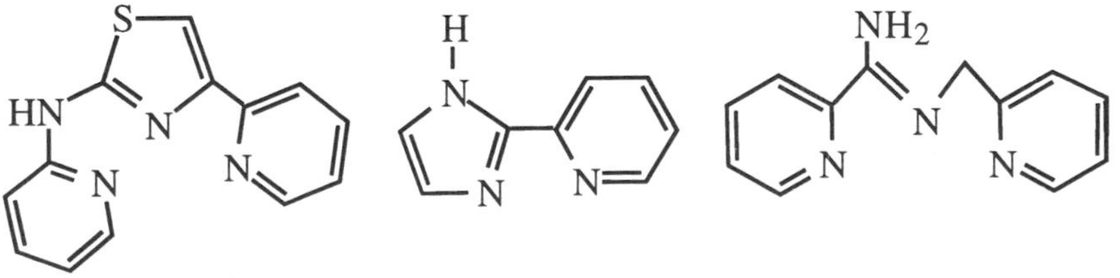

papth   pyimH   ppa

Figure 4.3. Examples of ligands that have iron(II) complexes with high-spin and low-spin forms.

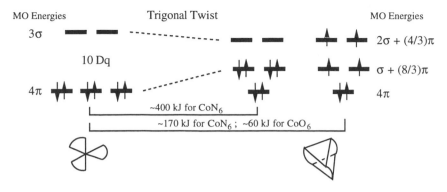

Figure 4.4. Metal *d* orbital energies for the trigonal twist rearrangement.

To calculate the ΔH^*, one must take into account electron repulsion–delocalization factors through the Racah parameters, B and C, in ligand field theory. The diagram gives a change of $6\sigma - 8\pi$ for formation of the transition state and $\Delta H^* = 6\sigma - 8\pi - 5B - 8C$ for a d^6 system. For Co(III), B = 7.14 kJ mol^{-1}, C = 44 kJ mol^{-1}, and $\Delta H^* \approx 180$ kJ mol^{-1} for Co(en)$_3{}^{3+}$. Oxygen donor ligand systems, with smaller σ values and π values > 0, should twist more readily than nitrogen donor systems.

For the dissociative mechanism, it was assumed that a square-based pyramid (C_{4v}) would form first and would rearrange to a trigonal bipyramid (D_{3h}) if some stereochemical change is observed. The orbital energies are given in Figure 4.5. The theory again predicts a spin-state change in the D_{3h} intermediate. As before, the diagram gives a change of $5\sigma - 6.5\pi$ for the high-spin transition state and for the Co(en)$_3{}^{3+}$, $\Delta H^* = 5\sigma - 6.5\pi - 5B - 8C \approx 85$ kJ mol^{-1}. A comparison of the ΔH^* values from the two mechanisms predicts that the rearrangement should go by a dissociative mechanism through a high-spin transition state.

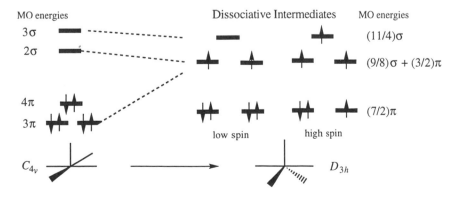

Figure 4.5. Metal *d* orbital energies for the dissociative rearrangement mechanism.

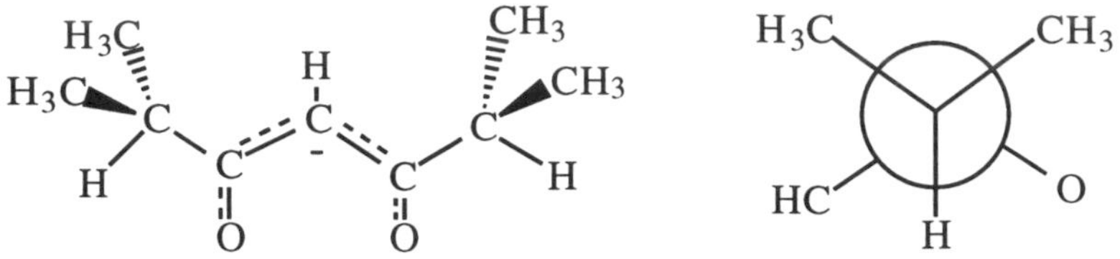

Figure 4.6. The diisobutyrylmethanide group with a chiral isopropyl carbon.

The theory has been extended to $Co(en)_2(A)X$ systems,[45] but the results are more complex because of the different ligand types. It was concluded by Vanquickenborne that $(\sigma_A - 2\pi_A)$ is the controlling factor, so that better π-donor A ligands will reduce the C_{4v} to D_{3h} energy barrier and therefore give more stereochemical change in a dissociative mechanism. This prediction is consistent with the fact that A ligands such as CN^- or NH_3 give hydrolysis with retention, while if A is Cl^- or OH^-, there is a considerable (~30 percent) change in configuration. The theory of Vanquickenborne predicts that the amount of rearrangement should increase with increasing temperature, but this aspect has not been tested experimentally.

For those interested in studying racemization, a major problem is the need to resolve the optical isomers. It is possible to avoid this problem by using a chelate with a chiral center and studying the system by NMR. An example, given in Figure 4.6, is diisobutyrylmethanide, where the isopropyl carbon is chiral and the view at the right down the C—C bond shows that the two CH_3 groups can never be magnetically equivalent. When inversion occurs in a tris chelate of this type, the two CH_3 groups are interchanged, and this is observed as a merging of the two NMR signals as the interconversion becomes fast on the NMR time scale; actually H–H coupling causes four peaks to collapse to two. The process is pictured in Figure 4.7 for one isopropyl group with the CH_3 groups designated as a and b. This method can be used for the H atoms of a $-CH_2CH_3$ substituent on an unsymmetrical chelate ring.

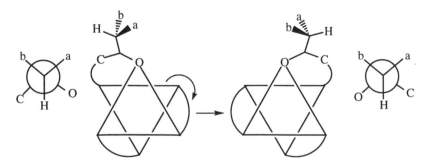

Figure 4.7. Interconversion of methyl groups a and b during a trigonal twist.

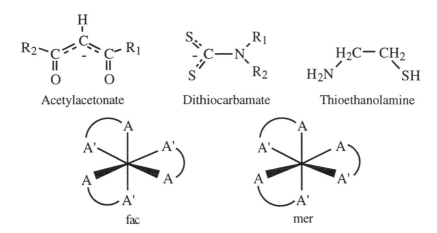

Figure 4.8. Examples of unsymmetrical chelating ligands and their geometrical isomers.

4.2.d Rearrangements in Unsymmetrical Chelates

There are a large number of unsymmetrical chelates that present the feature of having both geometrical and optical isomers, as shown by the examples in Figure 4.8. The fac and mer isomers each have optical isomers; thus, both racemization and geometrical isomerization can be observed in one system. This can serve to eliminate certain rearrangement processes.

For a dissociative mechanism, the different possible **D** transition states give different products, as in the following examples:

$$
\begin{aligned}
\text{Axial trigonal bipyramid} &\rightarrow \text{Isomerization} + \text{Inversion} \\
\text{Equatorial trigonal bipyramid} &\rightarrow \text{Isomerization} + \text{Inversion} \\
\text{Square-based pyramid} &\rightarrow \text{Some isomerization} + \text{Some inversion}
\end{aligned}
$$

Further possibilities of products from a dissociative mechanism have been given by Wilkins.[46]

The Al(III) and Ga(III) complexes of the unsymmetrical derivative of acetylacetonate (see Figure 4.8), with $R_1 = CF_3$ and $R_2 = 2\text{-}C_4H_4S$, isomerize and racemize by a common pathway.[47] A dissociative mechanism is suggested by the ten-fold increase in rate on changing the solvent from $(CDCl_2)_2$ to DMSO. The solvent effect criterion is a subjective one and an eight-fold change between H_2O and DMSO was taken to be minor for a tris-catecholate complex of gallium(III).[48]

The two intramolecular twist mechanisms give different results, as shown in Figure 4.9. The trigonal twist gives inversion without geometrical isomerization, but the rhombic bend gives both. Thus, if the reaction is shown to proceed without dissociation and gives inversion without isomerization, then a trigonal twist mechanism is established.

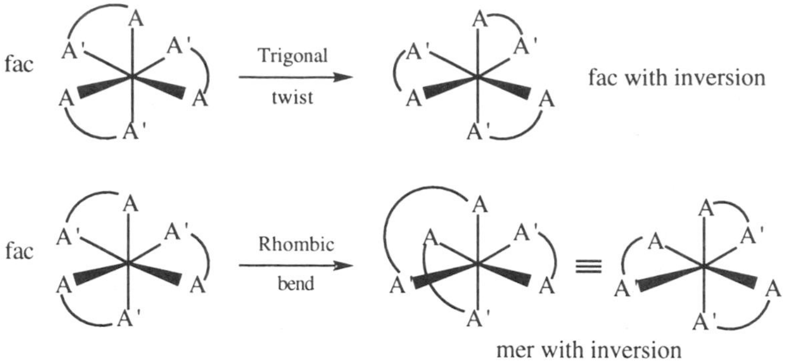

Figure 4.9. The products of the trigonal twist and rhombic bend for an unsymmetrical tris chelate.

The trigonal twist is the most restrictive in terms of the rearrangement results. This mechanism has been confirmed for the complexes of Al(III), Ga(III), and Co(III), with the following ligand:

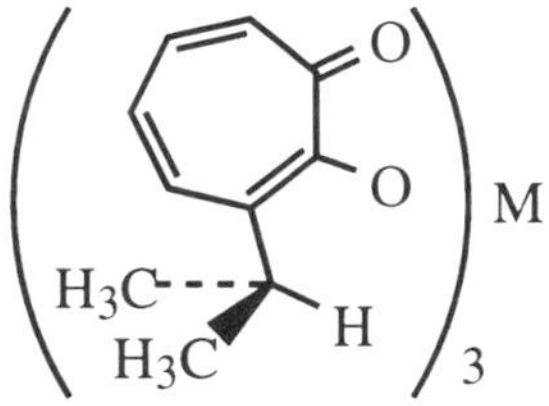

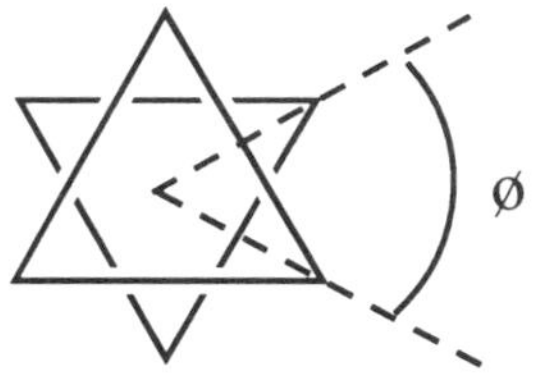

In these systems, the trigonal twist may be favored by ground-state distortions toward the trigonal transition state because of the small bite distance (2.5 Å) of the ligand.[49] This distortion is measured by the twist angle, Ø, which is 60° for an octahedron and 0° for the trigonal transition state. Several Fe(IV), Fe(III), and Ru(III) dithiocarbamates also use the trigonal twist mechanism.[50] It is fortunate that these systems rearrange by the one process that is established by NMR because it gives inversion without fac–mer rearrangement.

The structural features that may favor the trigonal twist relative to the rhombic bend have been discussed by Rodger and Johnson.[51] Their approach considers the ligand bite distance, b, and the hard-sphere contact distance, *l*, of the coordinating atoms, and they conclude that the trigonal twist is favored if b is much smaller than *l*. Relevant structural information on such systems has been reviewed by Keppert.[52] This approach does not take into account any differences in strain in the chelate backbone.

Instead of this mechanistic approach, a pure permutational analysis can be used. The possibilities are given by Holm[53] and in more detail by Musher[54] and Eaton et al.[55]

4.3 STEREOCHEMICAL CHANGE IN FIVE-COORDINATE SYSTEMS

Most of the effort in the area of five-coordinate systems has been on trigonal bipyramid systems such as PF_5 and $Fe(CO)_5$. The interconversion is between the axial, a, and equatorial, e, positions.

The axial–equatorial conversion is generally quite rapid for $M(L)_5$ systems, but a series of $M(P(OR)_3)_5$ complexes have rates that are accessible on the NMR time scale. Meakin and Jesson[56] found values for ΔH^* of 8 to 12 kcal mol^{-1} and showed that the process required the simultaneous interchange of two axial and two equatorial substituents. The *pseudorotation mechanism* suggested by Berry,[57] and shown in Scheme 4.4, is consistent with the observations.

Scheme 4.4

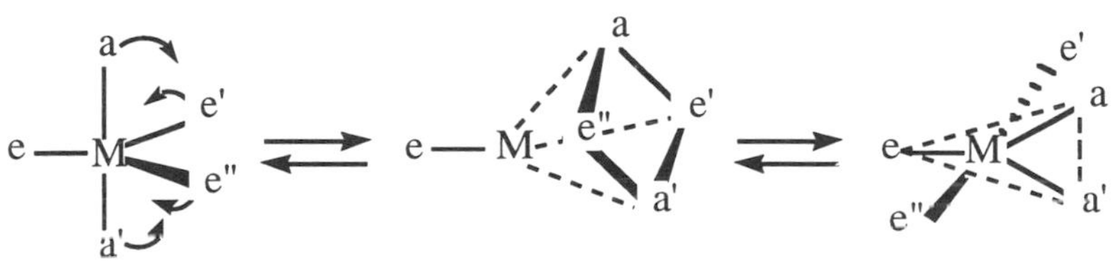

The initially equatorial e' and e" ligands become axial, whereas the initially axial a and a' become equatorial. The product appears to be rotated by 90° relative to the starting structure. Note that the intermediate structure is close to a square-based pyramid.

Ugi et al.[58] have suggested what is called the *turnstyle mechanism*, which is described in reaction (4.7). Since this process is also consistent with the observations, the two mechanisms cannot be distinguished experimentally. Theoretical arguments have favored the pseudorotation mechanism as providing the lower energy pathway for the interconversion.[59]

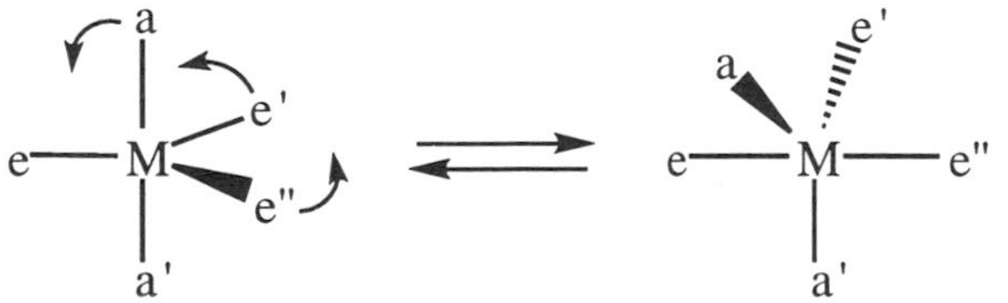

(4.7)

For some transition-metal hydrides of the type HMP_4, the ground-state structure is a distorted trigonal bipyramid and can be pictured as a tetrahedral arrangement of the P ligands around M with the H atom on one tetrahedral face. For such hydrides, a *tetrahedral jump mechanism* has been proposed[60] in which the H atom moves from one tetrahedral face to an edge and then to another face, as shown in Scheme 4.5.

Scheme 4.5

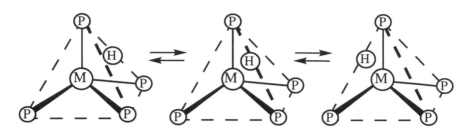

This mechanism is consistent with the NMR observations. If the H and the opposite P are considered as axial, then this process converts the axial P substituent to an equatorial position for each occurrence, and this is not the permutational equivalent of the pseudorotation process.

4.4 ISOMERISM IN SQUARE PLANAR SYSTEMS

Most examples in this area involve cis–trans isomerization of Pt(II) complexes. The area was reviewed by Anderson and Cross.[61] In coordinating solvents, the mechanism usually suggested involves solvolysis of one ligand followed by readdition of that ligand, both proceeding through the five-coordinate associative intermediate that seems consistent with most ligand substitution processes on Pt(II).

For Pt(II) complexes of the type $Pt(R)_2(L)_2$ and $Pt(Ar)_2(L)_2$ in non-coordinating solvents, a dissociative mechanism involving a 14-electron T-shaped intermediate has been suggested.[62] This pathway typically has a positive ΔS^* and shows kinetic retardation by free L in solution.

The anticancer agent cisplatin, cis-$Pt(NH_3)_2(Cl)_2$, has been extensively studied.[63] A metabolite of this drug is the bis-L-methionine complex, $Pt(L\text{-Met-}S,N)_2$, which forms an equilibrium mixture of cis and trans isomers in water. The trans to cis isomerization has $\Delta H^* = 95$ kJ mol^{-1} and $\Delta S^* = -18$ J mol^{-1} K^{-1}, and an intramolecular twist through a tetrahedral intermediate has been suggested.[64]

4.5 FLUXIONAL ORGANOMETALLIC COMPOUNDS

The mechanisms of geometrical and optical isomerization already discussed also apply to organometallic compounds, but these show some additional unique rearrangement processes. This area has been reviewed recently by Mann.[65,66]

4.5.a Iron Pentacarbonyl

Iron pentacarbonyl is a trigonal bipyramidal $M(L)_5$ complex for which the rearrangement mechanism has already been discussed. $Fe(CO)_5$

deserves special mention because of the amount of attention it has received. Sofar, the equatorial–axial exchange has proved to be too fast to measure by ^{13}C NMR, even down to $-120°C$.[60,67] This conclusion assumes that there is a significant chemical shift difference between the axial and equatorial ^{13}C nuclei; this has been questioned by Mahnke et al.[68] A solid-state NMR study[69] down to 100 K indicated some fluxional behavior with an energy barrier of ~1 kcal. More recent solid-state ^{13}C NMR results[70] indicate a shift difference of 182 Hz at 22.53 MHz and an interchange rate of $<10^2$ s^{-1} at $-38°C$. A study[71] of the polarized UV spectrum in a CO matrix at 20 K shows that there is slow interconversion at that temperature.

4.5.b Fluxional Ring Systems

The term *ring whizzers* was coined by Cotton to describe the phenomenon that was first studied quantitatively by Bennett et al.[72] in the compound $Fe(\eta^5\text{-}C_5H_5)(\eta^1\text{-}C_5H_5)(CO)_2$, whose structure follows:

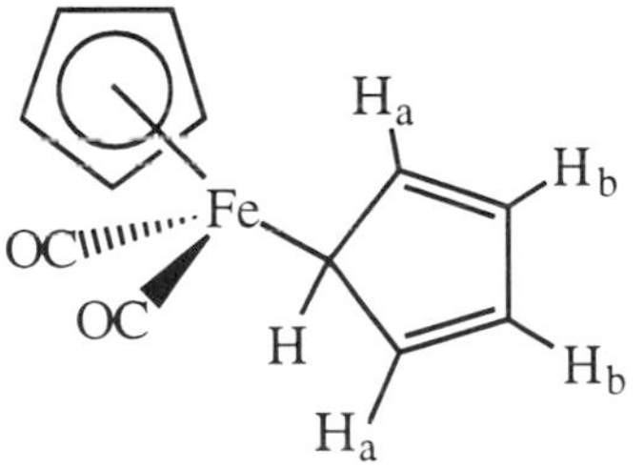

The ^1H NMR for this compound at 30°C in CS_2 shows only two peaks, one typical of the $\eta^5\text{-}C_5H_5$ hydrogens and one for all of the $\eta^1\text{-}C_5H_5$ hydrogens. If the latter is static, then it should show two peaks (due to a and b protons in the structure) plus the unique H on the same C as the Fe. At $-100°C$, the two peaks (+ coupling features) expected for the a and b sets become resolved. In addition, it was observed that the peak at lower field collapses more rapidly than the other one as the temperature is increased from $-100°C$.

These observations are consistent with the Fe moving around the $C_5H_5^-$ ring or the ring whizzing around the Fe, depending on your point of view. The mechanism could be a series of either 1,2 or 1,3 shifts of the Fe about the $\eta^1\text{-}C_5H_5$. If the shift were a random process, then all the peaks should collapse at the same rate. The two shift mechanisms are shown in Scheme 4.6, which starts at the center and goes through a 1,3 shift to the left and a 1,2 shift to the right. The change in magnetic type of the original H atoms is shown in Scheme 4.6 for each type of shift. For the 1,2 shift, the B' becomes a magnetically equivalent B, wheareas for the 1,3 shift, the A becomes a magnetically equivalent A'. Therefore, the mechanism is a 1,2 shift if the

B resonance is collapsing more slowly and a 1,3 shift if the A resonance is the one collapsing more slowly.

Scheme 4.6

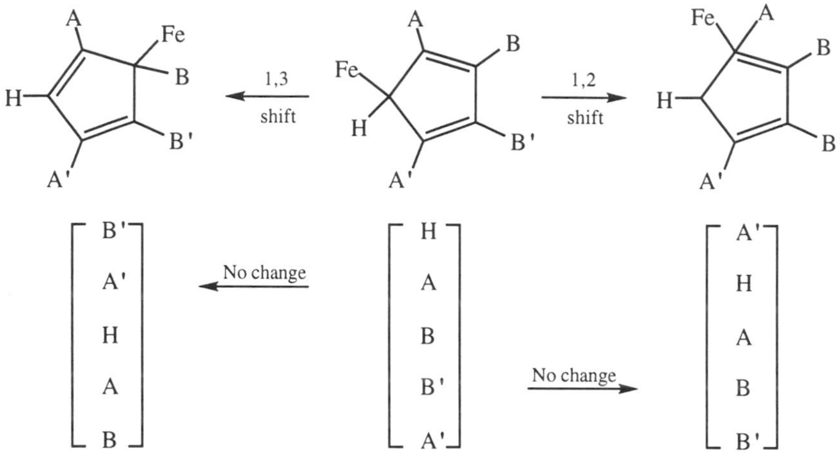

By analogy to other systems, Bennett et al. assigned the more slowly collapsing resonance to the B and B' hydrogens and concluded that the mechanism involved a 1,2 shift. However, the spectral assignment remained somewhat uncertain, and it is a common feature in this type of work that *the mechanism will rely directly on the validity of the spectral assignments for the low-temperature, nonexchanging state of the system.* To test their assignment of the spectrum and mechanism, Cotton et al.[73] prepared $Fe(\eta^5\text{-}C_5H_5)(\eta^1\text{-indenyl})(CO)_2$. They anticipated that a 1,2 shift, as shown in reaction (4.8), would be unfavorable in this system because of the loss of aromatic resonance energy in the product.

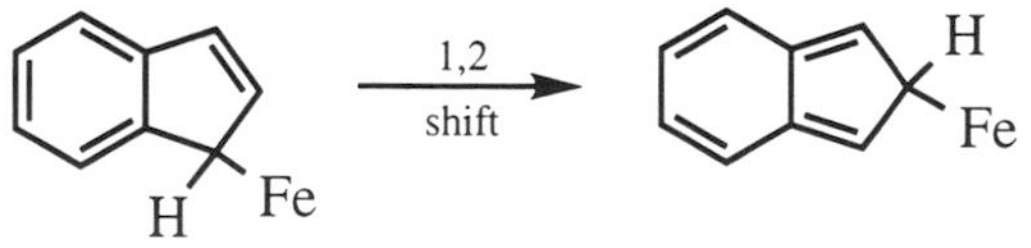

$$(4.8)$$

The compound shows no fluxional behavior, and all the expected ^1H resonances are observed in the room-temperature NMR spectrum.

4.5.c Symmetry Rules for Sigmatropic Shifts

As a result of the above and other work through the 1970s, it appeared that these fluxional processes prefer the 1,2 shift mechanism; this led to what has been termed "the principle of least motion" as being a major factor in determining the shift mechanism. Recent work has examined these processes in more detail and emphasized the orbital symmetry involved, following the Woodward–Hoffmann rules[74] for sigmatropic shifts in organic chemistry. The adaptation of these rules to inorganic transition-metal systems has had some success, but the inorganic systems have several potential complications that weaken the predictive power of the rules. The main problem is that p and d orbitals may be involved and their symmetry gives predictions opposite to those derived from organic systems where only sigma symmetry orbitals are needed. Another problem is that the inorganic systems may proceed by dissociation, so that the process is not concerted, as required for the extended Woodward–Hoffmann rules. The latter complication arises because C—H and C—C bonds are much stronger than metal–carbon bonds, so that dissociation is unlikely in the organic systems. Nevertheless, the current organometallic studies generally attempt to interpret results in terms of symmetry rules and focus on agreement and apparent exceptions to the rules.

A process is symmetry allowed if the migrating group can maintain proper overlap as it moves from the original position to make a new bond with an appropriate empty antibonding orbital, which produces the new species. In addition, the movement must be suprafacial rather than antarafacial, as described in the examples to follow.

A *1,3 sigmatropic shift* is symmetry allowed but occurs *antarafacially*. This means that the migrating group must move from one side of the molecule to the other, as shown by the orbital picture in Figure 4.10. The orbital diagram on the left in Figure 4.10 shows the unoccupied antibonding π^* orbital, which becomes the bonding π orbital in the product. To maintain proper symmetry overlap during the migration, the σ orbital must move antarafacially from the top to the bottom of the molecule. This movement is viewed as unlikely for organometallic systems, and the 1,3 process is not expected to be favorable.

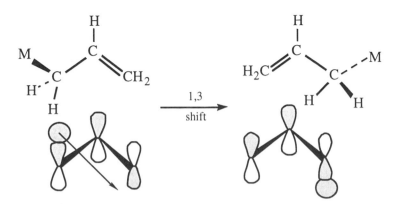

Figure 4.10. A structural and bonding representation of a 1,3 antarafacial sigmatropic shift.

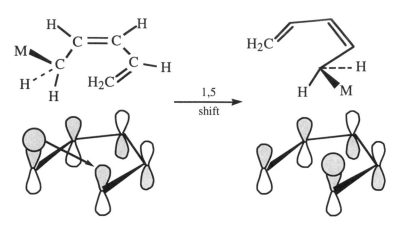

Figure 4.11. A structural and bonding representation of a 1,5 suprafacial sigmatropic shift.

A *1,5 sigmatropic shift* is symmetry allowed and occurs *suprafacially* with the migrating group staying on the same side of the molecule, as shown in Figure 4.11. This process should be facile.

Extensions to larger systems follow in an obvious way and predict that *1,7 shifts should be antarafacial and 1,9 shifts suprafacial*.

Mingos[75] extended these ideas to π-bonded cyclic systems by using a simple valence bond approach that greatly expanded the applicability of these rules in organometallic systems. Essentially, one draws the valence bond structure and moves electron pairs in the standard way of organic chemistry to obtain the product. If the rearrangement is symmetry allowed and suprafacial, then the fluxional process is expected to have a low energy barrier. For example, with the $Fe(\eta^1\text{-}C_5H_5)$ system discussed earlier, the process is shown by the following reaction:

$$Fe \quad \xrightarrow[\text{shift}]{1,5} \quad Fe \qquad (4.9)$$

Note that this corresponds to a 1,5 shift and would be predicted to be of low energy. We called this a 1,2 shift previously, but that is the same as a 1,5 shift for a C_5 ring.

An unexplained exception to the predictions is the $Re(\eta^1\text{-}C_7H_7)(CO)_5$ system,[76] which shows 1,2 (= 1,7) shifts with an energy barrier of ~80 kJ mol^{-1}. The symmetry rules predict that a 1,5 shift should be allowed, as shown in Figure 4.12. A minor 1,5 pathway has been observed with $Re(\eta^1\text{-}C_7H_7)(\eta^5\text{-}C_5H_5)(CO)_2$. Mann has suggested that the reaction

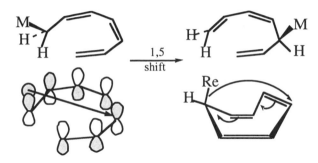

Figure 4.12. The predicted allowed shift for an η^1-C_7H_7 system.

may involve homolytic fission of the Re—C bond, or that the rules breakdown because of participation of metal p or d orbitals.

The delicate balance between dissociation and intramolecular pathways can be seen in the, albeit somewhat different, η^3-C_7H_6 complex of $Pt(PPh_3)_2$. In the initial study,[77] it was determined from saturation transfer and observation of scrambling of Pt isotopomers, that, at 80°C, the fluxionality was due to dissociation of $Pt(PPh_3)_2$ from the ring with $\Delta H^* = 26.8$ kcal mol^{-1} and $\Delta S^* = 15.1$ cal mol^{-1} K^{-1}, determined from 60°C to 80°C. Subsequent work[78] has confirmed the observations at 80°C, but shown that the process is entirely intramolecular at 60°C, possibly via a carbene intermediate. This study also showed that the bimetallic derivative $(OC)_3Mo(C_7H_6)Pt(PPh_3)_2$ shows fluxionality at lower temperature and proceeds by an intramolecular shift of $Pt(PPh_3)_2$.

The approach of Mingos can be used for higher levels of "hapticity". For example, an η^2-C_6H_6 system undergoes a favorable 1,5 shift, but an η^4-C_6H_6 system has an unfavorable 1,3 shift, as shown in Figure 4.13.

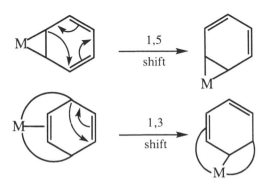

Figure 4.13. The allowed 1,5 shift for an η^2-C_6H_6 system and the forbidden 1,3 shift for an η^4-C_6H_6 system.

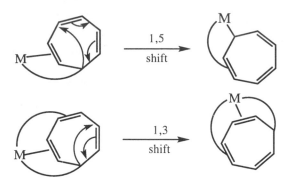

Figure 4.14. The allowed 1,5 shift for an η^3-C_7H_7 system and the forbidden 1,3 shift for an η^5-C_7H_7 system.

The η^3-C_7H_7 system is predicted to be easily fluxional by a 1,5 shift, but the η^5-C_7H_7 system should be static or highly hindered due to a forbidden 1,3 shift. Both shifts are shown in Figure 4.14. In fact, both of these systems are found to be highly fluxional.

The main exception to this prediction is $Fe(\eta^5$-$C_7H_7)(CO)_3^+$ and to a lesser extent $Mn(\eta^5$-$C_7H_7)(CO)_3$, with NMR coalescence temperatures of $-60°C$ and $20°C$, respectively.[79] The charge dependence of the activation parameters and the rather exceptional deviation from the rules has led to the proposal of a different mechanism involving a charge-separated intermediate instead of a concerted process.

For cyclooctatetraene systems, the prediction is that the η^2-C_8H_8 systems should be static because a 1,7 shift is required. The η^4-C_8H_8 systems should be fluxional by a 1,5 shift, but the η^6-C_8H_8 systems should be static because a 1,3 shift is required, as shown in Figure 4.15.

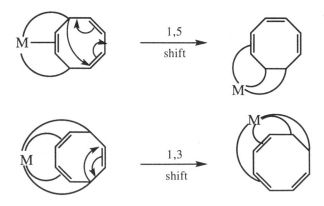

Figure 4.15. The allowed 1,5 shift for an η^4-C_8H_8 system and the forbidden 1,3 shift for an η^6-C_8H_8 system.

The η^6-C_8H_8 systems that have been studied,[80] $M(\eta^6$-$C_8H_8)(CO)_3$ (M = Cr, W), are fluxional, but with barriers >60 kJ mol^{-1} and with a mixture of shift types. This behavior has been attributed to formation of an η^4-C_8H_8 intermediate. Recently, a 1,5 shift has been identified in $Os(\eta^6$-$C_8H_8)(\eta^4$-1,5-cyclooctadiene).[81]

The general experimental observations are that the neutral complexes of η^2-C_6R_6, η^3-C_7H_7 and η^4-C_8H_8 are highly fluxional, whereas those of η^4-C_6R_6, η^5-C_7H_7, η^2-C_8H_8, and η^6-C_8H_8 are usually static or show high barriers (>80 kJ mol^{-1}).

The 1,5 shift in the η^4-C_8H_8 system predicts specific movements in the remaining ligands on the metal, as follows:

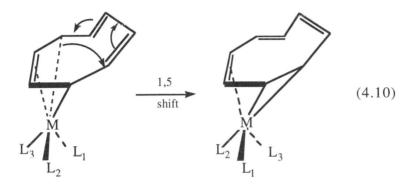

$$\xrightarrow[\text{shift}]{1,5} \tag{4.10}$$

These predictions have been tested for $Fe(\eta^4$-$C_8H_8)(CO)_2(i$-PrNC$)$ and are claimed to be confirmed.[82] The system is complicated by the fact there are two stable structural isomers with i-PrNC in positions L_1 or L_2 in the left-hand structure. However, for $Os(\eta^3$-$C_7H_7)(CO)_3(SnPh_3)$, a similar test is not consistent with the predictions, and the observations indicate that 1,2 shifts predominate.[83] The latter results seem more consistent with a slip mechanism, possibly involving a p orbital on the metal. The slip mechanism has been suggested in other cases, and one version is pictured in Scheme 4.7.

Scheme 4.7

4.5.d Rotations of π-Bonded Olefins

The bonding in π-bonded olefins consists of donation from the olefin π orbital to a metal σ orbital and back donation from a metal d to the olefin π* orbital, as shown in the following diagram:

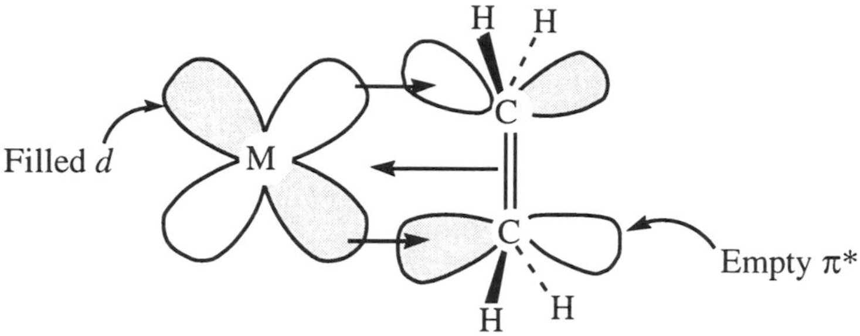

This amounts to at least a partial double bond and some restricted rotation of the olefin relative to the metal might be expected. The rotation could be about the M–olefin bond or about the C=C axis. These compounds normally have the C=C axis perpendicular to the square plane with $M(L)_3$ units and parallel to the trigonal plane with $M(L)_4$ units. Hoffmann and co-workers[84] have given an extensive theoretical analysis of the bonding and fluxional processes in these systems. They conclude that the structures of the square planar complexes are largely controlled by steric factors, while the trigonal bipyramidal complexes prefer ethylene in the trigonal xy plane because of better overlap with the hybridized d_{xy} orbital compared to the unhybridized d_{xz} orbital.

Cramer et al.[85] observed ethylene rotation in $M(\eta^5\text{-}C_5H_5)(C_2H_4)_2$ (M = Rh, Ir) complexes with the following structure:

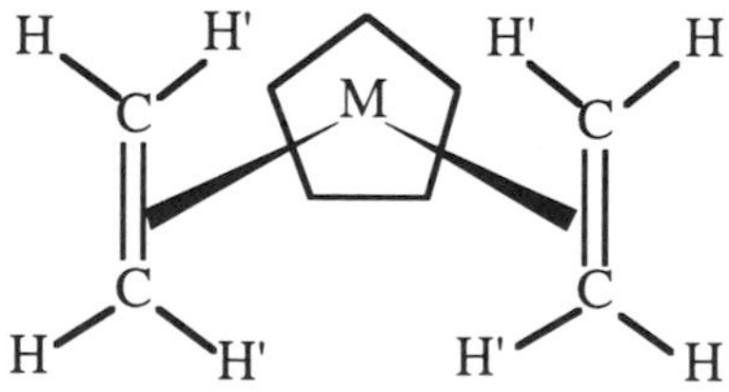

For M = Rh, the H and H' can be observed separately in the NMR spectrum at $-20°C$, but the peaks collapse to one at $57°C$. The analogous C_2F_4 compound shows no collapse of the ^{19}F spectrum up to $100°C$. The rotational barrier also is increased by CH_3 groups on the cyclopentadiene ring and by changing from Rh to Ir. These differences can be explained by steric and better back π-bonding factors, but they do not determine whether rotation is about the metal–olefin or C=C bond.

Rotation about the metal–olefin bond has been observed[86] in $Os(PPh_3)_2(CO)(NO)(C_2H_4)$, shown in the following diagram:

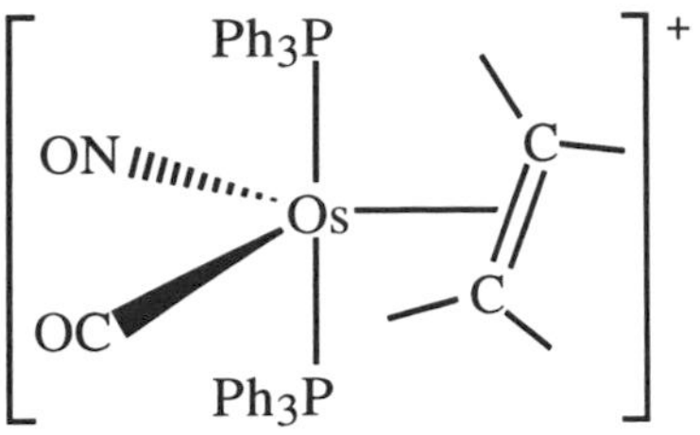

The two ends of the ethylene are different, and the ^{13}C NMR spectrum shows that they are static at $-80°C$, but interconversion is observed as the temperature is raised to $20°C$. Since the ^1H–^{31}P coupling is retained, the reaction does not involve dissociation of ethylene.

The theoretical work of Hoffmann suggests that the $(L)_4M$–ethylene systems should rearrange by a pseudorotation mechanism with the ethylene remaining in the trigonal plane. The $M(CO)_4(ethylene)$ compounds (M = Fe, Ru, Os) show axial–equatorial CO interchange.[87] Caulton and co-workers[88] have suggested that ethylene rotation may be facilitated by interaction with π^* orbitals of axial CO ligands in a transition state such as that in the following diagram. This is used to rationalize the nonfluxional character of systems that do not have a CO ligand, such as $Ir(PPhMe_2)_4(C_2H_4)^+$, $Ir(PPhMe_2)_3(CH_3CN)(C_2H_4)^+$, and $Ir(PPhMe_2)_3(CH_3)(C_2H_4)$.

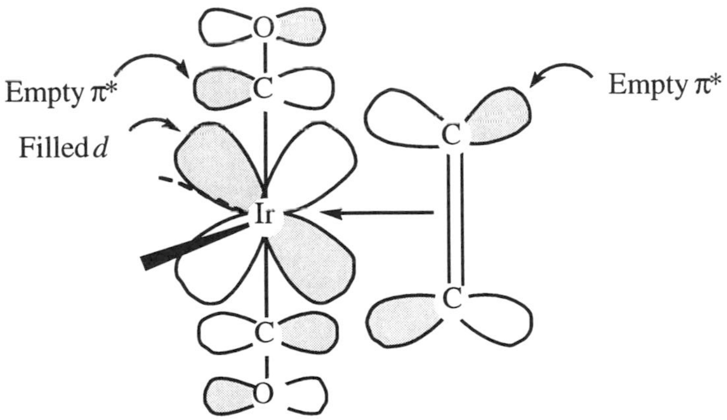

4.5.e Fluxional Allyl Complexes

The allyl group, $-C_3H_5^-$, can be bonded to a metal either as η^1-C_3H_5 or as η^3-C_3H_5, as shown in Figure 4.16. The former is normally observed with nontransition metals, and the latter is the predominant form with transition metals, but both forms have been isolated in some systems.[89]

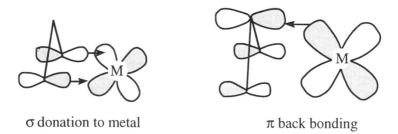

Figure 4.16. The η^1 and η^3 bonding modes of the allyl group.

In the η^1-C_3H_5 complexes, fluxional behavior involves movement of the metal from one end of the allyl ligand to the other. In η^3-C_3H_5 systems, the syn protons (H_1, H_4) and anti protons (H_2, H_3) may interchange and the C_1 and C_3 may be observed to interchange if other ligands make the two sides of the complex different. The bonding[90] and fluxional aspects[91] of these systems have been reviewed.

The η^1-C_3H_5 systems are predicted to be static by the symmetry rules, and this is the case for $Mn(CO)_5(\eta^1$-$C_3H_5)$. However, if the metal is coordinatively unsaturated (<18 valence electrons), then the systems are highly fluxional and are thought to proceed through an η^3-C_3H_5 intermediate, such as in $Pd(PPhMe_2)(NC_5H_4CO_2)(\eta^1$-$C_3H_5)$ and $Ti(NMe_2)_2(\eta^1$-$C_3H_5)$. A similar state can be attained by reversible loss of another ligand, and heterolytic bond cleavage also has been proposed for an Fe complex.[92]

The reason for restricted rotation in the η^3-C_3H_5 complexes can be understood from the following bonding pictures that depict the "sigma" donation from the ligand on the left and the back donation from the metal to the π^* orbital of the ligand on the right. Recent theoretical studies have examined the structures and reactions of these systems.[93]

σ donation to metal   π back bonding

Three processes have been suggested for the fluxionality observed in the η^3-C_3H_5 systems:

1. In the olefin rotation mechanism, the intermediate has a bond only with the C_1–C_2 end and the C_3 is free to rotate and interchange the syn and anti protons.

2. In the allyl flip mechanism, the intermediate is bonded only at C_1 and C_3, and the C_2 does an end-over flip. The syn and anti protons will interchange at both ends of the allyl ligand.
3. In the π–σ–π mechanism, the intermediate is η^1-σ bonded at one end (e.g., C_1) and the ligand is free to rotate about the C_1—C_2 bond. Each event causes syn–anti exchange at one end of the ligand.

The "π–σ–π" mechanism is now commonly accepted. Some of the evidence is from observations on the following system:[94]

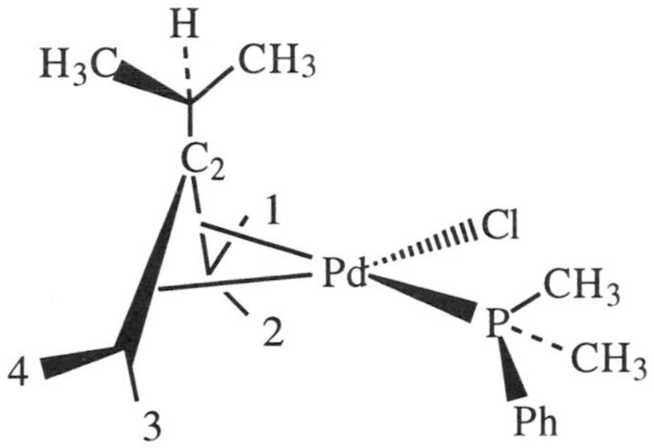

The 3 and 4 protons interchange at the same rate as the methyl protons on the phosphine and the methyl protons of the 2-isopropyl group. This is expected if the C_2 is moving above and below the P–Pd–Cl plane as in the "π–σ–π" mechanism.

4.5.f Bridge–Terminal Carbonyl Exchange

The phenomenon of bridge–terminal carbonyl exchange can occur in any dimetal system that can exist in CO-bridged and metal–metal bonded forms, such as $Co_2(CO)_8$:

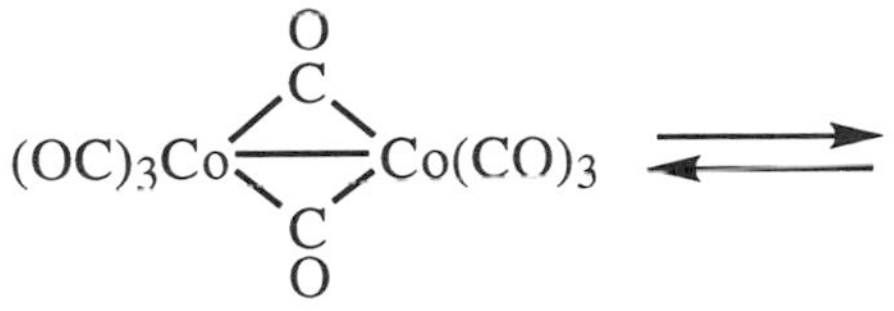

$$(OC)_3Co \rightleftharpoons Co(CO)_3 \quad\rightleftharpoons\quad (OC)_4Co-Co(CO)_4 \quad (4.11)$$

Another example[95] is $[Ru(\eta^5\text{-}C_5H_5)(CO)_2]_2$, where exchange is rapid on the NMR time scale down to $-100°C$.

A more complex and informative process has been observed with $[Fe(\eta^5\text{-}C_5H_5)(CO)_2]_2$, which has bridging and terminal CO ligands and cis and trans isomers. The temperature dependence of the cis–trans equilibrium adds a complication to the NMR analysis. Several detailed studies[96] have shown that the bridge–terminal CO exchange has a lower activation energy in the trans isomer. Furthermore, bridge–terminal exchange in the cis isomer occurs at the same rate as the cis–trans isomerization. These observations have been explained by Adams and Cotton[97] by the mechanism depicted in Scheme 4.8.

Scheme 4.8

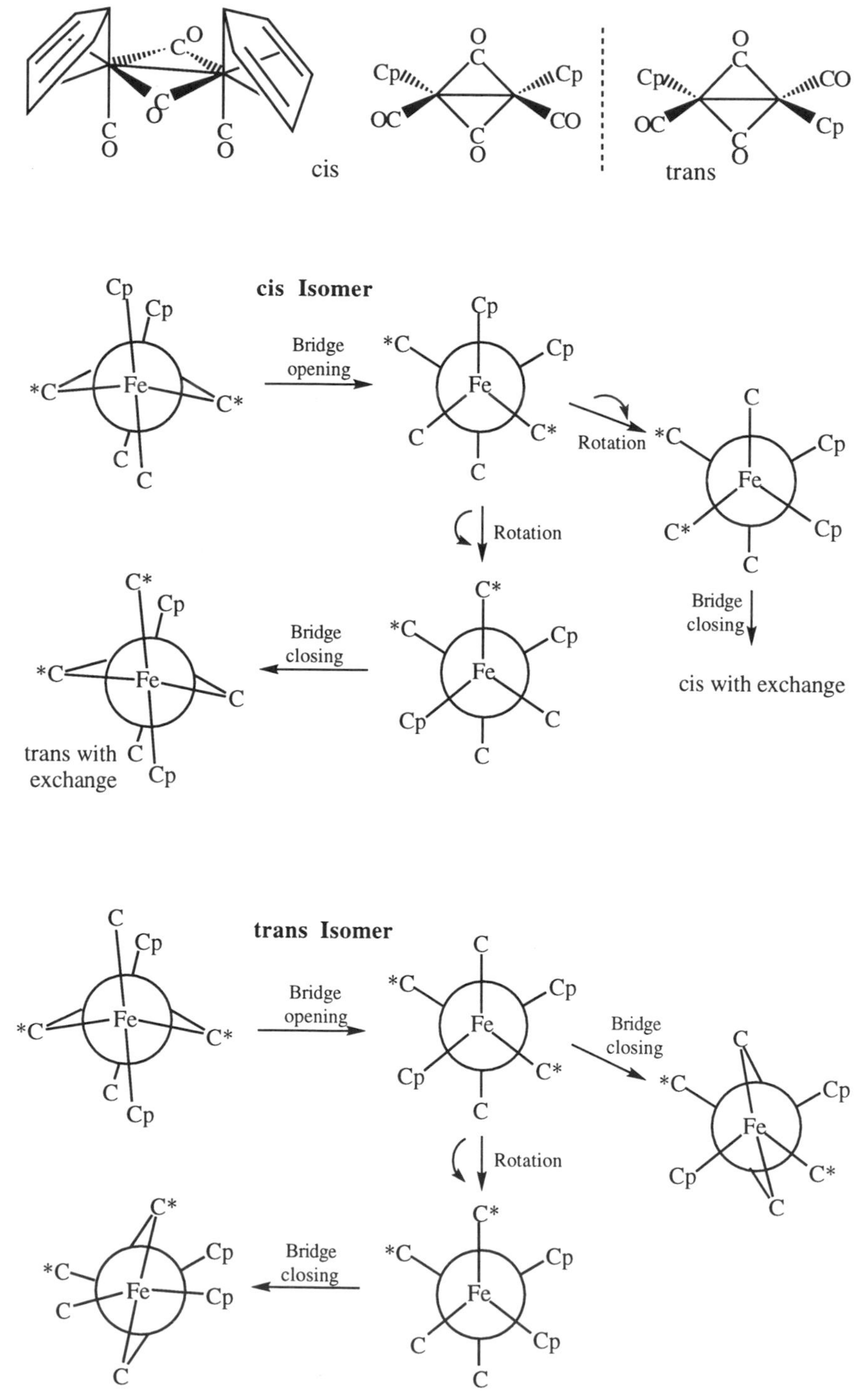

cis

trans

cis Isomer

Bridge opening

Rotation

Rotation

Bridge closing

cis with exchange

Bridge closing

trans with exchange

trans Isomer

Bridge opening

Bridge closing

Rotation

Bridge closing

The basis of the mechanism is a nonbridged intermediate, which in the cis case must undergo rotation to cause bridge–terminal exchange and isomerization. But the trans isomer can undergo exchange without rotation. It is necessary to remember that ring closing can only occur with two CO ligands on opposite sides of the molecule. This general mechanism seems to be consistent with observations on other systems. Thus, in $Fe_2(\eta^5\text{-}C_5H_5)_2(P(OPh)_3)(CO)_3$, the cis–trans isomerization and bridge–terminal exchange have the same rate because the phosphite ligand cannot occupy a bridging position.[98] Further examples with other metals are $Mn_2(\eta^5\text{-}C_5H_5)_2(NO)_2(CO)_2$, $Cr_2(\eta^5\text{-}C_5H_5)_2(NO)_4$,[99] and $Pt_2(\eta^5\text{-}C_5H_5)_2(CO)_2$.[100]

The fluxionality of $M_3(CO)_{12}$ systems (M = Fe, Ru, Os) and their derivatives has been extensively examined. The structures typically have a $(\mu\text{-}CO)_2$ bridge, but the unbridged (D_3) structure seems to be close in energy and has been characterized at 173 K in $FeRu_2(CO)_{12}$ by Braga et al.[101] Mann and co-workers[102] describe the fluxionality by a low-energy concerted bridge-opening bridge-closing mechanism around the Fe_3 plane and a higher-energy merry-go-round[103] in which the bridge opens and closes on different CO's. Johnson and co-workers[104] consider the process in terms of rotation of the M_3 plane and the $(CO)_{12}$ icosahedron.

References

1. Wilkins, R. G. *The Study of Kinetics and Mechanism of Reactions of Transition Metal Complexes*; Allyn and Bacon: Boston, 1974; Chapter 7.
2. Basolo, F.; Pearson, R. G. *Mechanisms of Inorganic Reactions*, 2nd ed.; Wiley: New York, 1967.
3. Holm, R. H.; O'Connor, M. J. *Prog. Inorg. Chem.* **1971**, *14*, 241.
4. Serpone, N.; Bickley, D. G. *Prog. Inorg. Chem.* **1972**, *17*, 391.
5. *Dynamic Nuclear Magnetic Resonance Spectroscopy*; Jackman, L. M.; Cotton, F. A., Eds.; Academic Press: New York, 1975; Chapters 8–12.
6. Beattie, J. K. *Acc. Chem. Res.* **1971**, *4*, 253.
7. McGarvey, J. J.; Wilson, J. *J. Am. Chem. Soc.* **1975**, *97*, 2531.
8. Knoch, R.; Elias, H.; Paulus, H. *Inorg. Chem.* **1995**, *34*, 4032, and references therein.
9. Ivin, K. J.; Jamison, R.; McGarvey, J. J. *J. Am. Chem. Soc.* **1972**, *94*, 1763; Campbell, L.; McGarvey, J. J.; Samman, N. G. *Inorg. Chem.* **1978**, *17*, 3378.
10. Beattie, J. K.; Moody, W. E. *J. Coord. Chem.* **1994**, *32*, 155.
11. Farina, R. D.; Swinehart, J. H. *Inorg. Chem.* **1972**, *11*, 645.
12. Balahura, R. J.; Lewis, N. A. *Coord. Chem. Rev.* **1976**, *20*, 109.
13. Burmeister, J. L. *Coord. Chem. Rev.* **1968**, *3*, 225.
14. Hubbard, J. L.; Zoch, C. R.; Elcesser, W. L. *Inorg. Chem.* **1993**, *32*, 3333.
15. Maher, J. M.; Lee, G. R.; Cooper, N. J. *J. Am. Chem. Soc.* **1982**, *104*, 6797.

16. Buckingham, D. A. *Coord. Chem. Rev.* **1994**, *135*, 587.
17. Buckingham, D. A.; Creaser, I. I.; Sargeson, A. M. *Inorg. Chem.* **1970**, *9*, 655.
18. Buckingham, D. A.; Clark, C. R.; Liddel, G. F. *Inorg. Chem.* **1992**, *31*, 2909.
19. Palmer, D. A.; van Eldik, R.; Kelm, H. *Inorg. Chim. Acta* **1978**, *30*, 83.
20. Basolo, F.; Baddley, W. H.; Weidenbaum, K. J. *J. Am. Chem. Soc.* **1966**, *88*, 1576.
21. Jackson, W. G.; Lawrance, G. A.; Lay, P. A.; Sargeson, A. M. *Aust. J. Chem.* **1982**, *35*, 1562.
22. Fairlie, D. P.; Angus, P. M.; Fenn, M. D.; Jackson, W. G. *Inorg. Chem.* **1991**, *30*, 1564.
23. Jackson, W. G.; Rahman, A. F. M. M. *Inorg. Chem.* **1990**, *29*, 3247.
24. Blackman, A. G.; Buckingham, D. A.; Clark, C. R.; Simpson, J. *J. Chem. Soc., Dalton Trans.* **1991**, 3031.
25. Jackson, W. G.; Cortez, S. *Inorg. Chem.* **1994**, *33*, 1921; Hubinger, S.; Hall, J. H.; Purcell, W. L. *Inorg. Chem.* **1993**, *32*, 2394.
26. Yeh, A.; Scott, N.; Taube, H. *Inorg. Chem.* **1982**, *21*, 2542.
27. Powell, D. W.; Lay, P. A. *Inorg. Chem.* **1992**, *31*, 3542.
28. Katz, N. E.; Fagalde, F. *Inorg. Chem.* **1993**, *32*, 5391.
29. Chou, M. H.; Brunshwig, B. S.; Creutz, C.; Sutin, N.; Yeh, A.; Chang, R. C.; Lin, C.-T. *Inorg. Chem.* **1992**, *31*, 5347.
30. Lay, P. A.; Harman, W. D. *Adv. Inorg. Chem.* **1991**, *37*, 219.
31. Damrauer, L.; Milburn, R. M. *J. Am. Chem. Soc.* **1971**, *93*, 6481.
32. Dalzell, B. C.; Eriks, K. *J. Am. Chem. Soc.* **1971**, *93*, 4298.
33. Gamsjäger, H.; Murmann, R. K. *Adv. Inorg. Bioinorg. Mech.* **1983**, *2*, 317.
34. Palmer, D. A.; Kelm, H. *J. Inorg. Nucl. Chem.* **1978**, *40*, 1095.
35. Odell, A. L.; Olliff, R. W.; Rands, D. B. *J. Chem. Soc., Dalton Trans.* **1972**, 752.
36. Lethbridge, J. W.; Glasser, L. S. D.; Taylor, H. F. W. *J. Chem. Soc. A* **1970**, 1862.
37. Lawrance, G. A.; Stranks, D. R. *Inorg. Chem.* **1977**, *16*, 929.
38. Evilia, R. F.; Young, D. C.; Reilley, C. N. *Inorg. Chem.* **1971**, *10*, 433; Ho, F. F.-L.; Reilley, C. N. *Anal. Chem.* **1969**, *41*, 1835.
39. Wilkins, R. G.; Williams, M. J. G. *J. Chem. Soc.* **1957**, 1763.
40. Basolo, F.; Hayes, J. C.; Neumann, H. M. *J. Am. Chem. Soc.* **1954**, *76*, 3807.
41. Lawrance, G. A.; Stranks, D. R. *Inorg. Chem.* **1978**, *17*, 1804; Ibid. **1977**, *16*, 929.
42. Beattie, J. K.; Binstead, R. A.; West, R. J. *J. Am. Chem. Soc.* **1978**, *100*, 3046.
43. McGarvey, J. J.; Lawthers, I.; Heremans, K.; Toftlund, H. *Inorg. Chem.* **1990**, *29*, 252.
44. Vanquickenborne, L. G.; Pierloot, K. *Inorg. Chem.* **1981**, *20*, 3673.
45. Vanquickenborne, L. G.; Pierloot, K. *Inorg. Chem.* **1984**, *23*, 1471.

46. Wilkins, R. G. *The Study of Kinetics and Mechanism of Reactions of Transition Metal Complexes*; Allyn and Bacon: Boston, 1974; p. 350.
47. Kite, K.; Orrell, K. G.; Sik, V. *Polyhedron*, **1995**, *14*, 2711.
48. Kersting, B.; Telford, J. R.; Meyer, M.; Raymond, K. N. *J. Am. Chem. Soc.* **1996**, *118*, 5712.
49. Eaton, S. S.; Eaton, G. R.; Holm, R. H. *J. Am. Chem. Soc.* **1973**, *95*, 1116.
50. Pignolet, L. H.; Duffy, D. J.; Que, L., Jr. *J. Am. Chem. Soc.* **1973**, *95*, 295; Palazzotto, M. C.; Duffy, D. J.; Edgar, B. L.; Que, L., Jr.; Pignolet, L. H. *J. Am. Chem. Soc.* **1973**, *95*, 4537.
51. Rodger, A.; Johnson, B. F. G. *Inorg. Chem.* **1988**, *27*, 3061.
52. Keppert, D. *Prog. Inorg. Chem.* **1977**, *23*, 1.
53. Holm, R. H. In *Dynamic Nuclear Magnetic Resonance Spectroscopy*; Jackman, L. M.; Cotton, F. A., Eds.; Academic Press: New York, 1975; p. 317.
54. Musher, J. I. *J. Chem. Educ.* **1974**, *51*, 94.
55. Eaton, S. S.; Hutchinson, J. R.; Holm, R. H. *J. Am. Chem. Soc.* **1972**, *94*, 6411.
56. Meakin, P.; Jesson, J. P. *J. Am. Chem. Soc.* **1973**, *95*, 7272.
57. Berry, R. S. *J. Chem. Phys.* **1960**, *32*, 933.
58. Ugi, I.; Marquarding, D.; Klusacek, H.; Gillespie, P. *Acc. Chem. Res.* **1971**, *4*, 288.
59. Strich, A. *Inorg. Chem.* **1978**, *17*, 942.
60. Meakin, P.; Muetterties, E. L.; Jesson, J. P. *J. Am. Chem. Soc.* **1972**, *94*, 5271.
61. Anderson, G. K.; Cross, R. J. *Chem. Soc. Rev.* **1980**, *9*, 185.
62. Alibrandi, G.; Scolaro, L. M.; Romeo, R. *Inorg. Chem.* **1991**, *30*, 4007; Minnitti, D. *Inorg. Chem.* **1994**, *33*, 2631.
63. Appleton, T. G.; Hall, J. R.; Ralph, S. F.; Thompson, C. S. M. *Inorg. Chem.* **1989**, *28*, 1989; Miller, S. E.; Gerard, K. J.; House, D. A. *Inorg. Chim. Acta* **1991**, *190*, 135; Mikola, M.; Arpalahti, J. *Inorg. Chem.* **1994**, *33*, 4439.
64. Murdoch, P. S.; Ranford, J. D.; Sadler, P. J.; Berners-Price, S. J. *Inorg. Chem.* **1993**, *32*, 2249.
65. Mann, B. E. In *Comprehensive Organometallic Chemistry*; Abel, E. W.; Stone, F. G. A.; Wilkinson, G., Eds.; Pergamon Press: London, 1982; Vol. 3, p. 89.
66. Mann, B. E. *Chem. Soc. Rev.* **1986**, *15*, 167.
67. Sheline, R. K.; Mahnke, H. *Angew. Chem., Int. Ed.* **1975**, *14*, 314.
68. Mahnke, H.; Clark, R. J.; Rosanske, R.; Sheline, R. K. *J. Chem. Phys.* **1973**, *60*, 2997.
69. Speiss, H. W.; Grosescu, R.; Haeberlen, U. *Chem. Phys.* **1974**, *6*, 226.
70. Hanson, B. E.; Whitmire, K. H. *J. Am. Chem. Soc.* **1990**, *112*, 974.
71. Burdett, J. K.; Gryzbowski, J. M.; Poliakoff, M.; Turner, J. J. *J. Am. Chem. Soc.* **1976**, *98*, 5728.
72. Bennett, M. J.; Cotton, F. A.; Davison, A.; Faller, J. W.; Lippard, S. J.;

Morehouse, S. M. *J. Am. Chem. Soc.* **1966**, *88*, 4371; Piper, T. S.; Wilkinson, G. *J. Inorg. Nucl. Chem.* **1956**, *3*, 104.

73. Cotton, F. A.; Musco, A.; Yagupsky, G. *J. Am. Chem. Soc.* **1967**, *89*, 6136.
74. Woodward, R. B.; Hoffmann, R. *J. Am. Chem. Soc.* **1965**, *87*, 2511.
75. Mingos, D. M. P. *J. Chem. Soc., Dalton Trans.* **1977**, 602.
76. Heinekey, D. M.; Graham, W. A. G. *J. Am. Chem. Soc.* **1982**, *104*, 915; Ibid. **1979**, *101*, 6115; *J. Organomet. Chem.* **1982**, *232*, 335.
77. Lu, Z.; Jones, W. M.; Winchester, W. R. *Organometallics*, **1993**, *12*, 1344.
78. Klosin, J.; Zheng, X.; Jones, W. M. *Organometallics*, **1996**, *15*, 3788.
79. Whitesides, T. H.; Budnik, R. A. *Inorg. Chem.* **1976**, *15*, 874.
80. Gibson, J. A.; Mann, B. E. *J. Chem. Soc., Dalton Trans.* **1979**, 1021.
81. Grassi, M.; Mann, B. E.; Spencer, C. M. *J. Chem. Soc., Chem. Commun.* **1985**, 1169.
82. Hails, M. J.; Mann, B. E.; Spencer, C. M. *J. Chem. Soc., Dalton Trans.* **1985**, 693.
83. Takats, J.; Kiel, G.-Y. *Organometallics* **1987**, *6*, 2009.
84. Albright, T. A.; Hoffmann, R.; Thibeault, J. C.; Thorn, D. L. *J. Am. Chem. Soc.* **1979**, *101*, 3801.
85. Cramer, R.; Kline, J. B.; Roberts, J. D. *J. Am. Chem. Soc.* **1969**, *91*, 2519.
86. Segal, J. A.; Johnson, B. F. G. *J. Chem. Soc., Dalton Trans.* **1975**, 677.
87. Takats, J.; Burke, M. R. *J. Am. Chem. Soc.* **1983**, *105*, 4092.
88. Lundquist, E. G.; Folting, K.; Streig, W. E.; Huffman, J. C.; Eisenstein, O.; Caulton, K. G. *J. Am. Chem. Soc.* **1990**, *112*, 855.
89. Ramdeehul, S.; Barloy, L.; Osborn, J. A. *Organometallics* **1996**, *15*, 5442; Werner, H.; Kühn. A.; Burschka, C. *Chem. Ber.* **1980**, *113*, 2291.
90. Mingos, D. M. P. In *Comprehensive Organometallic Chemistry*; Abel, E. W.; Stone, F. G. A.; Wilkinson, G., Eds.; Pergamon Press: London, 1982; Vol. 3, pp. 60–61.
91. Vrieze, K.; van Leeuwen, P. W. N. M. *Prog. Inorg. Chem.* **1971**, *14*, 1; Clarke, H. L. *J. Organomet. Chem.* **1974**, *80*, 155; Tsutsui, M.; Courtney, A. *Adv. Organomet. Chem.* **1977**, *16*, 241; Anderson, G. K.; Cross, R. J. *Chem. Soc. Rev.* **1980**, *9*, 185.
92. Rosenblum, M.; Waterman, P. *J. Organomet. Chem.* **1981**, *206*, 197.
93. Sakai, S.; Satoh, H.; Shono, H.; Ujino, Y.; *Organometallics* **1996**, *15*, 1713; Norrby, P. O.; Åkermark, B.; Haefner, F.; Hansen, S.; Blomberg, M. *J. Am. Chem. Soc.* **1993**, *115*, 4859.
94. van Leeuwen, P. W. N. M.; Praat, N. A. P.; van Diepen, M. *J. Organomet. Chem.* **1971**, *29*, 433.
95. Bullit, J. G.; Cotton, F. A.; Marks, T. J. *J. Am. Chem. Soc.* **1970**, *92*, 2155.
96. Adams, R. D.; Cotton, F. A. *J. Am. Chem. Soc.* **1973**, *95*, 6589; Gansow, O. A.; Burke, A. R.; Vernon, W. D. *J. Am. Chem. Soc.* **1972**, *94*, 2550.
97. Adams, R. D.; Cotton, F. A. In *Dynamic Nuclear Magnetic Resonance Spectroscopy*, Jackman, L. M.; Cotton, F. A., Eds.; Academic Press: New York, 1975; Chapter 12.

98. Cotton, F. A.; Kruczynski, L.; White, A. J. *Inorg. Chem.* **1974**, *13*, 1402.
99. Kirchner, R. M.; Marks, T. J.; Kristoff, J. S.; Ibers, J. A. *J. Am. Chem. Soc.* **1973**, *95*, 6602.
100. Boag, N. M.; Goodfellow, R. J.; Green, M.; Hessner, B.; Howard, J. A. K.; Stone, F. G. A. *J. Chem. Soc., Dalton Trans.* **1983**, 2585.
101. Braga, D.; Farrugia, L. J.; Gillion, A. L.; Greponi, F.; Tedesco, E. *Organometallics*, **1996**, *15*, 4684.
102. Adams, H.; Carr, A. G.; Mann, B. E.; Melling, R. *Polyhedron* **1995**, *14*, 2771; Adams, H.; Bailey, N. A.; Bentley, G. W.; Mann, B. E. *J. Chem. Soc., Dalton Trans.* **1989**, 1931.
103. Cotton, F. A.; Hunter, D. L. *Inorg. Chim. Acta* **1974**, *11*, L9.
104. Johnson, B. F. G.; Parisini, E.; Roberts, Y. V. *Organometallics* **1993**, *12*, 233; Johnson, B. F. G.; Roberts, Y. V. *J. Chem. Soc., Dalton Trans.* **1993**, 2945.

5

Reaction Mechanisms of Organometallic Systems

5.1 LIGAND SUBSTITUTION REACTIONS

The general principles discussed in Chapter 3 also apply to reactions of organometallic complexes. Because these systems do not have a wide range of structurally similar complexes with different metal atoms for comparative studies across the Periodic Table, comparisons are usually made down a particular group. However, there is a wide range of ligands available for studies of entering and leaving group effects. This area has been the subject of several recent reviews.[1-3] A major difference from the systems discussed in Chapter 3 is that many of these complexes are soluble in organic solvents, including hydrocarbons. This can minimize the complicating factor of solvent coordination, but these solvents often have quite low dielectric constants so that various types of preassociation are more probable.

5.1.a Metal Carbonyls

The metal carbonyl family of compounds is typical of the range of structures and reactivities of organometallic complexes. The rate of CO exchange was examined in early studies, and this work is the subject of a recent review.[4] The order of reaction rates is as follows:

$$V(CO)_6 > Ni(CO)_4 > Mo(CO)_6 > Cr(CO)_6 >$$
$$W(CO)_6 > Fe(CO)_5 > Mn_2(CO)_{10}$$

Where the rate law has been determined, the reaction is first order in the metal carbonyl and zero order in [CO]. This implies a **D** mechanism since a solvent intermediate is unlikely for the "noncoordinating" solvents. This mechanism also is probable for other ligand substitutions.

The main mechanistic exception to the above generalizations is $V(CO)_6$, which has an $\mathbf{I_a}$ mechanism for PR_3 substitution reactions.[5] This compound is unique in that it is the only 17-electron metal carbonyl and also is by far the most substitution labile. Some kinetic results for substitution on $V(CO)_6$ in hexane are given in Table 5.1.

Table 5.1. Rate Constants (25°C) and Activation Parameters for PR_3 Substitution on $V(CO)_6$ in Hexane

Entering Group	Cone Angle	k $(M^{-1} s^{-1})$	ΔH^* (kcal mol^{-1})	ΔS^* (cal mol^{-1} K^{-1})
PMe_3	118	132	7.6	−23.4
$P(n\text{-}Bu)_3$	132	50.2	7.6	−25.2
$P(OMe)_3$	107	0.70	10.9	−22.6
PPh_3	145	0.25	10.0	−27.8

The substitution rates have rather low ΔH^* values, and the negative ΔS^* values are typical of an associative process. The rates for various entering groups correlate with the basicity rather than the size, as measured by the cone angle. It has been suggested that formation of a 19-electron associative intermediate is much more favorable than a 20-electron intermediate from an 18-electron starting material.

Quantitative studies of CO exchange have been rather limited because of experimental difficulties, especially the low solubility of CO in most solvents. This is illustrated by work on $Ni(CO)_4$. The original study was incorrect because much of the exchange was occurring in the gas phase, where it is faster than in solution. Subsequent work[6] showed that the CO exchange and PPh_3 substitution occur at the same rate as expected for a **D** mechanism (k = 2.1 x 10^{-2} M^{-1} s^{-1} (30°C), ΔH^* = 24 kcal mol^{-1}, ΔS^* = 13.1 cal mol^{-1} K^{-1}). It seems surprising that a species with an initially small coordination number such as $Ni(CO)_4$ would react by a **D** mechanism, but it appears that 20-electron associative intermediates or transition states are less stable than the 16-electron dissociative ones because the extra electron pair must go into an antibonding orbital.

The exchange of CO on $Fe(CO)_5$ was originally reported to show two rates attributed to the axial and equatorial CO ligands. Of course, this is inconsistent with the very rapid fluxionality of $Fe(CO)_5$ discussed in Chapter 4. Later studies report[7] only one observable rate of CO exchange. The CO exchange on $Fe(CO)_5$ is too slow to measure relative to thermal decomposition. However, Basolo and co-workers[8] have estimated the rate for this exchange based on measured rates for PR_3 substitution in $Os(CO)_5$ and $Ru(CO)_5$.[9] The results are given in Table 5.2. The reactivity pattern down the group is similar to that for Group 6, discussed later, but $Fe(CO)_5$ is still surprisingly unreactive. Basolo has suggested that this could be because the {$Fe(CO)_4$} dissociative intermediate is high spin, consistent with observations of Poliakoff and Turner,[10] and therefore the process is spin forbidden. However, a spin-allowed dissociation to the higher-energy spin-paired species is possible.

Table 5.2. Rate Constants (50°C) and Activation Parameters for PR_3 Substitution on Group 5 Metal Carbonyls in Decalin

Compound	k (s^{-1})	ΔH^* (kcal mol^{-1})	ΔS^* (cal mol^{-1} K^{-1})
$Fe(CO)_5$	(6×10^{-11})	40	18
$Ru(CO)_5$	3.0×10^{-3}	27.6	15.2
$Os(CO)_5$	4.9×10^{-8}	30.6	1.33

Recent theoretical studies of Ziegler and co-workers[11] and by Ehlers and Frenking[12] have given first bond dissociation energies of 46, 33, 35 and 41, 28, 31 kcal mol^{-1}, respectively, for the Fe, Ru, and Os systems. These values are impressively similar to the ΔH^* values in Table 5.2, especially when one recognizes that they are the difference between the calculated energies of the $M(CO)_5$ ground state and the $M(CO)_4$ + CO products. The calculations of Ziegler indicate that the triplet state of $Fe(CO)_4$ is ~2 kcal mol^{-1} lower than the singlet, but both Ziegler and co-workers and Ehlers and Frenking assumed that dissociation occurred by the spin-allowed pathway to the singlet state. Ehlers and Frenking suggested that $Os(CO)_5$ proceeds by an I_a mechanism, but this is not consistent with the fact that the rate is independent of CO concentration.

Ziegler indicates that relativistic effects contribute about 3 and 10 kcal mol^{-1} to strengthen the bonds in $Ru(CO)_5$ and $Os(CO)_5$, respectively, and this difference largely accounts for the lower reactivity of $Os(CO)_5$. There is an implicit implication that the relativistic effects are relatively less important in the $M(CO)_4$ product than in the $M(CO)_5$ reactant.

The Group 6 $M(CO)_6$ compounds have been most widely studied with regard to CO exchange and substitution. The substitution reactions usually have the following two-term pseudo-first-order rate constant:

$$k_{exp} = k_1 + k_2 [L] \qquad (5.1)$$

The k_1 term is assigned to a **D** mechanism and the k_2 pathway to an I_a mechanism. Some results for CO exchange and $P(n\text{-}C_4H_9)_3$ substitution are given in Table 5.3. Clearly, the kinetic parameters for exchange and k_1 are quite similar, as expected if CO dissociation is rate controlling for both reactions. The parallel between f_{M-C} and ΔH^* and ΔH_1^* is expected for a dissociative process.

Any explanation for the lack of a smooth trend in ΔH^* down the group first requires a decision as to which member is out of line. It was argued by King[13] that $W(CO)_6$ has an especially strong M—C bond because the lanthanide contraction makes the covalent radius of W smaller than expected (Cr, 1.25 Å; Mo, 1.36 Å; W, 1.37 Å), with the

Table 5.3. Activation Parameters for CO Exchange and P(n-C$_4$H$_9$)$_3$ Substitution on M(CO)$_6$ in Decalin[a]

| Compound | Exchange | | Substitution | | | | |
	ΔH^*	ΔS^*	ΔH_1^*	ΔS_1^*	ΔH_2^*	ΔS_2^*	f_{M-C}[b]
Cr(CO)$_6$	38.7	18.5	40.2	22.6	25.5	−14.6	2.08
Mo(CO)$_6$	30.2	−0.4	31.7	6.7	21.7	−14.9	1.96
W(CO)$_6$	39.8	11.0	39.9	13.8	29.2	−6.9	2.36

[a] Activation enthalpies and entropies in kcal mol^{-1} and cal mol^{-1} K^{-1}, respectively; original sources are given in reference 11.

[b] Force constant for the M—C bond in mdyn Å^{-1}.

result that the $5d$ and maybe $4f$ orbitals give much better π back bonding. The recent theoretical studies[11,12] indicate that this argument is overly simplistic. According to Ziegler and co-workers, the π back bonding is more effective because relativistic effects raise the energy of the $5d$ π-type orbitals and reduce the energy gap with the π^* orbitals of CO so that back bonding is enhanced with W.

For phosphine substitution,[14] $\Delta V_1^* = 15$ cm^3 mol^{-1} for Cr(CO)$_6$ in cyclohexane and 10 cm^3 mol^{-1} for Mo(CO)$_6$ in isooctane. The values are positive, as expected for a dissociative process. For W(CO)$_6$, the ΔV_2^* of −10 cm^3 mol^{-1} is negative, as expected for an associative process. The negative values of ΔS_2^* in Table 5.3 also are consistent with an associative activation. The parallel between ΔH_1^* and ΔH_2^* indicates a significant amount of bond breaking in the associative transition state. Calculations by Ziegler and co-workers for an associative pathway indicate that this is only slightly less favorable than the dissociative one for Cr and Mo and is more favorable for W.

5.1.b Substituted Metal Carbonyls, M$_m$(CO)$_n$(X)$_x$

The CO exchange and substitution reactions on these systems are usually discussed in terms of the cis and trans effects of the heteroligand X and steric factors. The rate laws are similar to that described for the M(CO)$_6$ systems [Eq. (5.1)], but, depending on the system, the k$_1$ or k$_2$ path may dominate.

The Mn(CO)$_5$X (X = Cl or Br) systems have been the subject of a number of studies. Wojcicki and Basolo[15] originally reported that cis-CO exchange was much faster than trans-CO exchange. Later work[16] indicated that the two rates were within a factor of less than two of each other. However, the recent study of Atwood and Brown[17] finds that cis-CO exchange is more than 10 times faster than trans-CO exchange in

$Mn(CO)_5Br$ and $Re(CO)_5Br$. Exchange into the trans position is proposed to occur subsequent to cis exchange by a fluxional process in the intermediate. These results have been discussed in Chapter 1 with regard to the principle of microscopic reversibility.

A recurrent problem in this area and for other substitution reactions is the separation of the electronic and steric effects of the nonreacting ligands. In organometallic systems, the steric effect is usually measured by the Tolman cone angle, θ, but the electronic effect is a mixture of the σ-donor and π-acceptor abilities of the nonreacting ligands. Chen and Poë[18] have studied the reaction

$$Ru(CO)_4L + L' \xrightarrow{\ k_1\ } Ru(CO)_3(L)L' + CO \qquad (5.2)$$

and used the ^{13}C NMR chemical shift, δ, in $Ni(CO)_3L$ as a measure of the electronic factor for L to correlate the results with Eq. (5.3):

$$\log k_1 = \alpha + \beta_L\,\delta + \gamma_L\,\theta \qquad (5.3)$$

They also found that this approach correlates substitution kinetics for several other substituted metal carbonyls and for a methyl migration reaction. Giering and co-workers[19] have analyzed steric and electronic effects for a large number of systems and coined the expression "Qualitative Analysis of Ligand Effects", QUALE, for such studies. These studies measure the electronic effect by the parameter χ, derived from IR data for $Ni(CO)_3L$.[20]

The nonreacting ligand can have special properties that influence the substitution lability and mechanism. For example, $Cr(\eta^6\text{-arene})(CO)_3$[21] and $M(\eta^5\text{-}C_5H_5)(CO)_2$ compounds (M = Co, Rh, or Ir) react by an I_a mechanism.[22] This has been explained[23] as "slippage" of the $\eta^5\text{-}C_5H_5$ to an $\eta^3\text{-}C_5H_5$ form in the transition state, thereby liberating a metal orbital to accept an electron pair from the entering group. A similar argument has been used by Basolo and co-workers[24] to explain the unusually high associative substitution lability of $Rh(\eta^5\text{-indenyl})(CO)_2$, which is about 10^8 times more reactive than the ($\eta^5\text{-}C_5H_5$) analogue. This has come to be known as the *indenyl effect*. It is proposed that the intermediate formed by slippage is stabilized by the gain in resonance energy in the six-membered ring, as shown in Scheme 5.1.

Scheme 5.1

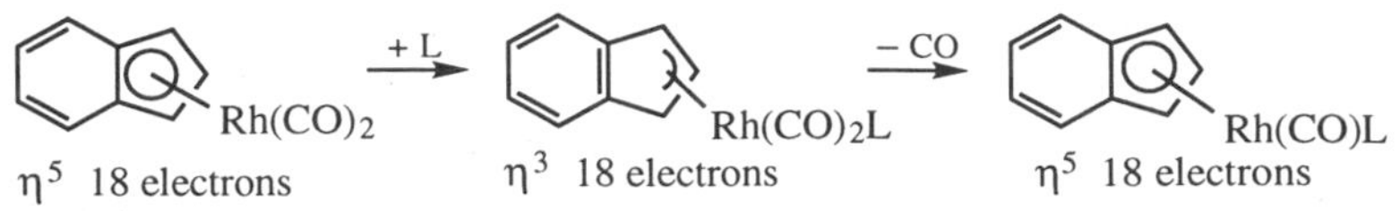

η^5 18 electrons η^3 18 electrons η^5 18 electrons

Similar rate enhancements and associative pathways were observed for indenyl complexes of iridium,[25] rhenium,[26] and manganese.[27] However, $Fe(\eta^5\text{-indenyl})(CO)_2I$[28] and $Ru(\eta^5\text{-indenyl})(PPh_3)_2I$[29] have a **D** substitution mechanism. The same mechanism is shown by the 19-electron radical $Fe(\eta^5\text{-indenyl})(CO)_3$ and it is 10^3 times less reactive than its $\eta^5\text{-}C_5H_5$ analogue.[30] It seems that factors other than slippage are contributing to the indenyl effect and the variation of the indenyl donor ability with ancillary ligands has been suggested in a recent theoretical study.[31]

The danger in making generalizations in these systems also is shown by CO exchange studies[32] on $V(\eta^5\text{-}C_5R_n)_2(CO)$ compounds. As expected, $V(\eta^5\text{-}C_5H_5)_2(CO)$ and $V(\eta^5\text{-}C_5Me_5)_2(CO)$ are quite labile and react by an I_a mechanism, k_2. This is consistent with either the 17- to 19-electron intermediate formation or slippage of the ring. The unexpected fact is that derivatives of $V(\eta^5\text{-}C_5H_7)_2(CO)$ are much more inert and show a significant dissociative reaction pathway, k_1. Some data for such systems are given in Table 5.4. Unexpectedly, the mixed complex $V(\eta^5\text{-}C_5H_5)(\eta^5\text{-}C_5H_7)(CO)$ is just slightly more reactive than $V(\eta^5\text{-}C_5H_7)_2(CO)$ and quite different from $V(\eta^5\text{-}C_5H_5)_2(CO)$. Basolo et al. attributed this great difference in reactivity to electronic factors related to the poorer donation of electrons to the metal from $\eta^5\text{-}C_5H_7$ than from $\eta^5\text{-}C_5H_5$. There is some evidence for this in the CO stretching frequencies, ν_{CO}, which are 1959 cm^{-1} in the former and 1881 cm^{-1} in the latter. The lower value indicates more back donation into the π^* orbital of CO and a stronger M—C bond. In any case, it is fair to say that these reactivity differences were quite unexpected.

Nitrosyl ligands tend to favor associative activation. For example, $Co(NO)(CO)_3$ and $Fe(NO)_2(CO)_2$ are isoelectronic with $Ni(CO)_4$ but have associative substitution mechanisms.[33] Basolo has suggested that if the NO is formally regarded as NO$^+$, then its stronger π-acceptor ability relative to CO can be rationalized, and NO$^+$ will withdraw more electron density from the metal and favor associative activation.

Table 5.4. Rate Constants (60°C) and Activation Parameters for CO Exchange on $V(\eta^5\text{-}C_5R_n)_2(CO)$ Systems in Decalin[a]

Compound	k_1	ΔH_1^*	ΔS_1^*	k_2	ΔH_2^*	ΔS_2^*
$V(\eta^5\text{-}C_5H_5)_2(CO)$	$\sim 10^{-4}$			8.0×10^2	~ 6	~ -30
$V(\eta^5\text{-}C_5Me_5)_2(CO)$	$\sim 10^{-4}$			2.6×10^2	8.9	-21
$V(\eta^5\text{-}C_5H_5)(\eta^5\text{-}C_5H_7)(CO)$	3×10^{-4}			5.7×10^{-3}		
$V(\eta^5\text{-}C_5H_7)_2(CO)$	8×10^{-6}	28.1	2	3.8×10^{-3}	22.7	-2

[a] k_1 (s^{-1}), k_2 (M^{-1} s^{-1}), ΔH^* (kcal mol^{-1}), and ΔS^* (cal mol^{-1} K^{-1}).

5.1.b.i Cis-Labilizing Effect

The cis-labilizing effect has been discussed previously, and the order of cis-labilizing influence is opposite to that of the trans effect, with π-donor ligands being the most cis-labilizing. The work of Darensbourg and co-workers[34] indicates that oxygen-donor ligands are especially effective for the cis labilization of CO ligands, and the use of phosphine oxides and acetate ion in the synthesis of specifically labeled metal carbonyls has been especially useful. More recently, they have demonstrated[35] the cis effect in $W(CO)_5F$ and found that 1,2-substituted benzene ligands can stabilize the coordinatively unsaturated species to the extent that they can be fully characterized.[36]

Rossi and Hoffmann[37] suggested that poor or non π acceptors prefer the equatorial site in a d^6 square pyramidal dissociative intermediate. The same conclusion was reached by Lichtenberger and Brown[38] for $Mn(CO)_5X$ systems. This prediction has been used in stereoselective labeling work, since it predicts that such heteroligands will minimize the amount of scrambling due to fluxionality in the intermediate. On the other hand, if the heteroligand is a better σ donor and π acceptor than CO, then the heteroligand should favor the axial position in the square pyramid and should minimize fluxionality. For example, this approach has been exploited[39] with CS as the heteroligand for the following synthesis:

$$
\begin{bmatrix}
& \overset{\displaystyle CO}{\underset{\displaystyle OC}{SC-W-I}}{}^{,CO}
\end{bmatrix}^{-}
\xrightarrow{Ag^+}
\left\{
\overset{\displaystyle CO}{\underset{\displaystyle OC}{SC-W}}{}^{,CO}_{CO}
\right\}
\xrightarrow{{}^{13}CO}
\overset{\displaystyle CO}{\underset{\displaystyle OC}{SC-W-{}^{13}CO}}{}^{,CO}_{CO}
\qquad (5.4)
$$

5.1.b.ii Heteroligand Replacement

For $M(CO)_5L$ systems, the kinetic order for replacement of the leaving group L in a first-order process is generally as follows:

$$py > AsPh_3 > CO \approx PPh_3 > P(OPh)_3 > P(OMe)_3 > P(n\text{-Bu})_3$$

High reactivity may be associated with poor π-acceptor ability, as for py and $AsPh_3$, or with poor σ-donor ability, as for CO. The phosphites are considered to be better π acceptors than the phosphines, but steric factors enter as a compensating factor. This type of dichotomy pervades the interpretation of reactivity patterns for these systems.

For $cis\text{-}Mo(CO)_4(L)_2$ complexes, steric effects seem to dominate the substitution reactions. For L = PPh_3, the rate of L displacement by CO is ~200 times larger than that for L = $PMePh_2$. The ground-state structures[40] show that the P—Mo—P angle is distorted from 90° to

104.6° in the former and only to 92.5° in the latter. The distortion of the structure is relieved in the dissociative transition state. It should be noted that the Mo—P bond lengths are very similar, 2.577 Å for PPh_3 and 2.555 Å for $PMePh_2$.

The replacement of the amine in $M(CO)_5(NHR_2)$ by phosphines is catalyzed by phosphine oxides. This has been ascribed[41] to a preassociation phenomenon involving hydrogen bonding of the phosphine oxide to the amine hydrogen, thereby weakening the M—N bond. With $OP(n-Bu)_3$ and for M = Mo and R = NHC_5H_{10}, the kinetics show a saturation effect with the concentration of oxide. In hexane at 34.5°C, the equilibrium constant for formation of the adduct is 600 M^{-1} and the following structure was proposed.

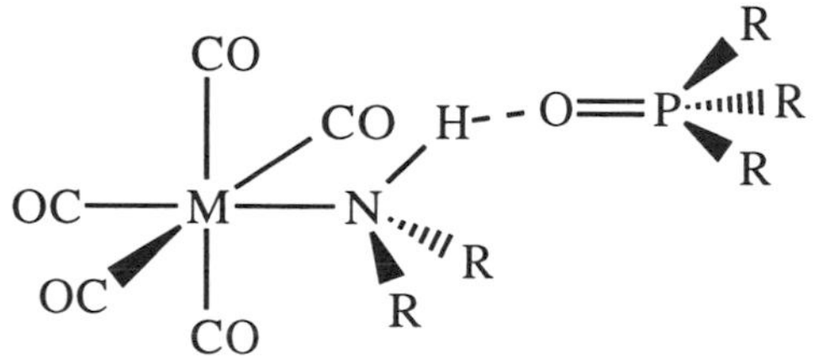

This is a particular case of general Lewis base catalysis of a substitution reaction. It may be troublesome for other studies because of the ease with which some phosphines are oxidized to phosphine oxides. This type of preassociation and catalysis will be favored by the low-polarity solvents used in this area and would probably be unobserved in more polar and hydroxylic solvents.

A coordinatively unsaturated species,[42] $W(CO)_3(PCy_3)_2$, has been found to undergo very rapid substitution,[43] with rate constants in the range of 10^3 to 10^6 M^{-1} s^{-1}, depending on the steric bulk of the entering group. The reactant is stabilized by an "agostic bond" to an H from a Cy ring. If these H atoms are replaced by D atoms, the rate of substitution of $P(OMe)_3$ increases by a factor of 1.15 at 25°C in toluene.

Chelate ring-opening processes also have been studied in these systems. Graham and Angelici[44] studied the reaction of $W(CO)_4(bpy)$ with phosphites and found that the pseudo-first-order rate constant is given by

$$k_{exp} = k_1 + k_2 \text{[phosphite]} \tag{5.5}$$

Memerring and Dobson[45] noted that the values of k_1 given by Graham and Angelici were not the same for different phosphites, and this is not consistent with a **D** mechanism for the k_1 path. They suggested that the reaction was proceeding through a ring-opened intermediate, as shown in Scheme 5.2, and that the apparent k_1 values are a composite of these two processes that resulted from an incomplete rate law in the original work. The **A** mechanism of the k_2 path is not shown in Scheme 5.2.

Scheme 5.2

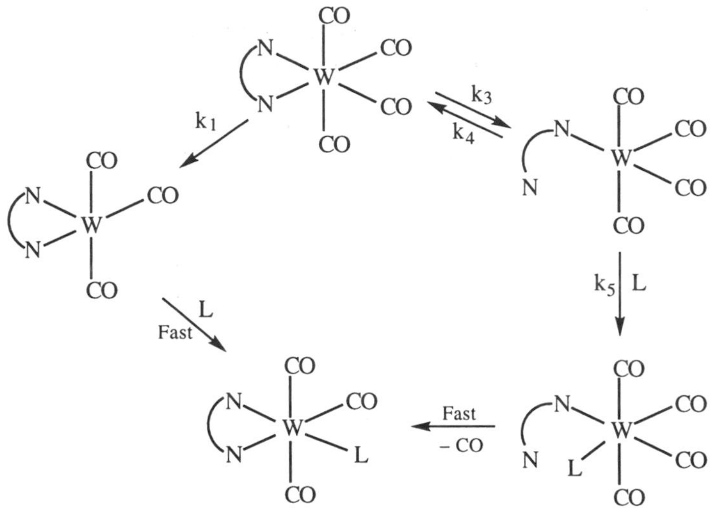

The experimental rate constant predicted from Scheme 5.2 is given by

$$k_{exp} = k_1 + k_2 [L] + \frac{k_3 k_5 [L]}{k_4 + k_5 [L]} \tag{5.6}$$

A detailed analysis confirmed this rate law and gave consistent k_1 values. More recent work is given by Dobson and co-workers[46] for other chelates and for systems in which the chelate is actually displaced. These reactions generally conform to the preceding mechanism. The volumes of activation have been measured[47] for displacement of sulfur-bonding chelates from $Cr(CO)_4(S \frown S)$ and the values of 14 cm^3 mol^{-1} are consistent with the **D** ring-opening process.

The mechanism of chelate ring closure in such systems bears directly on the ring-opening process because of microscopic reversibility. Eyring and van Eldik and co-workers[48] have studied a number of such reactions of Group 6 metal carbonyls. Recently, they examined ring-closing reactions of $Mo(CO)_5(phen)$ and substituted phenanthrolines and found rate constants varying from 10^4 to 10^{-2} s^{-1}. They concluded from ΔV^* values that slower reactions with bulky and less basic phenanthroline derivatives are controlled by release of CO, but the mechanism changes to I_a as the steric constraints are removed, and the phenanthroline becomes more basic.

5.1.c Metal–Metal Bonded Carbonyls and Clusters

Metal–metal bonded carbonyl systems have received considerable attention and have been the subject of some controversy in recent years. The reactions may proceed by a normal substitution reaction or by homolytic cleavage of the M—M bond, as shown in Scheme 5.3, to give radicals that then undergo rapid substitution before recombining to form the products. The homolysis is a possibility because of the weakness of the M—M bonds.

Scheme 5.3

$$(OC)_nM—M(CO)_n + L \longrightarrow (OC)_nM—M(CO)_{n-1}L + CO$$

$$(OC)_nM—M(CO)_n \underset{}{\overset{Homolysis}{\rightleftharpoons}} 2\ \{\bullet M(CO)_n\}$$

$$\{\bullet M(CO)_n\} \longrightarrow \{\bullet M(CO)_{n-1}L\} + CO$$

$$\{\bullet M(CO)_n\} + \{\bullet M(CO)_{n-1}L\} \longrightarrow (OC)_nM—M(CO)_{n-1}L$$

The first study on $Mn_2(CO)_{10}$ by Wawersik and Basolo[49] indicated a **D** mechanism for PR_3 substitution because the rate was inhibited by free CO. However, later work by Poë and co-workers[50] claimed to find no CO inhibition, and the homolytic cleavage mechanism in Scheme 5.3 was proposed. It is important to note that these were primarily studies of the decomposition of $Mn_2(CO)_{10}$ either in the absence or presence of O_2 and mainly in decalin, at temperatures of 115°C to 180°C. Poë and co-workers found that the rate is half-order in $[Mn_2(CO)_{10}]$ under an argon atmosphere and proposed the following mechanism:

$$Mn_2(CO)_{10} \underset{k_{-1}}{\overset{k_1}{\rightleftharpoons}} 2\ \{Mn(CO)_5\} \overset{k_2}{\longrightarrow} Products \qquad (5.7)$$

If the dimetal and monometal species are represented by M_2 and M, respectively, and a steady state is assumed for M, then the rate law can be developed as follows:

$$2\,k_1\,[M_2] - 2\,k_{-1}\,[M]^2 - k_2\,[M] = 0 \qquad (5.8)$$

and

$$\frac{d\,[M_2]}{d\,t} = k_1\,[M_2] - k_{-1}\,[M]^2 = \frac{k_2\,[M]}{2} \qquad (5.9)$$

The first equation can be solved for [M] in terms of $[M_2]$ to give

$$[M] = \frac{k_2}{4\,k_{-1}}\left[-1\pm\left(1+\frac{16\,k_{-1}\,k_1}{(k_2)^2}[M_2]\right)^{1/2}\right]$$

$$= \frac{k_2}{4\,k_{-1}}\left[-1+(1+a)^{1/2}\right] \tag{5.10}$$

and the positive root is chosen because [M] must be positive. If the term $1+(1+a)^{1/2}$ is multiplied and divided into Eq. (5.10), then one obtains

$$[M] = \frac{k_2}{4\,k_{-1}}\left(\frac{a}{1+(1+a)^{1/2}}\right) \tag{5.11}$$

Substitution for [M] in Eq. (5.9) gives

$$-\frac{d\,[M_2]}{dt} = \frac{(k_2)^2}{8\,k_{-1}}\left(\frac{a}{1+(1+a)^{1/2}}\right) \tag{5.12}$$

If $a^{1/2} \gg 1$, which requires that $a \gg 1$, then Eq. (5.12) yields

$$-\frac{d\,[M_2]}{dt} = \frac{(k_2)^2}{8\,k_{-1}}(a)^{1/2} = \frac{k_2}{2}\left(\frac{k_1\,[M_2]}{k_{-1}}\right)^{1/2} \tag{5.13}$$

Therefore, the rate predicted by this mechanism can be half-order in $[M_2]$. It remains questionable as to whether the assumption about the magnitude of a is reasonable. More recent work[51] has shown that the recombination is nearly diffusion controlled ($k_{-1} \approx 10^9$ M^{-1} s^{-1}).

Later studies have questioned the observations of Poë and co-workers and especially their relevance to the mechanism for CO substitution on $Mn_2(CO)_{10}$. Sonnenberger and Atwood[52] studied the reaction of $(OC)_5Mn$—$Re(CO)_5$ with PR_3 and found no $Mn_2(CO)_9(PR_3)$ or $Re_2(CO)_9(PR_3)$ products. They expected the latter to form by radical recombination, but this expectation has been questioned by Poë on the basis of probable lifetimes and reactivities of the $\{\bullet M(CO)_5\}$ intermediate. Atwood has argued that the time dependence of the product distribution also is inconsistent with a radical mechanism, since $(OC)_5Mn$—$Re(CO)_4(PR_3)$ and $(R_3P)(OC)_4Mn$—$Re(CO)_5$ increase in concentration and then decay as the disubstituted species is formed. Muetterties and co-workers[53] found that $^{185}Re_2(CO)_{10}$ does not give isotopic scrambling with $^{187}Re_2(CO)_{10}$ under thermal decomposition conditions, even when CO is added to suppress the decomposition. They also found that $Mn_2(^{12}CO)_{10}$ and $Mn_2(^{13}CO)_{10}$ show no CO scrambling in octane at 120°C over time periods in which there is substantial exchange with free CO. In addition, the half-times for CO exchange and PPh_3 substitution are 45 and 46 minutes, respectively, indicative of the same rate-controlling process for both reactions. The various details of this problem are discussed in an exchange of notes between Poë[54] and Atwood.[55]

There have been a number of studies of substitution reactions on metal carbonyl clusters derivatives of the general form $M_m(CO)_n(L)_p$. These include examples of Co_4,[56] Rh_4,[57] Ir_4,[58] Ru_3,[59] and Os_3.[60] The substitution reactions generally have the two-term rate law given by Eq. (5.1). Studies of the derivatives indicate that the dissociative pathway, k_1, is enhanced by the number and steric bulk of the L hetero-ligand. Poë and Brodie[61] have recently correlated such reactions for $Ru_3(CO)_{11}L$ and $Ru_3(CO)_{10}(L)_2$ systems by Eq. (5.3) and found a strong correlation of $\log k_1$ with the cone angle of the spectator ligand L. Basolo and Shen[62] have studied anion activation in $Os_3(CO)_{11}X^-$ systems with X = Cl, Br, I, NCO and conclude that the k_1 pathway is facilitated by rearrangement to a bridging X intermediate prior to substitution.

The associative pathway for CO substitution on $Os_3(\mu_2-H)_2(CO)_{10}$ has been extensively studied and analyzed.[63] The rate constant, k_2, shows a steric threshold of ~147° and decreases for cone angles above this value. The ΔV^* values are negative and correlate with the cone angle function $\tan^2(\theta/2)$, because the latter is related to the ligand volume. The reactivities of $Ru_6C(CO)_{17}$[64] and $Ru_5C(CO)_{15}$[65] have been correlated to entering group size and nucleophilicity, and the latter system shows adduct formation for ligands with $\theta \leq 133°$. Basolo and co-workers[66] have observed adduct intermediates with $Ru_3(\mu_3-\eta^2-(2-PhNpy)(CO)_9)^+$ which they ascribe to opening of the μ_3 bridge to a μ_1 form.

5.1.d Radical Pathways for Ligand Substitution

Although the question of radical pathways in the metal–metal bonded systems remains doubtful, there are examples of organometallic radical reactions, and they may be more prevalent than originally expected.

Absi-Halabi and Brown[67] found that substitution on $Cl_3Sn—Co(CO)_4$ shows properties characteristic of a radical process. The reaction is catalyzed by light and inhibited by radical traps, such as O_2 and galvinoxyl. The proposed process is given in Scheme 5.4, followed by an assortment of radical recombination reactions and electron transfers to the reactant.

Scheme 5.4

$$Cl_3Sn—Co(CO)_4 + L \longrightarrow Cl_3(L)Sn—Co(CO)_4$$

$$Cl_3(L)Sn—Co(CO)_4 \longrightarrow \{Cl_3(L)Sn^\bullet\} + \{^\bullet Co(CO)_4\}$$

$$\{^\bullet Co(CO)_4\} + L \longrightarrow \{^\bullet Co(L)(CO)_3\} + CO$$

Byers and Brown[68] observed that $Re(CO)_5H$ is inert toward phosphine substitution at 60°C in hexane in the dark under an N_2 atmosphere. However, any sort of radical initiator gave complete substitution in a few hours, and the reactive species was proposed to be $\{\bullet Re(CO)_5\}$. On the other hand, $Mn(CO)_5H$ appears to undergo normal substitution[69] without evidence for a radical process. However, Sweany and Halpern[70] reported that the hydrogenation of $PhMeC=CH_2$ by $Mn(CO)_5H$ proceeds by a radical pathway, based on the observation of chemically induced dynamic nuclear polarization, CIDNP, in the proton NMR of the methyl styrene. The radicals initially formed by hydrogen atom abstraction by methyl styrene from $Mn(CO)_5H$ can either recombine or diffuse apart and undergo further reaction. Bullock and Samsel[71] have suggested a similar mechanism for the reactions of several metal carbonyl hydrides with α-cyclopropylstyrene. This work also provides relative rates for H-atom abstraction and cyclopropyl ring opening by the organic radical.

The mechanism of substitution on the 17-electron radical species has been the subject of recent work. For $\{\bullet Mn(CO)_5\}$, Herrinton and Brown[72] have shown that the substitution is a bimolecular process, with the rate constants for different entering groups given in Table 5.5. It also was concluded from this study that decomposition of $\{\bullet Mn(CO)_5\}$ by loss of CO must have a rate constant <90 s^{-1}. The reactions of CO with the substituted radicals $\{\bullet Mn(CO)_3(PR_3)_2\}$ are second order, with rate constant values of 42 and 0.32 M^{-1} s^{-1} (24°C in hexane)[73] for R = *n*-Bu and *i*-Bu, respectively. Poë and co-workers[74] have shown that substitution on $\{\bullet Re(CO)_5\}$ is a second-order process.

Trogler and co-workers[75] studied the reactions of $\{\bullet Fe(CO)_3(PR_3)_2{}^+\}$ radicals and found that CO substitution is a second-order process whose rate depends on the steric bulk of the entering group. The rate constants correlate with the pK_a of the pyridine nucleophiles. For pyridine reacting with $\{\bullet Fe(CO)_3(PPh_3)_2{}^+\}$ (25°C in CH_2Cl_2), the parameters are k = 13.6 M^{-1} s^{-1}, $\Delta H^* = 9.8$ kcal mol^{-1}, $\Delta S^* = -21$ cal mol^{-1} K^{-1}; these values are typical of the other systems.

Table 5.5. Rate Constants (24°C) for the Reaction of $\{\bullet Mn(CO)_5\}$ with Various Entering Groups in Hexane

Entering Group	k (M^{-1} s^{-1})
PPh_3	1.7×10^7
$AsPh_3$	6.5×10^4
$P(n\text{-Bu})_3$	1.0×10^9
$P(i\text{-Pr})_3$	6.7×10^7
$P(O\text{-}i\text{-Pr})_3$	3.1×10^7

Theoretical aspects of the substitution on 17-electron systems, including an analysis of the geometries of the radicals and the direction of nucleophilic attack, have been discussed recently by Therien and Trogler.[76] The prevalence of associative attack is consistent with the observations of Basolo and co-workers[5] on various 17-electron V(0) species.

5.2 INSERTION REACTIONS

5.2.a CO "Insertion"

The classic example of a CO insertion reaction is

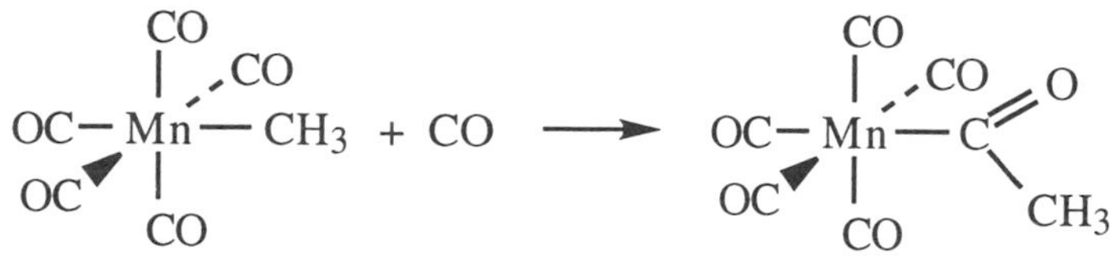

$$
\begin{array}{c}
\text{CO} \\
| \quad ,\!\text{CO} \\
\text{OC—Mn—CH}_3 + \text{CO} \quad \longrightarrow \quad \text{OC—Mn—C} \overset{\text{O}}{\underset{\text{CH}_3}{\diagup}} \\
\text{OC} \quad | \\
\text{CO}
\end{array}
\qquad (5.14)
$$

The CO insertion can be an important step in carbon–carbon bond-forming reactions that are catalyzed by organometallic complexes. At first sight, this appears to be an insertion of the entering CO into the Mn—CH$_3$ bond and the name insertion has continued to be used. However, phosphines and other nucleophiles (L) bring about an analogous transformation, as shown in the following reaction:

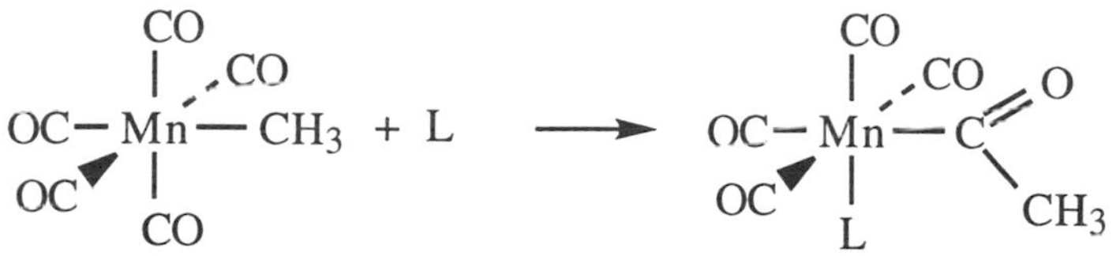

$$
\begin{array}{c}
\text{CO} \\
| \quad ,\!\text{CO} \\
\text{OC—Mn—CH}_3 + \text{L} \quad \longrightarrow \quad \text{OC—Mn—C} \overset{\text{O}}{\underset{\text{CH}_3}{\diagup}} \\
\text{OC} \quad | \\
\text{CO} \quad\quad\quad\quad \text{L}
\end{array}
\qquad (5.15)
$$

This could still be an insertion of a coordinated CO into the M—CH$_3$ bond, but it might also be migration of the CH$_3$ to a coordinated CO. The classic IR study of Noack and Calderazzo[77] using ^{13}C-labeled CO showed that the reaction is actually -CH$_3$ migration. The original work has been confirmed by ^{13}C NMR studies by Flood et al.[78]

The basis of the mechanistic conclusion for this system relies on the product distribution from the reverse reaction (decarbonylation of CO), as shown in Scheme 5.5. If the mechanism is CH$_3$ migration, then the products should be 25 percent with the labeled CO trans to CH$_3$, 50 percent with the labeled CO cis to the CH$_3$, and 25 percent with no label, as shown in the diagram. If the reaction goes by insertion of a coordinated CO, then the products should be 75 percent with labeled

CO in the cis position and 25 percent with no label. The results of both studies give a *product distribution consistent with CH₃ migration.*

Scheme 5.5

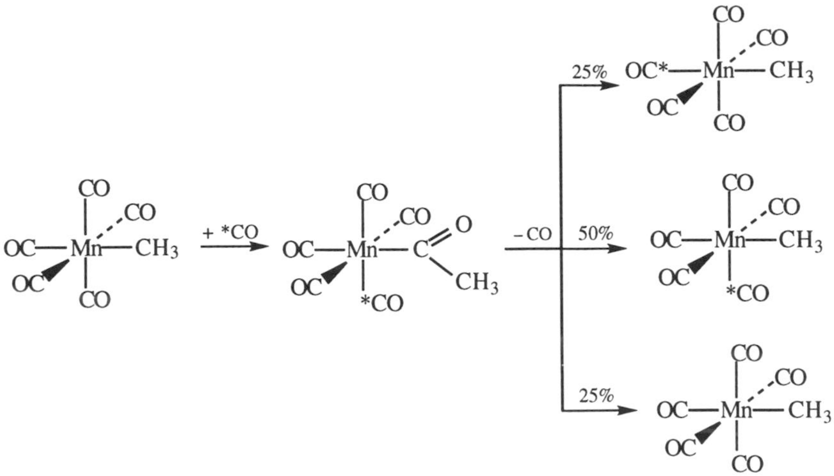

For many other metal carbonyls the mechanistic details are not known, and it is assumed that they are also proceeding by the methyl migration mechanism. The results of Wright and Baird[79] on the reaction of ^{13}CO with $Fe(CO)_2(PMe_3)(CH_3)I$ are consistent with methyl migration, but the interpretation is complicated by I^- dissociation. Jablonski et al.[80] found, from the ^{13}C NMR spectrum at 203 K, that $Fe(CO)_2(PPh_2Me)(CH_3)I$ exists in equilibrium with the η^2-acyl form, and the temperature dependence of formation of the latter gave values of $\Delta H° = -5.4$ kJ mol^{-1} and $\Delta S° = -6.5$ J mol^{-1} K^{-1}. Saturation transfer NMR experiments at 223 K showed that the acyl carbon exchanged specifically into the site cis to the CH_3 of its isomer.

The CO insertion in optically active $Fe(\eta^5-C_5H_5)(CO)(PR_3)(CH_3)$ was studied by Flood and Campbell,[81] who expected the products to reflect which group migrates, as shown in Scheme 5.6. The results indicated that in nitromethane and acetonitrile the products are consistent with methyl migration, while in dimethylsulfoxide, *N,N*-dimethylformamide, propylene carbonate, and hexamethylphosphoramide, the products imply CO migration. However, the possible intervention of the η^2-acyl intermediate, shown in braces in Scheme 5.6, makes the interpretation less than definitive with regard to the migrating group.

Brunner and co-workers[82] have shown that methyl migration occurs in the system studied by Flood if the reaction is catalyzed by BF_3. This catalysis is consistent with BF_3 complexing with the oxygen of CO, thereby decreasing the electron density on the C and promoting CH_3 migration.

Scheme 5.6

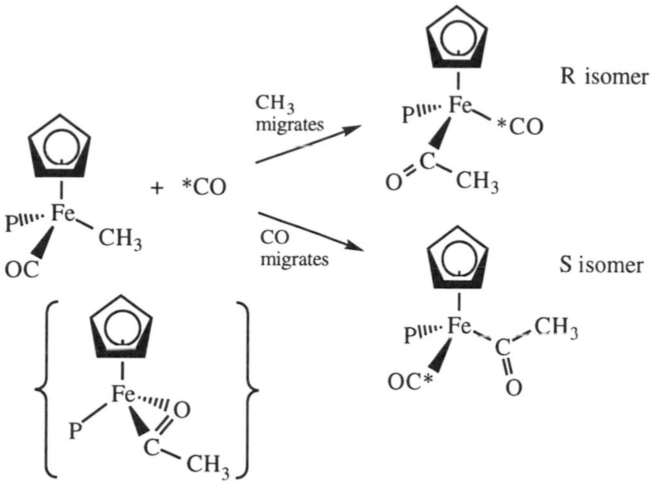

By using an unsymmetrical phosphine chelate on Pt(II) and ^{31}P NMR, it has been shown[83] that Ph migration occurs through identification of the isomers of the insertion products in Scheme 5.7.

Scheme 5.7

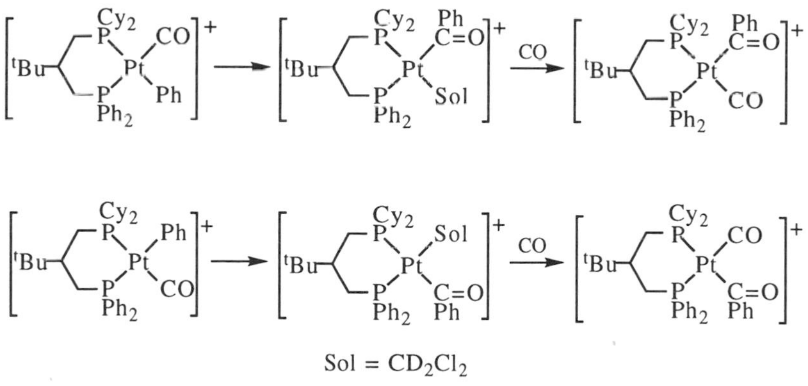

Sol = CD_2Cl_2

For the analogous CH_3 derivative of Pd(II), CH_3 migration was found. Qualitative rate observations indicated that the least stable isomer, based on the trans effect, is the most reactive in these systems.

The kinetics for these reactions with $Mn(CO)_5(CH_3)$ are consistent with a **D** mechanism through formation of an unsaturated intermediate, as shown in Scheme 5.8. The intermediate may be stabilized by solvent coordination, by an η^2-acyl interaction and/or by an agostic interaction with an H of the migrating group.

Scheme 5.8

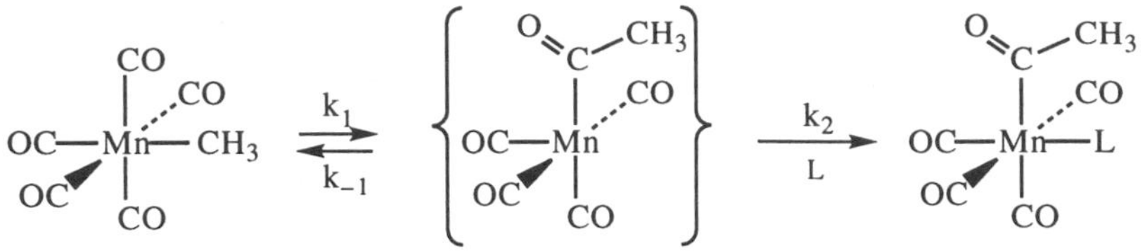

The pseudo-first-order rate constant ($[L] \gg [Mn]$) is given by

$$k_{exp} = \frac{k_1 k_2 [L]}{k_{-1} + k_2 [L]} \tag{5.16}$$

If L is CO, it is found that k_{exp} shows a direct dependence on [CO]. This is interpreted to mean that $k_2[CO] \ll k_{-1}$, because of the small CO concentration imposed by the low solubility of CO in most solvents. If L is pyridine,[84] then $k_{exp} = k_1$, apparently because $k_2[L] \gg k_{-1}$.

The reaction rate shows a substantial dependence on the nature of the solvent,[85] with faster reactions in more polar solvents. The solvent effect could be due either to better solvation of the polar intermediate or to direct coordination of the solvent.

Boese and Ford[86] have studied the photolysis of $Mn(CO)_5(C(O)CH_3)$ to generate the intermediate shown in Scheme 5.8. They reason that the photochemical and thermal reaction intermediates are the same because, for L = $P(OMe)_3$ at 25°C in THF, they give the same values of k_{-1}/k_2 of $6.6 \pm 1.4 \times 10^{-3}$ and $5.5 \pm 1.5 \times 10^{-3}$ M, respectively. In C_6H_{12}, the reaction is first order in $[P(OMe)_3]$, but this is not consistent with the condition in Eq. (5.16) that $k_2[L] \gg k_{-1}$, because the measured values are $k_{-1} = 9$ s^{-1} and $k_2 = 1.4 \times 10^6$ M^{-1} s^{-1}. The authors suggest that there is another pathway involving direct attack of L on Mn in alkane solvents, analogous to that proposed in the following example.

For $Mo(Cp)(CO)_3(CH_3)$[87] reacting with PPh_3, the rate is first order in $[PPh_3]$ in benzene but independent of $[PPh_3]$ in tetrahydrofuran. Recently, Wax and Bergman[88] have studied this system in a series of methyl-substituted tetrahydrofuran solvents that were chosen because of their similar polarities. Their results are consistent with Scheme 5.9. The pseudo-first-order rate constant is given by

$$k_{exp} = \frac{k_1 [PMePh_2]}{\dfrac{k_{-1}}{k_2} + [PMePh_2]} + k_3 [PMePh_2] \tag{5.17}$$

In THF and 3-MeTHF, the kinetics show saturation, and one can calculate k_{-1}/k_2; but in 2-MeTHF and 2,5-Me$_2$THF saturation is not

Scheme 5.9

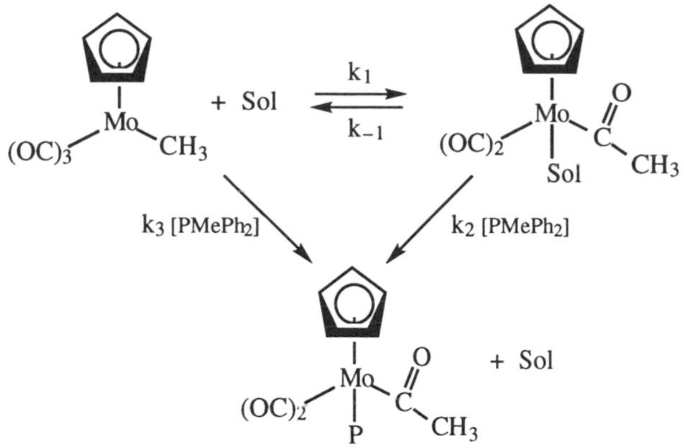

observed, indicating that k_{-1}/k_2 is much smaller. It is assumed that the latter solvents are much more weakly coordinated, so that k_2 is larger. The kinetic results are summarized in Table 5.6. Studies in mixed THF and 2,5-Me$_2$THF also show that the k_1 path is first order in THF, as expected. The k_1 and k_2 steps appear to be associative because of the first-order dependence on the entering group. This might be caused by slippage of the η^5-C$_5$H$_5$ to an η^3-C$_5$H$_5$ form to allow for coordination of the entering group. It may be noted that the k_1 values show the trend

Table 5.6. Methyl Migration Rate Constants (59.9°C) for Mo(Cp)(CO)$_3$(CH$_3$) in Various Solvents

Solvent	$10^4 \times k_1$ (s^{-1})	k_{-1}/k_2 (M)	$10^4 \times k_3$ ($M^{-1} s^{-1}$)
	7.78	0.0104	1.73
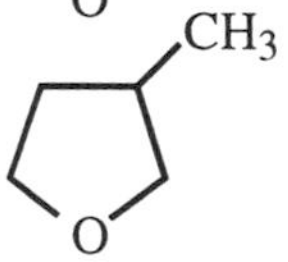	6.46	0.0082	1.86
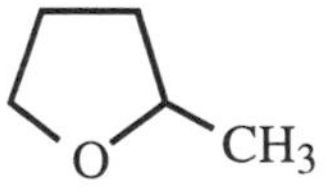	1.48		1.95
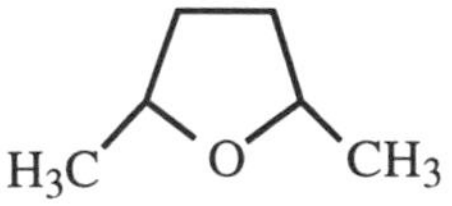	0.23		1.67

expected for steric inhibition of the entering group with associative activation. The k_3 values are relatively constant, as expected if the general solvent effects are not large. The earlier work of Butler et al.[87] found the rate with PPh_3 in THF to be independent of $[PPh_3]$, which implies that k_3 and k_2 are much smaller with this phosphine. This is consistent with a steric effect of the entering group.

Insertion reactions on $R—Fe(CO)_4^-$ have been studied kinetically by Collman et al.[89] This work shows the importance of ion pairs in these reactions, a factor that must be remembered especially when charged species are involved. The system can be described by Scheme 5.10, which includes ion pairs and ion triplets with the reactant and product and where $Z^+ = Na^+$, Li^+, or $(Ph_3P)_2N^+$.

Scheme 5.10

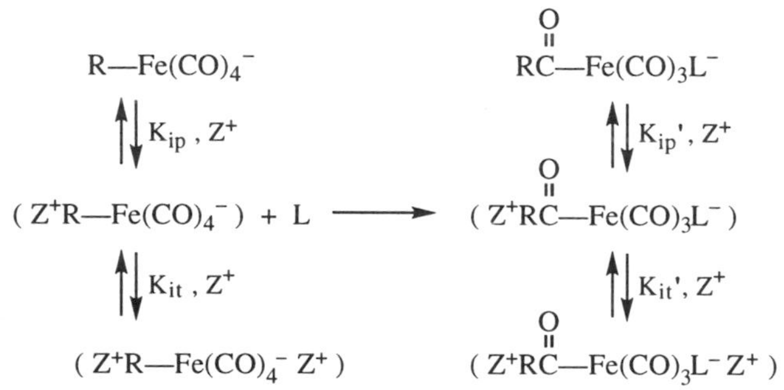

$$
\begin{array}{ccc}
R—Fe(CO)_4^- & & \overset{\overset{O}{\parallel}}{R}C—Fe(CO)_3L^- \\
\Big\updownarrow K_{ip},\ Z^+ & & \Big\updownarrow K_{ip}',\ Z^+ \\
(Z^+R—Fe(CO)_4^-)\ +\ L & \longrightarrow & (Z^+\overset{\overset{O}{\parallel}}{R}C—Fe(CO)_3L^-) \\
\Big\updownarrow K_{it},\ Z^+ & & \Big\updownarrow K_{it}',\ Z^+ \\
(Z^+R—Fe(CO)_4^-\ Z^+) & & (Z^+\overset{\overset{O}{\parallel}}{R}C—Fe(CO)_3L^-\ Z^+)
\end{array}
$$

The values of K_{ip}, K_{ip}', K_{it}, and K_{it}' are 1.1×10^4, 2×10^6, 1.7×10^2, and 1×10^2 M, for Na^+ in THF. The rate is first order in $[L]$ and in the reactant ion pair concentration. An increase in the solvent polarity reduces the rate, presumably because of less ion pair formation. The ion pair may be the more reactive species because of cation interaction with the CO ligands ($Fe—CO\cdots Na^+$), which will favor methyl migration.

5.2.b Sulfur Dioxide Insertion

Sulfur dioxide insertion has been found to proceed via the O-bonded sulfinate intermediate, which rearranges to the S-bonded product,[90] as shown in Scheme 5.11.

Scheme 5.11

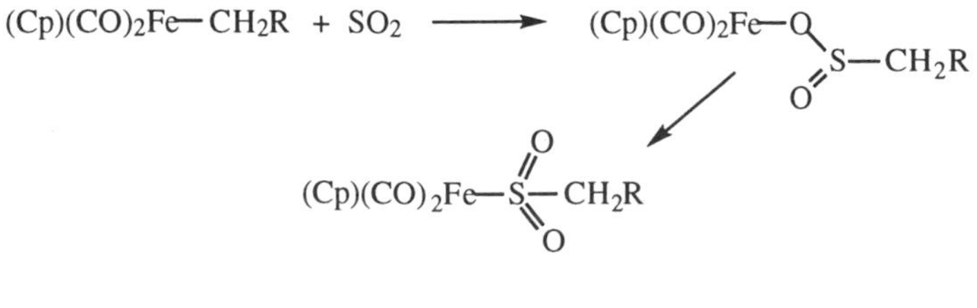

$$
(Cp)(CO)_2Fe—CH_2R\ +\ SO_2\ \longrightarrow\ (Cp)(CO)_2Fe—O\underset{O}{\overset{}{\diagdown}}S—CH_2R
$$

$$
(Cp)(CO)_2Fe—\overset{\overset{O}{\parallel}}{\underset{\underset{O}{\parallel}}{S}}—CH_2R
$$

It has been observed[91] that this reaction proceeds with inversion at the alkyl carbon bonded to M, and the mechanism is believed to involve electrophilic attack of SO_2 at this C followed by rearrangement to the O-bonded isomer. An analogous pathway has been found[92] for SO_2 insertion into $W(CO)_5(Y(CH_3)_3)^-$, where Y = Si or Sn.

5.2.c Carbon Dioxide Insertion

Carbon dioxide is an abundant industrial raw material,[93] and the carbon dioxide insertion, shown in (5.18), is a potentially important C—C bond-forming process.[94]

$$(L)_nM\text{—}R \;+\; CO_2 \longrightarrow \begin{array}{c}(L)_nM\text{—}O\\ \diagdown\\ C\text{—}R\\ \diagup\diagup\\ O\end{array} \qquad (5.18)$$

Sakakai and Musashi[95] have reported a theoretical analysis of such a reaction for the $Cu(PH_3)_2R + CO_2$ system (R = H, CH_3, OH).

The kinetics and stereochemistry of CO_2 insertion reactions have been investigated by Darensbourg and co-workers.[96] The rate is increased by more electron-rich metal complexes and shows much less sensitivity to the nature of the carbon center than the CO insertions. The reactions of $W(CO)_4(L)R^-$ (L = CO, phosphine, phosphite; R = Me, Et, Ph) have been studied in THF. The kinetics for $W(CO)_5(CH_3)^-$ show a first-order dependence on $[CO_2]$ and metal complex. The rate is increased by addition of Na^+, presumably due to ion pair formation. The kinetic work used $(Ph_3P)_2N^+$ as the counter ion, which should at least minimize ion pair effects because of the size and charge delocalization in this cation. For $W(CO)_4(P(OMe)_3)(CH_3)^-$, the activation-parameter values are $\Delta H^* = 42.7$ kJ mol^{-1} and $\Delta S^* = -181$ J mol^{-1} K^{-1}.

The stereochemistry at the carbon bonded to W was investigated using the threo-ligand isomer of $(OC)_5W$—CHD–CHDPh$^-$ and was found to proceed with retention of stereochemistry at the α carbon, as shown by

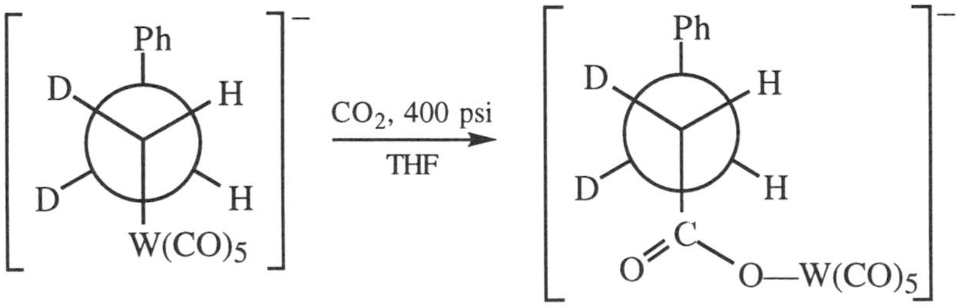

$$\qquad (5.19)$$

threo   threo

The mechanism proposed to be consistent with the observations is shown in Scheme 5.12.

Scheme 5.12

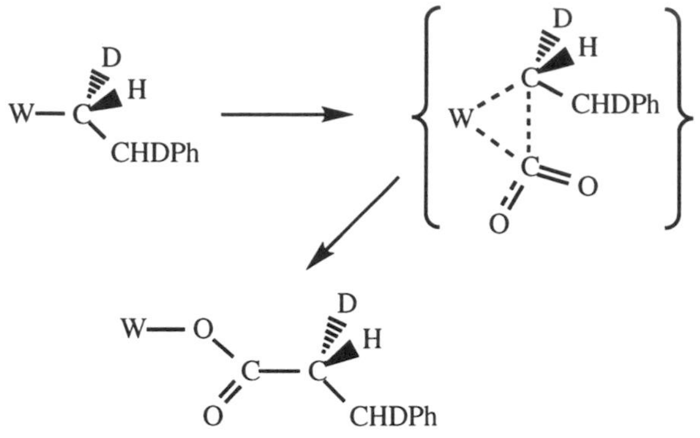

Carbon dioxide also can undergo insertion into metal—heteroatom bonds. The reversible reaction with alkoxides is shown in (5.20).[97,98]

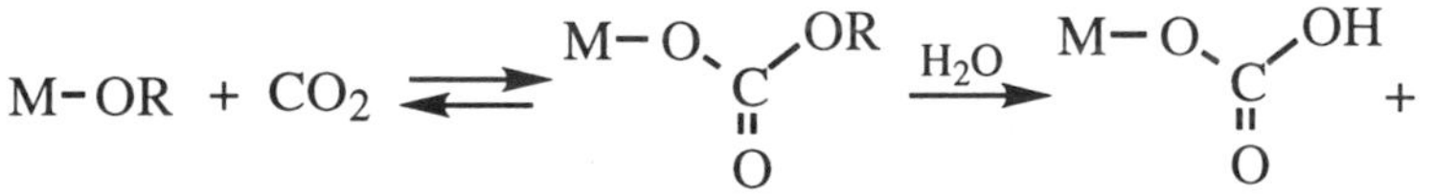

$$M-OR + CO_2 \rightleftharpoons M-O\underset{O}{\overset{}{\underset{\parallel}{C}}}OR \xrightarrow{H_2O} M-O\underset{O}{\overset{}{\underset{\parallel}{C}}}OH + HOR \quad (5.20)$$

The carbonate ester product is easily hydrolyzed to the carbonate and sometimes to further metal derivatives because of the cis-labilizing and chelating ability of the carbonate ligand. M—NHR species react to give the O-bonded amide. It is generally believed that these reactions occur by electrophilic attack of CO_2 on the atom coordinated to the metal.

5.3 OXIDATIVE ADDITION REACTIONS

5.3.a General Considerations and Mechanisms

Oxidative addition reactions usually involve a coordinatively unsaturated 16-electron metal complex or a five-coordinate 18-electron species, and take the following general form:

$$(L)_nM + XY \longrightarrow (L)_nM(X)(Y) \quad (5.21)$$

If the X and Y ligands in the product are considered to be formally -1, then the metal center has increased its formal oxidation state by $+2$, and this is the origin of the name *oxidative addition*. The reverse reaction is called *reductive elimination*. Some specific oxidative addition reactions are given in (5.22). The stereochemistry of the products can be controlled by subsequent isomerization reactions and is not always indicative of the immediate oxidative addition product.

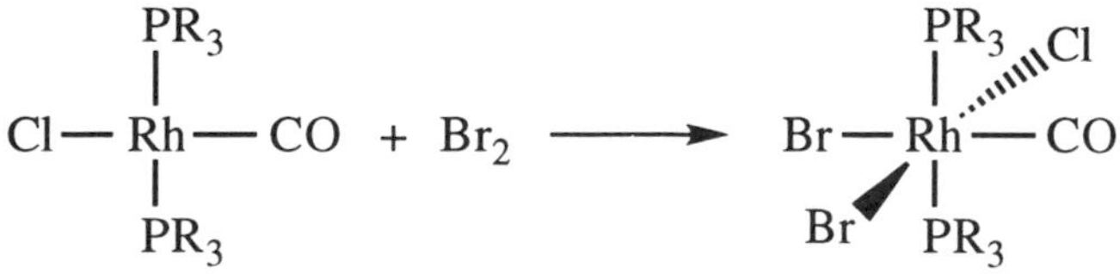

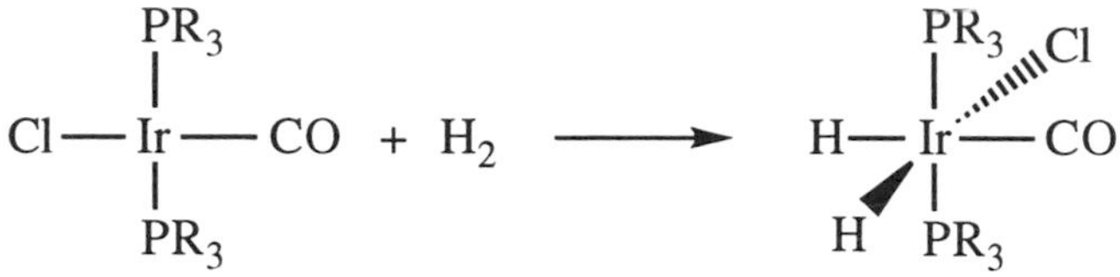

$$Cl-\underset{\underset{PR_3}{|}}{\overset{\overset{PR_3}{|}}{Rh}}-CO \;+\; Br_2 \;\longrightarrow\; Br-\underset{\underset{PR_3}{|}}{\overset{\overset{PR_3}{|}}{Rh}}\overset{\text{\tiny{}Cl}}{-}CO$$

$$Cl-\underset{\underset{PR_3}{|}}{\overset{\overset{PR_3}{|}}{Ir}}-CO \;+\; H_2 \;\longrightarrow\; H-\underset{\underset{PR_3}{|}}{\overset{\overset{PR_3}{|}}{Ir}}\overset{\text{\tiny{}Cl}}{-}CO$$

$$(5.22)$$

$$Cl-\underset{\underset{PR_3}{|}}{\overset{\overset{PR_3}{|}}{Ir}}-CO \;+\; HBr \;\xrightarrow{CH_3CN}\; H-\underset{\underset{PR_3}{|}}{\overset{\overset{PR_3}{|}}{Ir}}\overset{\text{\tiny{}Cl}}{-}CO \;+\; H-\underset{\underset{PR_3}{|}}{\overset{\overset{PR_3}{|}}{Ir}}\overset{\text{\tiny{}Cl}}{-}Br$$

$$Fe(CO)_4{}^{2-} \;+\; CH_3Br \;\longrightarrow\; \left[OC-\underset{\underset{CO}{|}}{\overset{\overset{CH_3}{|}}{Fe}}\overset{\text{\tiny{}CO}}{-}CO \right]^{-} \;+\; Br^{-}$$

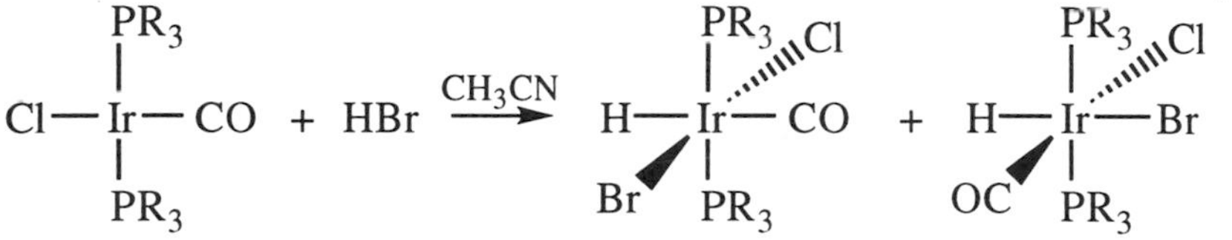

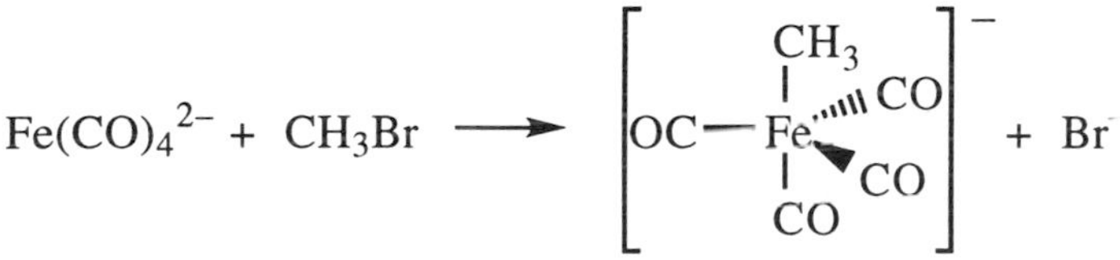

Oxidative addition reactions can also produce less obvious products, as shown in the following examples:

$$2\ Co(CN)_5{}^{3-} \;+\; CH_3I \;\longrightarrow\; Co(CN)_5(CH_3)^{3-} \;+\; Co(CN)_5I^{3-}$$

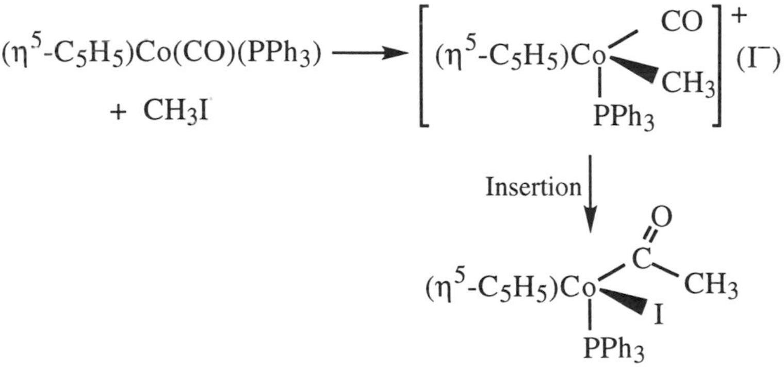

$$(\eta^5\text{-}C_5H_5)Co(CO)(PPh_3) \;\longrightarrow\; \left[(\eta^5\text{-}C_5H_5)Co\underset{\underset{PPh_3}{|}}{\overset{\overset{CO}{\diagup}}{\diagdown}}CH_3 \right]^{+} (I^-)$$
$$+\; CH_3I$$

$$\text{Insertion} \Big\downarrow \qquad\qquad (5.23)$$

$$(\eta^5\text{-}C_5H_5)Co\underset{\underset{PPh_3}{|}}{\overset{\overset{\displaystyle C \!=\! O}{\diagup}}{\diagdown}}\!\!\!\!\underset{I}{\overset{}{CH_3}}$$

The rate law for these reactions is usually first order in the metal complex and XY concentrations. The reaction mechanism depends on the nature of XY and can be broken into three generally recognized categories, as shown in Scheme 5.13.

Scheme 5.13

Ionic

$$XY \rightleftharpoons X^+ + Y^-$$

$$(L)_nM + Y^- \longrightarrow \{(L)_nM(Y)\}^- \xrightarrow{X^+} (L)_nM(Y)(X)$$

Nucleophilic Attack (S_N2)

$$(L)_nM + XY \longrightarrow \{(L)_nM\text{---}X\text{--}Y\} \longrightarrow (L)_nM(Y)(X)$$

Free Radical (Single Electron Transfer, SET)

$$(L)_nM + XY \rightleftharpoons \{(L)_nM\bullet\}^+ + \{\bullet XY\}^-$$

$$\{\bullet XY\}^- \longrightarrow \{\bullet X\} + Y^-$$

Possible Radical Termination Steps

$$\{(L)_nM\bullet\}^+ + \{\bullet X\} \longrightarrow \{(L)_nM\text{---}X\}^+ \xrightarrow{Y^-} (L)_nM(Y)(X)$$

$$\{(L)_nM\bullet\}^+ + Y^- \longrightarrow \{(L)_nM\text{---}Y\} \xrightarrow{\bullet X} (L)_nM(Y)(X)$$

$$(L)_nM + \{\bullet X\} \longrightarrow \{(L)_nM\bullet\}^+ + X^-$$

$$\{(L)_nM\bullet\}^+ + XY \longrightarrow \{(L)_nM\text{---}X\}^+ + \{\bullet Y\}$$

For species that are known to ionize, such as HI and HBr (and less obvious examples, such as $SnCl_4$ and acyl halides), the ionic mechanism is the most probable and will be especially favored in more polar solvents. The nucleophilic mechanism occurs with many organic halides and requires the availability of an unshared electron pair on the $(L)_nM$ species. The free radical electron-transfer mechanism is an obvious candidate when XY is an oxidizing agent such as Cl_2 and Br_2. The other possible reactions, noted in Scheme 5.13 for the radical path, take account of the fact that the concentration conditions are such as to favor reaction of the radicals with reagents rather than with other radicals, unless the latter are reasonably persistent.

A fourth alternative, not widely encountered in organometallic systems, is the atom transfer mechanism in Scheme 5.14.

Scheme 5.14

Atom Transfer

$$(L)_nM + XY \longrightarrow (L)_nM\text{---}X + \bullet Y$$

$$(L)_nM + \bullet Y \longrightarrow (L)_nM\text{---}Y$$

If $(L)_nM$—X and $(L)_nM$—Y are stable products, the implication is that $(L)_nM$ is an odd-electron (e.g., 17-electron) organometallic complex. This mechanism is observed for bis(dimethylglyoxime)cobalt(II) and $Co(CN)_5^{3-}$ reacting with organic halides, but only the products are true organometallic complexes.

5.3.b Oxidative Addition of H_2

In view of the fact that the H—H bond energy is 105 kcal mol^{-1}, this reaction is unexpectedly facile with 16-electron complexes. The ΔH^* values are typically in the range of 5 to 10 kcal mol^{-1} and the ΔS^* values are in the range of -20 to -50 cal mol^{-1} K^{-1}. The reaction has always been observed to give the *cis*-dihydride product. The rate is increased by more electron-donating ligands on the metal. Some typical data are given in Table 5.7. The reaction shows a relatively modest deuterium isotope effect of ~1.2, which indicates that the H—H bond is largely intact in the transition state. A recent theoretical analysis[99] of the addition of H_2 to $Ir(CO)(PH_3)_2X$ indicates that π-acceptor X ligands give more stable oxidative addition products.

There has been a great deal of interest in this reaction because of its importance in catalytic hydrogenation reactions. Crabtree and co-workers[100] discussed the stereochemistry of addition to d^8 complexes. Recently, complexes containing η^2-H_2 have been isolated[101] and characterized. One such complex is $W(CO)_3(PR_3)_2(H_2)$, which has a pentagonal bipyramid structure with an H—H bond length of 0.82 Å and a W—H length of 1.89 Å. Numerous further examples have been identified and the field has been reviewed.[102] The theoretical aspects of

Table 5.7. Rate Constants (30°C) and Activation Parameters for the Oxidative Addition of H_2 to $Ir(CO)(PR_3)_2X$ in Benzene

X	R	k (M^{-1} s^{-1})	ΔH^* (kcal mol^{-1})	ΔS^* (cal mol^{-1} K^{-1})
Cl	Ph	0.93 [a]	10.8	−23
Br	Ph	14.3 [a]	12.0	−14
I	Ph	>100 [a]		
Cl	p-OCH$_3$Ph	0.66 [b]	6.0	−39
Cl	p-CH$_3$Ph	0.53 [b]	4.3	−45
Cl	p-ClPh	0.16 [b]	9.8	−28
Cl	p-FPh	0.25 [b]	11.6	−22

[a] Halpern, J.; Chock, P. B. *J. Am. Chem. Soc.* **1966**, *88*, 3511.

[b] Ugo, R.; Pasini, A.; Fusi, A.; Cenini, S. *J. Am. Chem. Soc.* **1972**, *94*, 7364.

the bonding and stability have been discussed,[103,104] with the conclusion that the stability is strongly dependent on the π-acceptor properties of the other ligands on the metal, especially CO in the preceding example. This is consistent with the observation that the isoelectronic $W(PR_3)_5$ complexes react with H_2 to give the classical dihydride product because back donation from the metal to the antibonding σ orbital of H_2 promotes splitting of the H—H bond and formation of the dihydride.

The dihydrides, $M(P(OR)_3)_5(H)_2$ (M = Cr and W),[105,106] have been found to be readily fluxional and this seems to be a general property of many hydride and hydride-η^2-H_2 systems. Several recent papers have discussed the nature of this fluxionality and the mechanistic possibilities in, for example, $Ir(Cp^*)(PR_3)(H)_3{}^+$,[107] $Os(BH_4)(PR_3)_2(H)_3$,[108] and $Ir(P^iPr_3)_2(H)_2(H_2)X$.[109]

The reaction of H_2 with Wilkinson's compound, $Rh(PPh_3)_3Cl$, is of special significance because of the applications of this complex as a catalyst and is discussed in Section 5.5.b.

5.3.c Oxidative Addition of Organic Halides

The available evidence indicates that oxidative addition of organic halides quite often proceeds by nucleophilic attack of the metal center on the halogen-bearing carbon, as shown in Scheme 5.15.[110]

Scheme 5.15

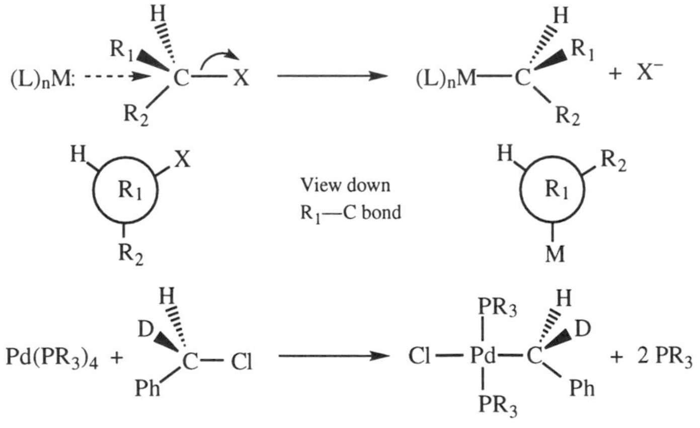

In several examples, it has been shown that the reaction proceeds with inversion of configuration at C, as predicted by the mechanism. The general example implies that addition of X^- is a subsequent process, and it is usually too fast to be observed. The fact that these reactions are not stereoselective is consistent with this. In the example in Scheme 5.15, the alkyl group and Cl^- are in a trans configuration in the product, and this would be difficult to rationalize if the addition were a concerted process.

Table 5.8. Rate Constants[a] (25°C) and Activation Parameters for the Reaction of $Ir(CO)(PR_3)_2Cl + CH_3I$ in Benzene

R	k $(M^{-1} s^{-1})$	ΔH^* (kcal mol^{-1})	ΔS^* (cal mol^{-1} K^{-1})
Ph	3.3×10^{-3}	7.0	−47
EtPh$_2$	1.2×10^{-2}	9.8	−34
p-CH$_3$Ph	3.3×10^{-2}	13.8	−20
p-FPh	1.5×10^{-4}	17.0	−20
p-ClPh	3.7×10^{-5}	14.9	−28

[a] Chock, P. B.; Halpern, J. *J. Am. Chem. Soc.* **1966**, *88*, 3511; Ugo, R.; Pasini, A.; Fusi, A.; Cenini, S. *J. Am. Chem. Soc.* **1972**, *94*, 7364.

The rates of these reactions are first order in each reactant, and some typical kinetic data for $Ir(CO)(PR_3)_2Cl$ are given in Table 5.8. The rate is much more affected by the ligand environment than is H$_2$ addition and increases with increasing polarity of the solvent. Decreasing the ligand basicity or increasing the steric bulk decreases the rate.[111] The order of reactivity of various halides is generally as follows:

$$CH_3 > CH_2CH_3 > CH(R)_2 > cyclohexyl > adamantyl$$

The S$_N$2 mechanism usually is observed with methyl, benzyl, and allyl halides and α haloethers. But other saturated alkyl halides, vinyl and aryl halides, and α haloesters show characteristics of a radical pathway with $Ir(CO)(PMe_3)_2Cl$.[112] Direct EPR evidence has been obtained for radicals in the reaction of $Zr(\eta^5\text{-}C_5H_5)_2(PMePh_2)$ with butyl chlorides.[113] The reaction of a gold(I) dimer, $(Au(CH_2)_2PPh_2)_2$, has been suggested[114] to proceed by a single electron transfer, SET, mechanism, based on the parallel between the rates and the reducibility of the organic halide. The reaction of alkyl halides with $Pt(PPh_3)_2$ is proposed to proceed by a halogen atom abstraction mechanism.[115]

5.4 REACTIONS OF ALKENES

Alkenes and alkynes are capable of normal substitution reactions like other nucleophiles, but this is rarely an associative process with 18-electron systems because alkenes and alkynes are poor nucleophiles. Such systems require prior dissociation of a ligand to allow coordination of the alkene, which can then form strong complexes because of the back π bonding from the metal to the π* orbital of the C—C multiple bond. The bonding π electrons have relatively low nucleophilicity. However, in coordinatively unsaturated systems, prior dissociation is not a problem. The expected initial reactions are shown in Scheme 5.16.

Scheme 5.16

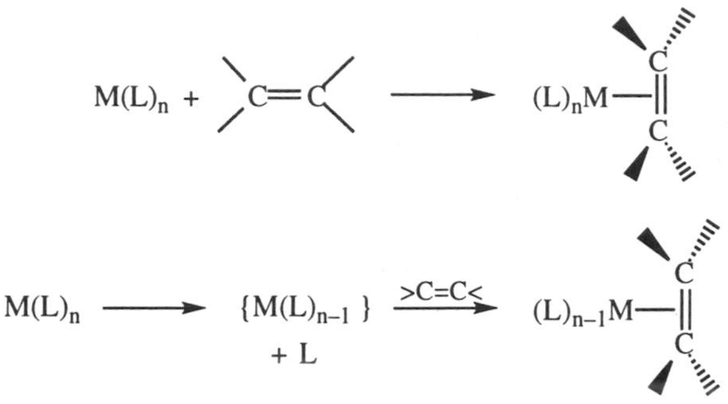

These reactions may be followed by rearrangement with H-atom migration to a σ-bonded form, as shown in (5.24).

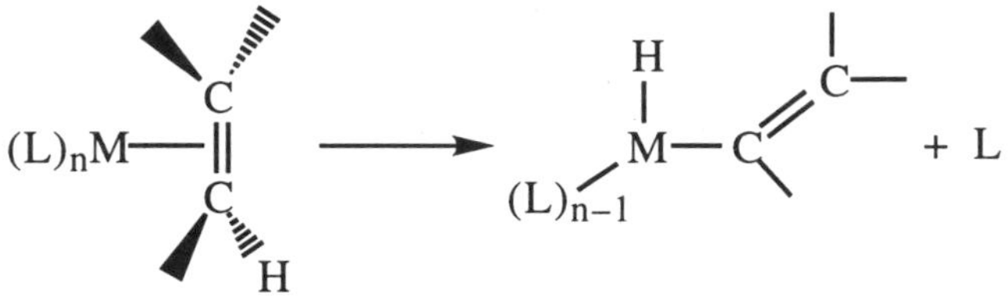

$$(5.24)$$

The latter process would seem to be a straightforward rearrangement subsequent to π-complex formation. But the work of Stoutland and Bergman[116] indicates that the process is more complex. The thermal decomposition of $Ir(\eta^5\text{-}C_5Me_5)(PMe_3)(C_6H_{11})H$ yields C_6H_{12} and the unsaturated reactive species $Ir(\eta^5\text{-}C_5Me_5)(PMe_3)$, which reacts with ethylene to give the kinetic products shown in Scheme 5.17.

Scheme 5.17

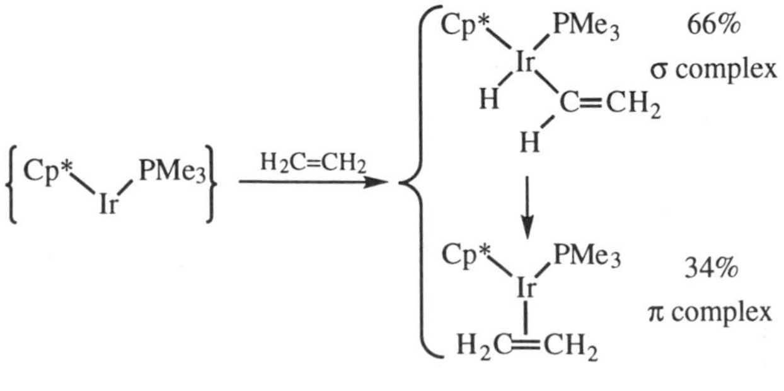

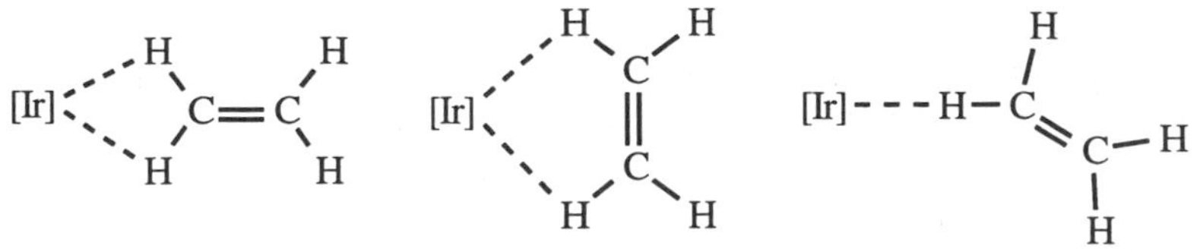

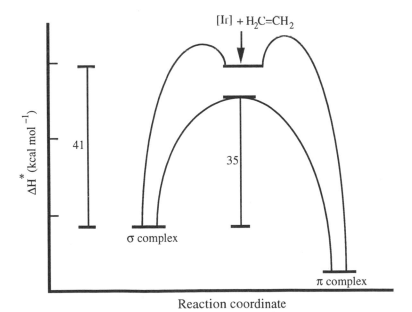

Figure 5.1. Possible transition states for the addition of ethylene to coordinatively unsaturated iridium.

The σ complex will isomerize to the thermodynamically more stable π complex ($\Delta H^* = 34.6$ kcal mol^{-1}, $\Delta S^* = 2.6$ cal mol^{-1} K^{-1}), but the latter is the minor kinetic product. The implication is that the two products are forming directly from two slightly different transition states. Stoutland and Bergman suggested the possible transition states shown in Figure 5.1, which involve initial bonding between the metal and hydrogen substituents on the ethylene. A perplexing feature of the observations is that the same product ratio is obtained from $H_2C=CD_2$ and *cis*-(HD)C=C(HD). It was expected that an isotope effect acting on the proposed transition states would cause a change in product distribution for the deuterated reactants.

A combination of calorimetric measurements and bond energy estimates leads to the reaction coordinate diagram in Figure 5.2.

Figure 5.2. Reaction coordinate diagram for the addition of ethylene to coordinatively unsaturated iridium.

This diagram is different from the normal one in that the reaction starts at the high-energy intermediate; this will then react to give the two products in amounts that depend on the relative heights of the energy barriers to the left and right. It should be noted that the difference in energy barriers must be less than 1 kcal mol^{-1} in order to explain the approximately two-to-one product distribution. The difference is exaggerated in Figure 5.2, in order to show that the barrier leading to the π complex must be higher to account for the product distribution.

Baker and Field[117] have observed rather similar behavior with Fe(P$_2$(CH$_2$)$_2$(R)$_4$)$_2$(H)$_2$, which gives mainly the cis σ complex after photolysis in the presence of ethylene. The cis σ complex isomerizes to trans and eventually yields the more stable π complex. Graham and co-workers[118] have observed that the trifluoromethylpyrazolylborate, HBPf$_3$, π complex of iridium, Ir(η^2-HBPf$_3$)(CO)($^2\eta$-C$_2$H$_4$), converts to the more stable σ complex, Ir(η^2-HBPf$_3$)(CO)(H)(-C$_2$H$_3$), at 100°C. However, the σ complex of rhodium isomerizes to the more stable π complex at 25°C. Clearly, there is a delicate balance between the stabilities of these species.

Grant and co-workers[119] studied the dissociation of ethylene from Cr(CO)$_5$(C$_2$H$_4$) in the gas phase in the presence of CO and C$_2$H$_4$. The rate of formation of Cr(CO)$_6$ has the usual dissociative rate law, with a competition ratio of 1.1 for C$_2$H$_4$ to CO. The activation parameters for the dissociation are $\Delta H^* = 24$ kcal mol^{-1} and $\Delta S^* = 15$ cal mol^{-1} K^{-1}. The ΔH^* is indicative of the Cr—(C$_2$H$_4$) bond energy.

An unusual reaction of olefins with Co(η^5-C$_5$H$_5$)(NO)$_2$ has been studied by Becker and Bergman.[120] They examined the direct reaction shown in (5.25) and the olefin exchange described in Scheme 5.18.

$$(5.25)$$

The direct reaction is first order in the metal complex and first order in olefin, and the second-order rate constant depends on the olefin, as indicated by some of the data in Table 5.9. Strain in the olefin appears to increase its reactivity, as shown for norbornene and cyclopentene. There may be some steric effects of olefin substituents, but these effects may be attenuated by better electron donation from the substituents. It was found also that the reaction rate with 2,3-dimethyl-2-butene was insensitive to the polarity of the solvent, with relative values of 1:0.6:0.8 in cyclohexane, THF, and methanol, respectively. This seems to rule out an ionic or polar transition state or intermediate, and the authors favor a concerted cycloaddition mechanism.

Table 5.9. Rate Constants (20°C) for the Reaction
of Olefins with $Co(Cp)(NO)_2$ in Cyclohexane

Olefin	k_2 (M^{-1} s^{-1})
	130
	3.9
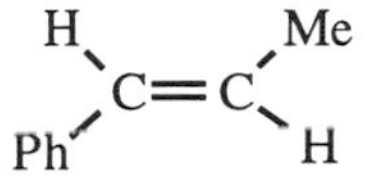	0.84
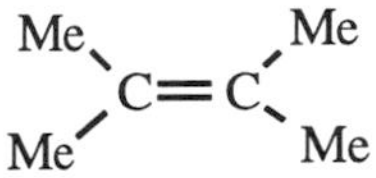	0.25

The exchange of one olefin adduct with another has reaction kinetics
that are consistent with the pathway shown in Scheme 5.18.

Scheme 5.18

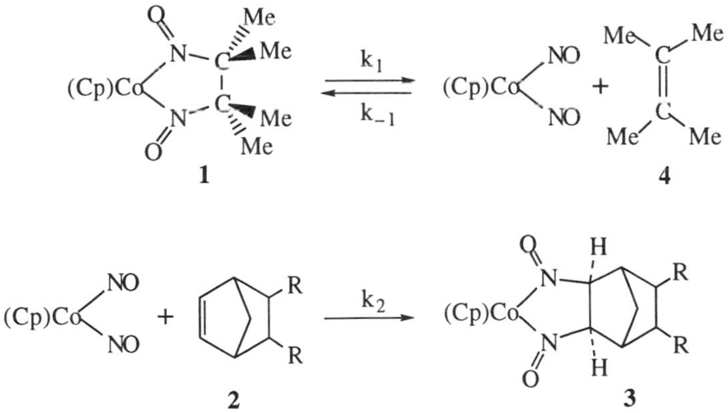

If a steady-state condition is applied to $Co(Cp)(NO)_2$, then

$$k_{exp} = \frac{k_1 [1] [2]}{\left(\dfrac{k_{-1}}{k_2}\right)[4] + [2]}$$
(5.26)

The kinetic results give $k_1 = 4.3 \times 10^{-4}$ s^{-1} ($E_a = 29.4$ kcal mol^{-1}) and
$k_2/k_{-1} = 115$ for $R = CO_2Me$ at 75°C in toluene.

The exchange of one olefin for another is stereospecific, with no isomerization of the olefin. This observation indicates that the exchange does not proceed through an intermediate with a C—C single bond, or at least that such an intermediate is not persistent enough to allow rotation about the C—C bond, as indicated in Scheme 5.19. This also would be consistent with a concerted mechanism (no intermediate) for the addition and dissociation reactions.

Scheme 5.19

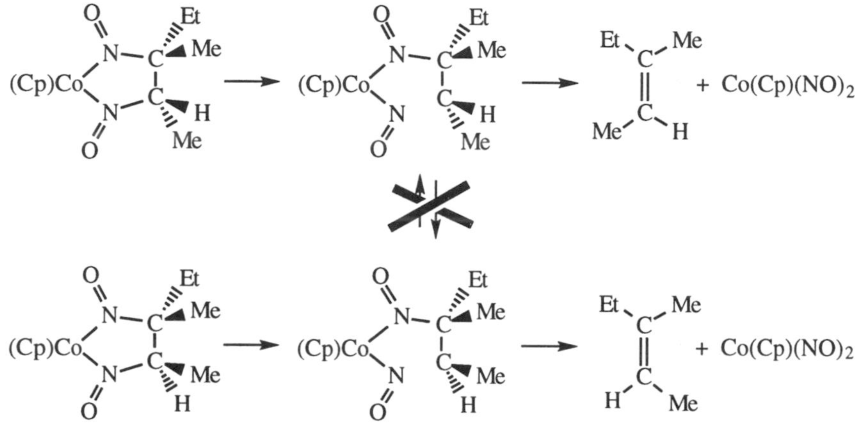

Since Co(Cp)(NO)$_2$ can achieve an 18-electron configuration by considering one NO as a three-electron donor and the other as a one-electron donor, this reaction could be considered as a formal analogue of 1,3-dipolar cycloadditions in organic chemistry. It is also possible for the NO to switch from a three- to a one-electron donor in the transition state, thereby leaving the Co unsaturated and able to form a transient π complex before rearranging to the observed product. Slippage of the C$_5$ ring is another possibility, but the rate only changes by a factor of about three when the C$_5$ is changed from C$_5$H$_5$ to C$_5$H$_4$CO$_2$Me to C$_5$(Me)$_5$.

5.5 CATALYTIC HYDROGENATION OF ALKENES

The addition of H$_2$ to a >C=C< system is thermodynamically favorable but generally difficult to achieve. In the laboratory, chemists use catalysts such as platinum black and Rainey nickel to make the reaction proceed at a reasonable rate. The kinetic barrier for this reaction can be understood in terms of simple orbital symmetry diagrams. The basic principle is that in the activated state the electrons must flow in such a way as to make and break the appropriate bonds. In the case of hydrogenation, we want to break the H—H σ and C—C π bonds and make two C—H bonds. The electrons must flow from an occupied

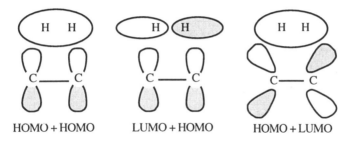

HOMO + HOMO   LUMO + HOMO   HOMO + LUMO

Figure 5.3. The highest occupied molecular orbitals and their combinations with the lowest unoccupied molecular orbitals in H_2 and CH_2CH_2.

orbital on one molecule to an unoccupied orbital on the other, that is, from the highest occupied orbital, HOMO, on one species to the lowest unoccupied orbital, LUMO, on the other.

For H_2, the HOMO is the σ-bonding orbital and the LUMO is the corresponding σ-antibonding orbital. For an alkene, the HOMO is the π-bonding orbital and the LUMO is the corresponding π-antibonding orbital. To stabilize the activated complex, the appropriate HOMO and LUMO must produce good overlap; that is, the signs of the radial parts of the wave functions should be the same in the overlap region. In addition, electron flow from HOMO to LUMO should break and make the required bonds. The first diagram in Figure 5.3 shows that the overlap is correct for the two HOMO orbitals in this system, but there can be no useful electron flow between two occupied orbitals. The second and third diagrams show the HOMO–LUMO combinations that would give the appropriate electron flow, but these do not produce net overlap; thus, the transition state will not be stabilized. Such a reaction is said to be *symmetry forbidden*. A catalyst must somehow overcome this symmetry restriction.

5.5.a General Mechanisms

The two generally recognized routes by which an organometallic complex can catalyze the hydrogenation of alkenes are referred to as the *olefin route* and the *hydride route*, as shown in Scheme 5.20. Both pathways start from a coordinatively unsaturated (16-electron) metal complex, indicated as $M(L)_4$ in the Scheme. This species might be formed by ligand dissociation, as shown, or by any process that can generate unsaturation.

The first step in the olefin route is addition of the olefin to form a π complex, while the first step in the hydride route is oxidative addition of H_2 to form a dihydride complex. These steps are followed by ligand dissociation and addition of either H_2 or olefin to generate the key dihydride-olefin species in the center of the Scheme.

Scheme 5.20

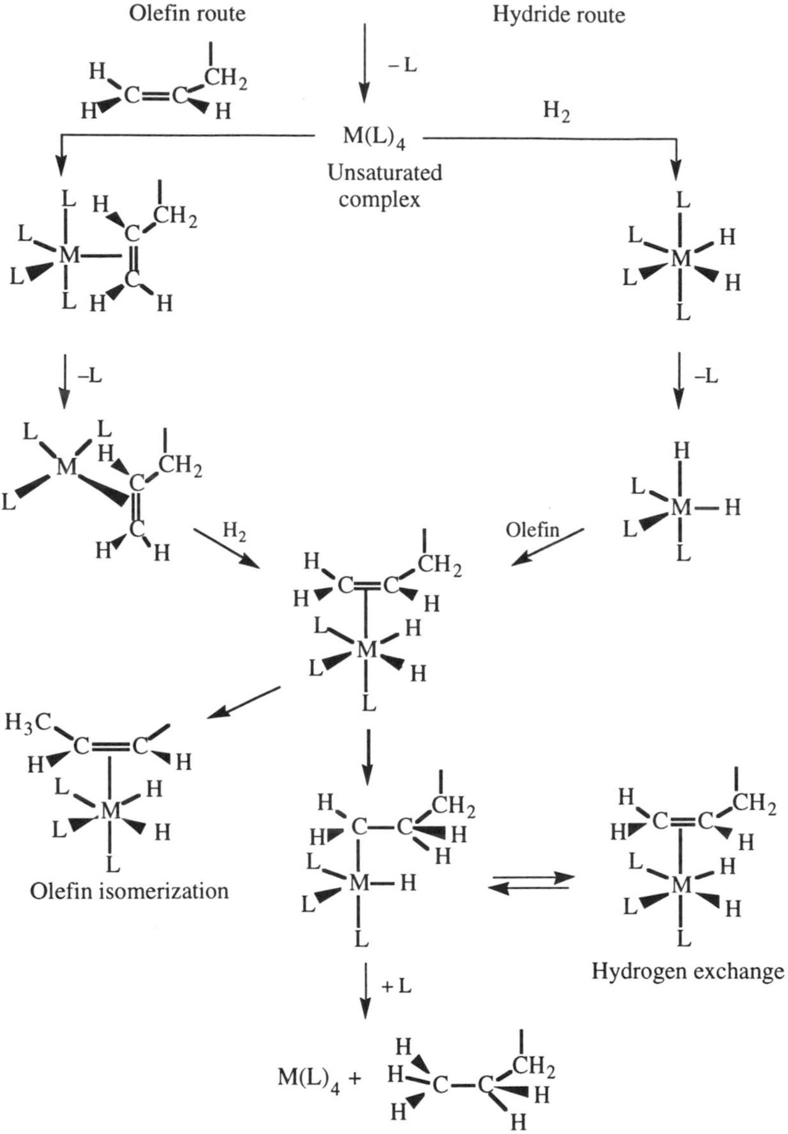

5.5.b Hydrogenation by Wilkinson's Catalyst: Rh(PPh₃)₃Cl

The catalytic properties of $Rh(PPh_3)_3Cl$ were first reported by Wilkinson and co-workers.[121] The nature of the species present in hydrocarbon solvents was the subject of controversy until the work of Arai and Halpern,[122] which indicated some PPh_3 dissociation, shown by

$$Rh(PPh_3)_3Cl \rightleftharpoons Rh(PPh_3)_2Cl + PPh_3 \qquad (5.27)$$

However, Tolman and co-workers[123] found that phosphine liberation was accompanied by dimer formation, which has $K = 2.4 \times 10^4$ M in benzene at 25°C. The reaction and suggested structure for the dimeric product are given by

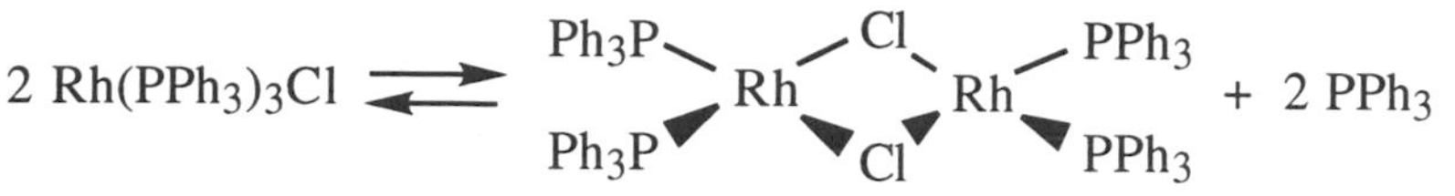

$$2 \ Rh(PPh_3)_3Cl \rightleftharpoons \qquad + \ 2 \ PPh_3 \quad (5.28)$$

Halpern and Wong[124] acknowledged the correctness of Tolman's interpretation and studied the kinetics of the hydrogenation of $Rh(PPh_3)_3Cl$ in the presence of excess PPh_3 to suppress the formation of dimer, with the conclusions that the reaction pathways are as shown in Scheme 5.21.

Scheme 5.21

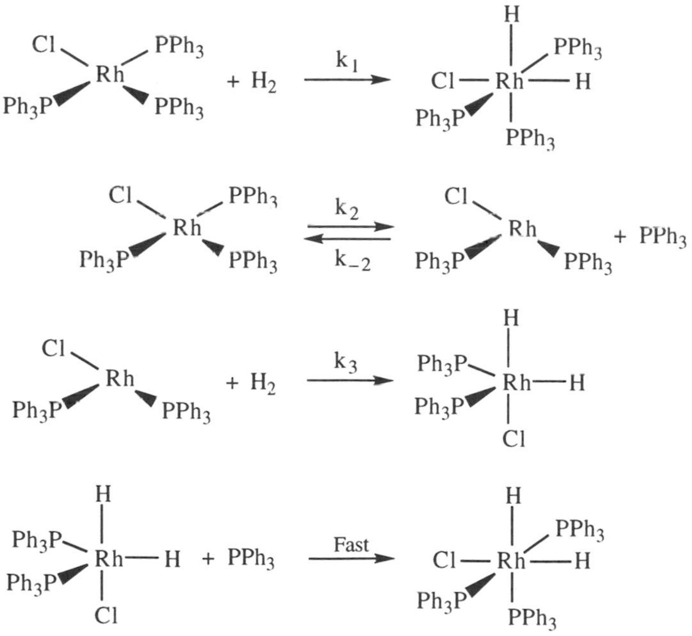

With $[H_2]$ and $[PPh_3] >> [Rh]$ at 25°C in benzene, the kinetics indicate that the reaction proceeds by parallel paths involving oxidative addition to Wilkinson's catalyst and to the species with one phosphine dissociated, as shown in Scheme 5.21. The rate law for Scheme 5.21 gives the following expression for the rate of reaction, if a steady state is assumed for $Rh(PPh_3)_2Cl$:

$$Rate = \left(k_1 + \frac{k_2 k_3}{k_{-2}[PPh_3] + k_3[H_2]} \right)[H_2][Rh(PPh_3)_3Cl] \quad (5.29)$$

The variation of the rate with $[PPh_3]$ and $[H_2]$ gives $k_1 = 4.8$ M^{-1} s^{-1}, $k_2 = 0.71$ s^{-1}, and $k_{-2}/k_3 = 1.1$ (25°C in benzene). At low $[PPh_3]$, the pseudo-first-order rate constant $k_{exp} = k_1[H_2] + k_2$, and for typical H_2 concentrations of ~2 x 10^{-3} M (at 1 atm), the k_2 path is dominant.

Separate studies on the dimer gave a rate that is first order in [dimer] and $[H_2]$ and independent of $[PPh_3]$, with $k = 5.4$ M^{-1} s^{-1}. Tolman et al.[125] showed that the dimer reacts with H_2 and with ethylene to give the following dihydride and ethylene complexes, but it does not react with cyclohexene.

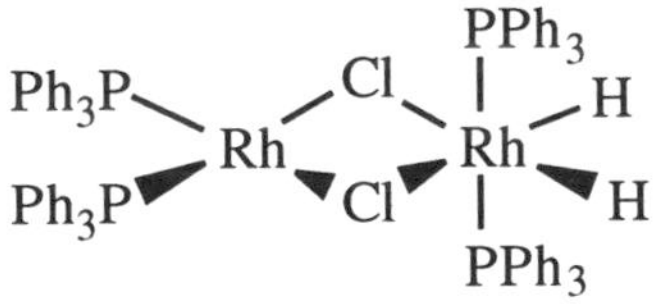

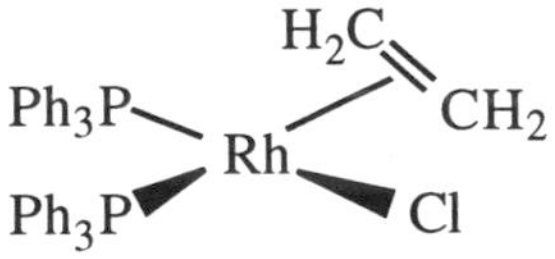

Halpern et al.[126] found that the kinetics of the hydrogenation of cyclohexene by the rhodium dihydride are consistent with Scheme 5.22, with $K_5 = 3.4$ x 10^{-4} and $k_6 = 0.2$ s^{-1} at 25°C in benzene.

Scheme 5.22

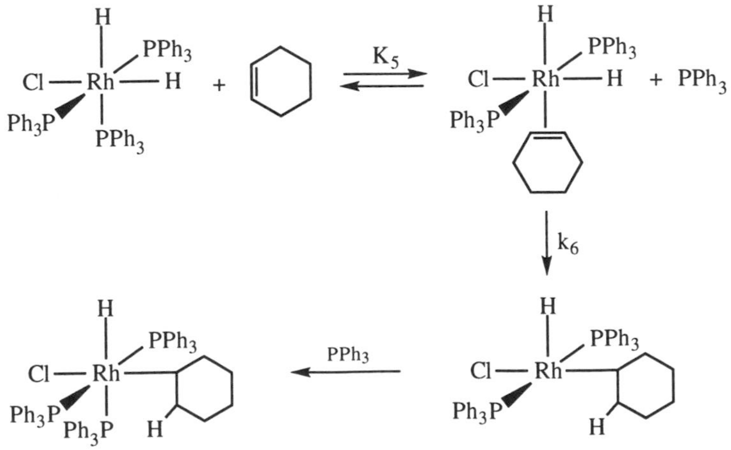

The above examples are considered to be consistent with the hydride route for hydrogenation with Wilkinson's catalyst. However, the hydride route is not universally observed even for closely related catalysts.

Halpern and co-workers[127] studied the diphos system and found that the kinetics were consistent with the reaction sequence in Scheme 5.23.

Scheme 5.23

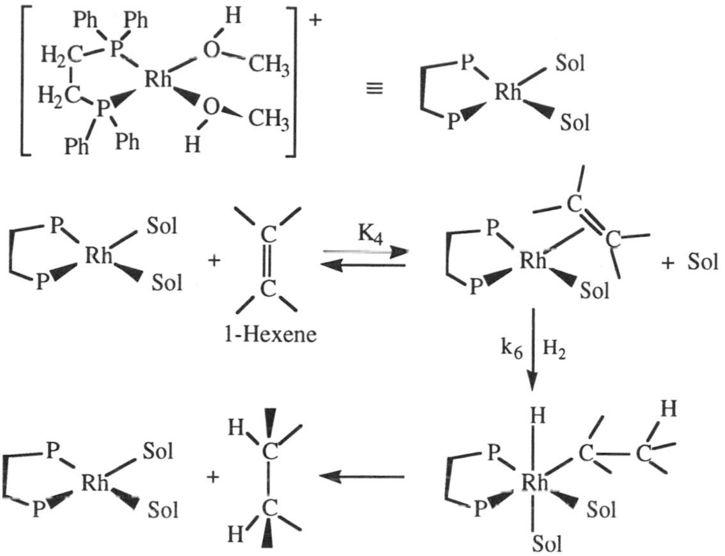

The reaction follows the "olefin route" (Scheme 5.20), and the kinetics give $K_4 = 1.6$ M^{-1} and $k_6 = 0.18$ atm^{-1} s^{-1} at 25°C. Halpern has rationalized the difference caused by using a chelating phosphine as due to the instability of dihydride that forms by oxidative addition of H_2. The trans configuration of the hydride and phosphine ligands is thought to be unstable because of the trans effect of the phosphine.

5.5.c Asymmetric Hydrogenation

An important commercial application of these types of catalysts is in the production of L-dopa for the treatment of Parkinson's disease. The key to this application is the stereoselectivity shown by the phosphine chelate 2S,3S bis(diphenylphosphino)butane, called chiraphos. Halpern and co-workers[128] have shown that the system is unusual in that the most stable olefin adduct does not lead to the major or desired product. The system also follows the "olefin route" like the chelate in Scheme 5.23. The kinetic components of various steps have been studied by Landis and Halpern,[129] who concluded that the minor species generates the major portion of the product because of the greater reactivity of the intermediate with H_2. The mechanistic steps and stereochemical results are described in Scheme 5.24. A recent study[130] has indicated that the H_2 pressure dependence of the products may result from rate-limiting transfer of H_2 between gas and solution phases.

Scheme 5.24

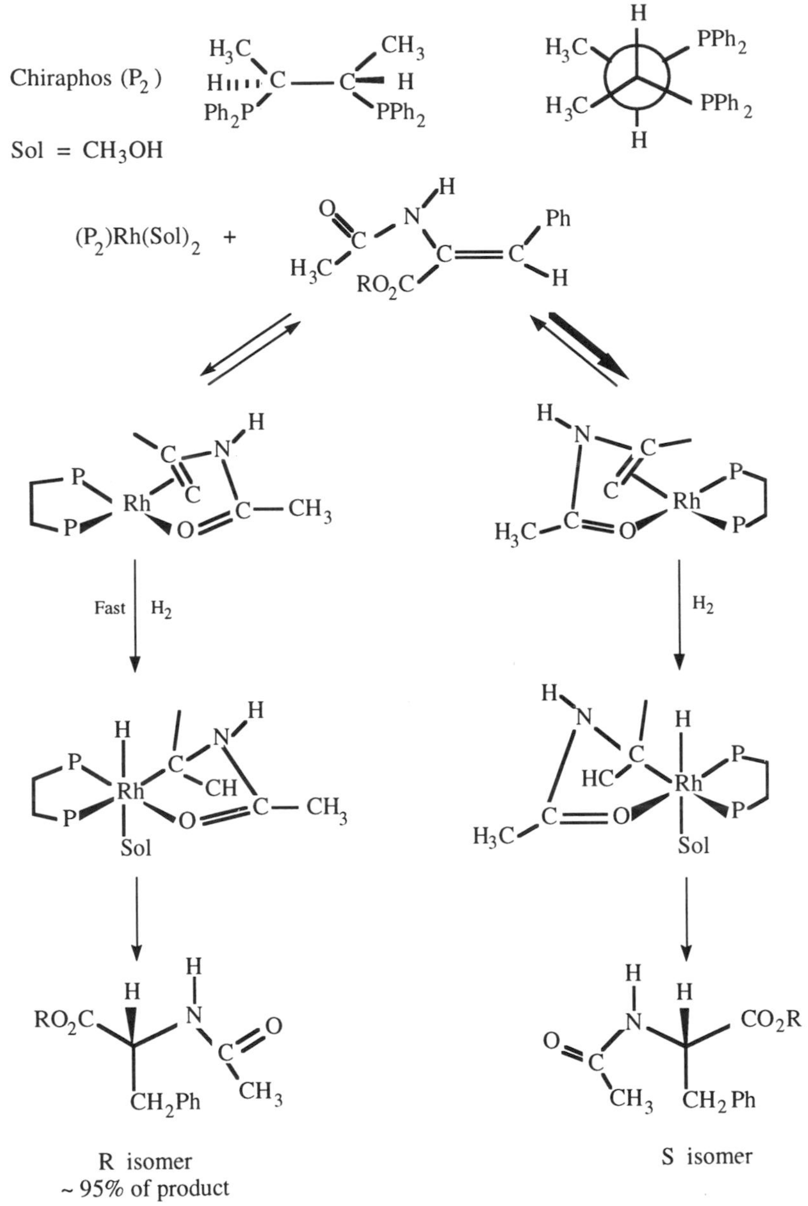
Chiraphos (P_2)
Sol = CH_3OH
$(P_2)Rh(Sol)_2$ +
Fast
H_2
H_2
R isomer
~ 95% of product
S isomer

The field of asymmetric hydrogenation has been described in a recent review[131] and a book.[132] There are a number of examples of such reactions in which the mechanistic details are uncertain. Recently, Takahashi et al.[133] have reported the highly selective hydrogenation of 3-(aryloxy)-2-oxo-1-propylamine derivatives with a chiral phosphine of

Rh(I). Chan and Osborn[134] found an Ir(III) diphosphine-monohydride that gives enantioselective hydrogenation of imines. Garland and Blaser[135] have reported good enantioselectivity for the hydrogenation of ethyl pyruvate on a heterogeneous catalyst of Pt on Al_2O_3 modified with 10,11-dihydrocinchonidine.

5.5.d Carbon–Hydrogen Bond Activation

Carbon–hydrogen bond activation is formally the oxidative addition of a hydrocarbon to a metal complex, as shown in (5.30). It is a potentially important reaction because it represents the initial step in a possible route to functionalize hydrocarbons.

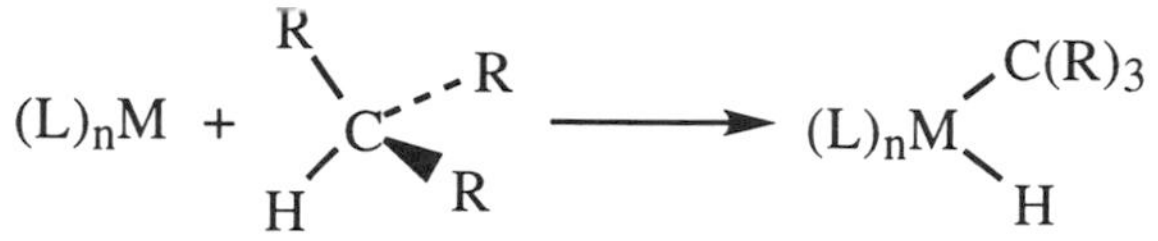

$$
(L)_nM \; + \; \underset{H}{\overset{R}{\underset{\big|}{C}}}\!\!\!\!-\!\!R \quad \longrightarrow \quad (L)_nM\underset{H}{\overset{C(R)_3}{\big\langle}} \tag{5.30}
$$

Until the discovery of this process in 1982,[136,137] it was thought to be difficult at best, and perhaps impossible under moderate conditions, because of the low reactivity of hydrocarbons and the high strength (95–100 kcal mol^{-1}) of the C—H bond. However, the first observations of this type of reaction indicated that it could occur under relatively mild conditions, as shown in the following examples:

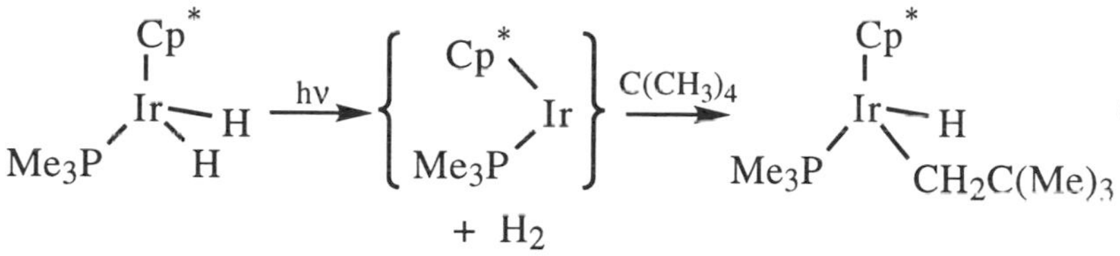

$$
\underset{Me_3P}{\overset{Cp^*}{\underset{}{}}}\!\!Ir\!\!\underset{H}{\overset{H}{}} \;\xrightarrow{h\nu}\; \left\{ \underset{Me_3P}{\overset{Cp^*}{}}\!Ir \right\} \;\xrightarrow{C(CH_3)_4}\; \underset{Me_3P}{\overset{Cp^*}{}}\!Ir\!\underset{CH_2C(Me)_3}{\overset{H}{}} \qquad (5.31)
$$
$$
+ \; H_2
$$

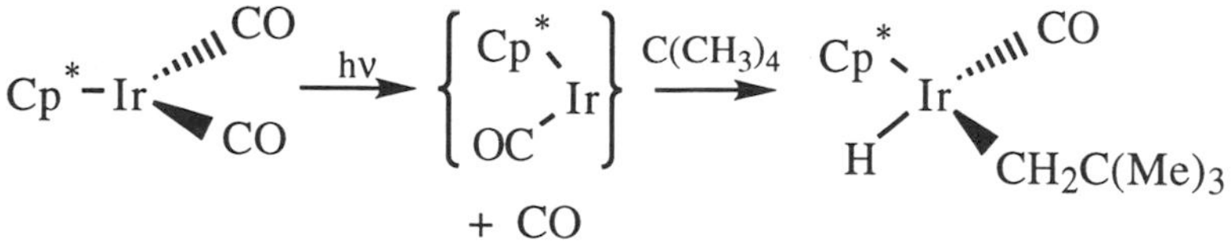

$$
Cp^*\!-\!Ir\underset{CO}{\overset{CO}{}} \;\xrightarrow{h\nu}\; \left\{ \underset{OC}{\overset{Cp^*}{}}\!Ir \right\} \;\xrightarrow{C(CH_3)_4}\; \underset{H}{\overset{Cp^*}{}}\!Ir\underset{CH_2C(Me)_3}{\overset{CO}{}} \qquad (5.32)
$$
$$
+ \; CO
$$

In both cases, a highly reactive unsaturated species is assumed to be generated by photolysis and reacts with the neopentane solvent as shown. Cyclohexane reacted similarly in both cases. Since the original discovery, many more metal complexes have been found to undergo oxidative addition of hydrocarbons,[138–141] and the area has been reviewed recently.[142,143]

Ghosh and Graham[144] have shown that Rh(HBPz$^*_3$)(CO)$_2$, where HBPz$^*_3$ is tris(dimethylpyrazolyl)borato, photolyzes under mild conditions with elimination of CO and oxidative addition of hydrocarbons. In this system, the cyclohexyl hydride complex exchanges with methane, as shown in (5.33), to give the methyl hydride complex.

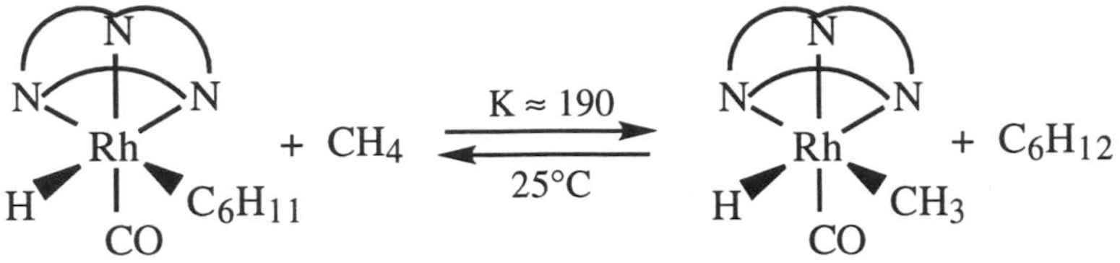

$$\text{(5.33)}$$

Ghosh and Graham[145] have used $Rh(HBPz^*_3)(CO)(C_2H_4)$ to prepare the first generation of functionalized products from such reactions, as shown in Scheme 5.25.

Scheme 5.25

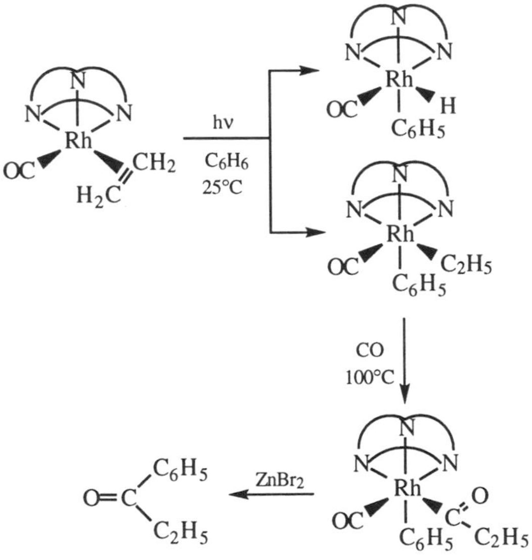

Tanaka and co-workers[146] have found that the irradiation of $Rh(PR_3)_2(CO)Cl$ in hydrocarbon solvents under 1 atm of CO can yield aldehyde and alcohol derivatives of the hydrocarbon in a photoassisted catalytic reaction. They attribute the reaction to photochemical dissociation of CO followed by C—H activation and then CO insertion. The observation that the catalysis is improved if the irradiation wavelength is below the absorbance maximum of $Rh(PR_3)_2(CO)Cl$ indicates that photochemistry may be involved in other steps in addition to the decarbonylation. This system has been the subject of a theoretical study.[147]

The energetics of the C—H activation process have been of concern since its discovery. Calorimetric, equilibrium constant, and kinetic information have been used to obtain estimates of the M—C and M—H bond energies. In general, these bonds have been found to be stronger than originally anticipated. For the simple oxidative addition of a hydrocarbon, the enthalpy change can be estimated as

$$\Delta H_{rxn} = BE(C\text{—}H) - [BE(M\text{—}H) + BE(M\text{—}C)] \qquad (5.34)$$

It is commonly assumed that the reaction will proceed if it is exothermic (i.e., entropy effects are assumed to be small, so that $\Delta H_{rxn} \approx \Delta G_{rxn}$). Thus, the bond energies can be used to predict whether an oxidative addition of a hydrocarbon is probable. Some bond energies[148] for different metal centers are given in Table 5.10. The C—H bond energies, in kcal mol^{-1}, are 96 for C_6H_{12}, 110 for C_6H_6, and 98 for CH_4, and the C—C bond energy is 83 kcal mol^{-1} in alkanes. For the oxidative addition of methane to $Th(Cp)_2$, the data in Table 5.10 can be used to estimate that $\Delta H_{rxn} \approx 98 - (97.5 + 82) \approx -82$ kcal mol^{-1}, and therefore the reaction should be quite exothermic. However, it is of greater interest and relevance to estimate the relative stabilities of the reactants and products for overall reactions such as

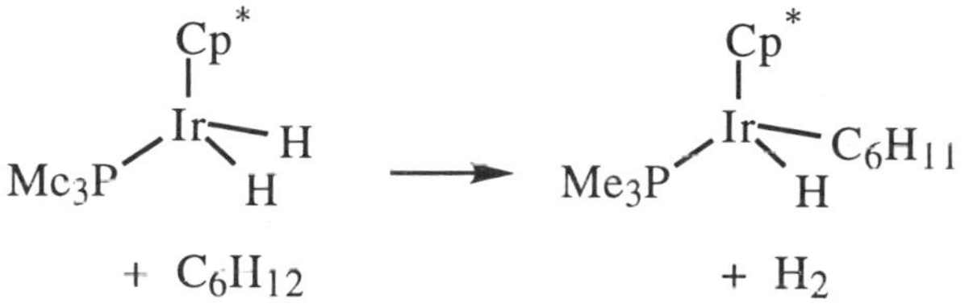

$$(5.35)$$

whose enthalpy change is given by

$$\begin{aligned} \Delta H_{rxn} &= BE(C\text{—}H) + BE(Ir\text{—}H) - BE(Ir\text{—}C) - BE(H\text{—}H) \\ &= 96 + 74 - 51 - 104 = 15 \text{ kcal mol}^{-1} \end{aligned} \qquad (5.36)$$

Table 5.10. Bond Energies for M—X in $(L)_nM(X)_2$ Systems (kcal mol^{-1})

X	$(Me_3P)(Cp^*)Ir(X)_2$	$(Cp)_2W(X)_2$	$(Cp)_2Th(X)_2$	$Cl(L)_2(CO)Ir(X)_2$
H	74	73	97.5	60
Cl	90	83		71
Br	76	71.5		53
I	64	64		35
CH_3			82	35
C_6H_5	81		91	
C_6H_{11}	51		75	

A notable feature in Table 5.10 is the strength of the M—C_6H_5 bond. This implies that benzene can be expected to replace other oxidatively added hydrocarbons in an exchange reaction. Since the C—C bond in alkanes is weaker than the C—H bond, one might anticipate C—C bond activation, as in the reaction

$$(Cp)_2Th\begin{smallmatrix}H\\H\end{smallmatrix} + H_3C—CH_3 \longrightarrow (Cp)_2Th\begin{smallmatrix}CH_3\\CH_3\end{smallmatrix} + H_2 \quad (5.37)$$

whose enthalpy change is given by

$$\begin{aligned}\Delta H_{rxn} &= BE(C—C) + 2\,BE(Th—H) - 2\,BE(Th—C) - BE(H—H)\\ &= 83 + (2 \times 97.5) - (2 \times 82) - 104 = 10 \text{ kcal mol}^{-1} \quad (5.38)\end{aligned}$$

For comparison, the corresponding C—H oxidative addition gives

$$\begin{aligned}\Delta H_{rxn} &= BE(C—H) + BE(Th—H) - BE(Th—C) - BE(H—H)\\ &= 98 + 97.5 - 82 - 104 = 9.5 \text{ kcal mol}^{-1} \quad (5.39)\end{aligned}$$

Thermodynamics allows no clear choice between these two possibilities, and kinetic factors must be at work to favor C—H activation.

Mechanistic studies on these reactions have been limited because of the necessity to generate the unsaturated metal center photochemically and the need for an appropriate "inert" solvent also is a problem.

Marx and Lees[149] studied quantum yields, Φ, for the photochemical reaction of Ir(Cp)(CO)$_2$ with C_6H_6 in hexafluorobenzene at 20°C and found that Φ shows a saturation effect with increasing [C_6H_6] and is independent of [CO]. They concluded that the rate-controlling process is not CO dissociation and suggest that it is an η^5 to η^3 slippage, which then allows complexation of benzene prior to oxidative addition. Drolet and Lees[150] proposed a similar slippage mechanism for the photolysis of Rh(Cp)(CO)$_2$ in the presence of PPh$_3$ in decalin at 10°C. Their conclusion is based on the observation that the quantum yield for CO replacement is first order in [PPh$_3$] and unaffected by 9 x 10^{-3} M CO.

Bergman and co-workers[151] studied the reaction of Rh(Cp*)(CO)$_2$ with C_6H_{12} by IR laser flash kinetics in inert gas solvents. Their conclusions are different from those of Lees and co-workers. At −31°C in xenon, irradiation gives a new species that Bergman and co-workers assign as Rh(Cp*)(CO) or its solvated form, Rh(Cp*)(CO)Xe. A similar but more reactive species is observed in krypton and when cyclohexane is added to the solution, the Rh(Cp*)(CO)Kr undergoes oxidative addition with a rate that shows saturation behavior described by

$$k_{exp} = \frac{\alpha\,[C_6H_{12}]}{[C_6H_{12}] + \beta} \quad (5.40)$$

If the mechanism involves rate-controlling dissociation of the solvent followed by cyclohexane addition, then α should be independent of the nature of the alkane. However, if C_6D_{12} is used, α decreases by a factor of about 20. Therefore, Bergman and co-workers have proposed the mechanism in Scheme 5.26.

Scheme 5.26

$$Rh(Cp^*)(CO)Kr + C_6H_{12} \underset{\longleftarrow}{\overset{K_{eq}}{\longrightarrow}} Rh(Cp^*)(CO)(C_6H_{12}) + Kr$$

$$\Big\downarrow k_1$$

$$Rh(Cp^*)(CO)(C_6H_{11})H$$

From the preceding expression for k_{exp}, $\alpha = k_1$ and $\beta = [Kr]/K_{eq}$. The nature of the cyclohexane precursor complex is uncertain, but studies with neopentane have allowed the IR spectra of the krypton and neopentane complexes to be resolved.[152]

Bergman and co-workers[153] analyzed the energy profile for the gas-phase reaction of methane and $Rh(Cp)(CO)$. It was estimated that the formation of the precursor complex, $Rh(Cp)(CO) \cdot HCH_3$, is exothermic by ~10 kcal mol^{-1}, the C—H activation energy is ~4.5 kcal mol^{-1}, and the overall exothermicity is ~15 kcal mol^{-1}. It also has been found[154] that $Co(Cp)(CO)$ gives no C—H activation and does not even form a precursor complex, but reacts with its parent to give $Co_2(Cp)_2(CO)_3$. There have been a number of theoretical studies of these systems,[155] and Siegbahn[156] has extended and compared the theoretical results. This analysis indicates that $Co(Cp)(CO)$ is different and unreactive because it has a triplet ground state that cannot easily cross to the singlet state due to the substantial differences in bond lengths in the two states.

The differences in reactivity of the Fe and Ru species $M((R_2PCH_2)_2)_2$, generated by photolysis of the corresponding dihydride,[157] might originate from the factors described by Siegbahn. The Fe derivative has a significantly different electronic spectrum and shows a much greater range of reactivities with nucleophiles, compared to the Ru analogue. This is not a problem for all first-row transition metals because $Mn(Cp)(CO)_3$ forms a complex with n-heptane.[158]

Perutz and co-workers[159] observed the time-resolved photochemistry of $Rh(Cp)(CO)_2$ and $(Rh(Cp)(CO))_2(\mu\text{-}CO)$ in the 300-nm region. The results are summarized in Scheme 5.27, where RH is a solvent such as C_6H_{12} or C_6H_6. The structure of $(Rh(Cp)(\mu\text{-}CO))_2$ was assigned as shown on the basis of its polarized IR spectrum. The unusual aspects of these observations are the high dimerization rate and the rapid formation of the hydride species of cyclohexane within 400 ns (the product with benzene could be $\eta^2\text{-}C_6H_6$).

Scheme 5.27

$$Rh(Cp)(CO)_2 \xrightarrow{\text{hv}} Rh(Cp)(CO) + CO$$

$$Rh(Cp)(CO) + RH \underset{k = 2.7 \times 10^3 \, s^{-1}}{\overset{k \geq 3 \times 10^5 \, M^{-1} s^{-1}}{\rightleftharpoons}} Rh(Cp)(CO)(R)H$$

$$Rh(Cp)(CO)(R)H + L \xrightarrow{k \approx 3 \times 10^8 \, M^{-1} s^{-1}} Rh(Cp)(CO)L + RH$$

$$Rh(Cp)(CO) + Rh(Cp)(CO)(R)H \xrightarrow{k \approx 10^{10} \, M^{-1} s^{-1}} [Rh(Cp)(\mu\text{-}CO)]_2 + RH$$

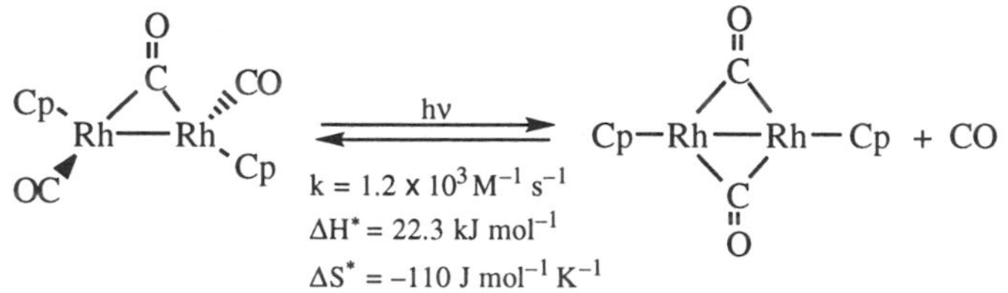

Studies of the reverse reaction, reductive elimination, have also been used to shed light on the overall reaction mechanism. Norton and co-workers[160] studied the reductive elimination of methane from $W(Cp)_2(CH_3)H$ and found that H exchange into the CH_3 group occurs more rapidly than CH_4 elimination. Such exchange has been observed in other systems, and some rate constants at 48°C in 10 percent CD_3CN–90 percent C_6D_6 are given in Scheme 5.28. The CD_3CN serves to trap the $W(Cp)_2$ as $W(Cp)_2(\eta^2\text{-}CD_3CN)$, but the kinetics are unaffected by the concentration of CD_3CN.

Scheme 5.28

$$W(Cp)_2(CH_3)D \underset{k = 5 \times 10^{-6} \, s^{-1}}{\overset{k = 2.5 \times 10^{-5} \, s^{-1}}{\rightleftharpoons}} W(Cp)_2(CDH_2)H$$

$$W(Cp)_2(CH_3)H \xrightarrow[\substack{\Delta H^* = 24.5 \text{ kcal mol}^{-1} \\ \Delta S^* = -5.6 \text{ cal mol}^{-1} K^{-1}}]{k = 8.4 \times 10^{-6} \, s^{-1}} \{W(Cp)_2\} + CH_4$$

Elimination has an inverse isotope effect, H/D, of 0.75 at 72.6°C. The authors propose that the exchange and reductive elimination proceed through a σ complex that may revert to reactant with exchange or give elimination, as shown in Scheme 5.29. The isotope effect was ascribed to greater stability of the σ complex $W(Cp)_2(D\text{—}CH_3)$ compared to $W(Cp)_2(H\text{—}CH_3)$ in a preequilibrium before elimination of methane.

Scheme 5.29

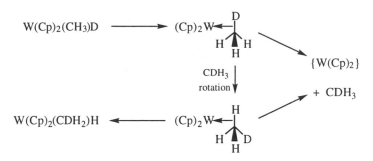

Bergman and co-workers[161] studied the exchange reaction in Scheme 5.30 and found that the kinetics are consistent with the mechanism shown.

Scheme 5.30

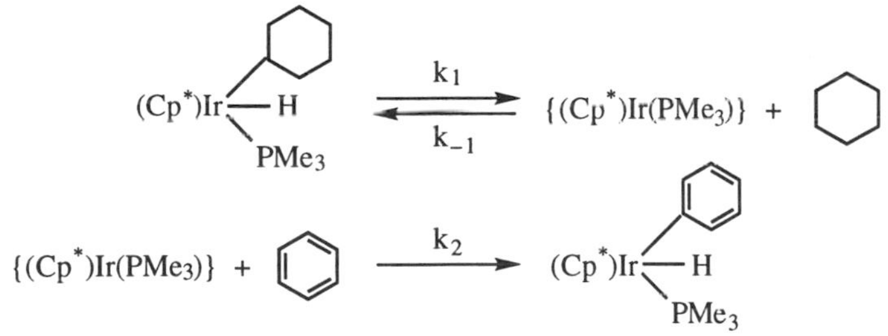

$$
(Cp^*)Ir\!\!\begin{array}{c} \diagup H \\ \diagdown PMe_3 \end{array} \quad \underset{k_{-1}}{\overset{k_1}{\rightleftharpoons}} \quad \{(Cp^*)Ir(PMe_3)\} \;+\; \bigcirc
$$

$$
\{(Cp^*)Ir(PMe_3)\} \;+\; \hexagon \quad \overset{k_2}{\longrightarrow} \quad (Cp^*)Ir\!\!\begin{array}{c} \diagup H \\ \diagdown PMe_3 \end{array}
$$

The pseudo-first-order rate constant is given by Eq. (5.41), and the reaction rate is unaffected by PMe_3.

$$
k_{exp} = \frac{k_1 [C_6H_6]}{\left(\dfrac{k_{-1}}{k_2}\right)[C_6H_{12}] + [C_6H_6]} \tag{5.41}
$$

The reaction was confirmed to be intramolecular and not to involve H transfer to Cp^*, since $Ir(C_6H_{11})H + C_6D_6$ yields only C_6H_{12} with no isotope scrambling. This seems to eliminate radical mechanisms involving Ir—C bond homolysis. Ring slippage is compatible with the rate law only if the oxidative addition of C_6H_{12} to the intermediate is competitive with the η^3 to η^5 change to give product; Bergman and co-workers believe that this is unlikely. There is an inverse isotope effect on k_1 (H/D $\approx$ 0.7) and the values of the activation parameters for k_1 are $\Delta H^* = 35.6$ kcal mol^{-1} and $\Delta S^* = 10$ cal mol^{-1} K^{-1}. Isotope exchange was observed in $Ir(C_6H_{11})D$, with the α position of C_6H_{11} becoming deuterated. All these observations are similar to those of Norton and co-workers, and an analogous σ complex would explain the results.

The Shilov reaction is a remarkable example of C—H activation because substitutionally inert aqueous Pt(II) activates alkanes to form alcohols. The reactive species may be *trans*-Pt(Cl)$_2$(OH$_2$)$_2$. Bercaw and co-workers[162] have suggested that this involves oxidative addition of the alkane to give a Pt(IV) alkyl hydride. A recent theoretical study[163] has suggested that the reaction may proceed via a transient Pt(II)-alkane complex that transfers a proton to a coordinated chloride and then to water and forms a Pt(II)-alkyl complex as the key intermediate.

5.6 HOMOGENEOUS CATALYSIS BY ORGANOMETALLIC COMPOUNDS

There are a number of processes that appear to be catalyzed by organometallic complexes. The reactions generally involve C—C bond formation and often involve CO and H$_2$. In the industrial application, the catalyst is often added as a metal salt, and it is assumed that this is transformed to active organometallic species under the reducing conditions of the process. In most cases, there may be several organometallic species present, and the nature of the catalytic mechanism is inferred from the known chemistry of simpler systems and the overall rate law.

5.6.a Hydroformylation or "Oxo" Reaction

The overall process is described by

$$RCH{=}CH_2 \xrightarrow[\text{CO}]{H_2} RCH_2CH_2{-}C\!\!\begin{array}{c}\nearrow O \\ \searrow H\end{array} \xrightarrow{H_2} RCH_2CH_2CH_2OH \quad (5.42)$$

In 1938, Roelen discovered that the reaction is catalyzed by cobalt salts, and, more recently, rhodium catalysts have been developed. The main commercial product is *n*-butanol from propylene, and the major problems are to avoid branched-chain alcohols and alkanes. The reaction was reviewed by Pruett.[164] With cobalt as the catalyst, the reaction is done at 100°C to 200°C and at 100 to 500 psi of H$_2$ + CO. Linear product is favored by lower CO pressures and by "modified cobalt catalysts" that contain phosphines in the reaction mixture. The rate law is given by

$$\text{Rate} = k\,[\text{Co}]\,[\text{Alkene}]\,[\text{H}_2]\,[\text{CO}]^{-1} \quad\quad (5.43)$$

It is accepted generally that the important cobalt species is Co(CO)$_4$H, formed by the following reaction:

$$Co_2(CO)_8 + H_2 \rightleftharpoons 2\,Co(CO)_4H \quad\quad (5.44)$$

The mechanism in Scheme 5.31 was first proposed by Heck and Breslow[165] and is still thought to be essentially correct. The hydride shift in the third step followed by CO "insertion" are common steps proposed in many such processes. The first dissociation of CO is assumed to be the source of the $[CO]^{-1}$ term in the rate law.

Scheme 5.31

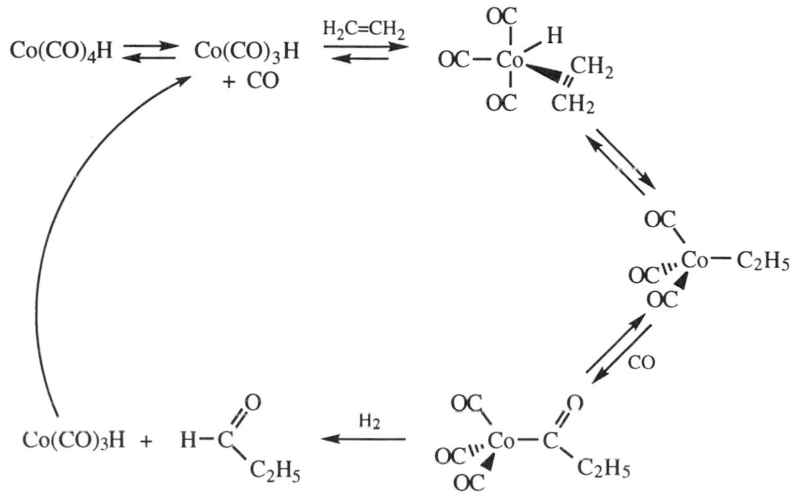

Wilkinson and co-workers[166] discovered that rhodium catalysts are effective in this reaction. It is thought that the key intermediate may be $Rh(PPh_3)_2(CO)_2H$, which may react by dissociative loss of PPh_3 or by direct formation of a π-olefin complex. The reaction could proceed by a hydride route, as proposed in some hydrogenation processes.

5.6.b Reppe Synthesis

The Reppe synthesis, shown in (5.45), has a formal resemblance to hydroformylation, with H_2 replaced by H_2O.

$$RCH=CH_2 + 2\,H_2O + 3\,CO \longrightarrow R(CH_2)_2CH_2OH + 2\,CO_2 \quad (5.45)$$

The most effective catalysts for the reaction are iron species often derived from $Fe(CO)_5$. Pettit and co-workers[167] found that the catalyst is most effective at pH >11. This has been rationalized by the known reactions shown in (5.46) and assuming that $H_2Fe(CO)_4$ is the active species that interacts with alkenes.

$$Fe(CO)_5 + OH^- \longrightarrow HFe(CO)_4^- \underset{}{\overset{H_2O}{\rightleftharpoons}} H_2Fe(CO)_4 \quad (5.46)$$
$$+ CO_2 \qquad\qquad + OH^-$$

5.6.c Fischer–Tropsch Reaction

The Fisher–Tropsch reaction involves the conversion of coal to hydrocarbons as follows:

$$C + H_2O \rightleftharpoons CO + H_2 \longrightarrow \text{Hydrocarbon products} \quad (5.47)$$

The process often is considered in conjunction with reaction (5.48), the water gas shift reaction, since this increases the H_2 content of the reactant mixture. The reaction is attractive because of the conversion of cheap materials, coal and water, to valuable hydrocarbons.

$$H_2O + CO \rightleftharpoons H_2 + CO_2 \quad (5.48)$$

The process was developed by Fischer and Tropsch in 1923 using heterogeneous iron catalysts, and catalyst improvements have continued in a rather unsystematic way ever since. The reaction became uneconomical with the advent of cheap oil after World War II but came into vogue again in the late 1970s as oil prices rose. Only South Africa has operating plants using the process, with the industrial conditions being 200°C to 300°C and a total pressure of ~25 atm.

It has been found by Muetterties and co-workers[168] and by Ford and co-workers[169] that polynuclear carbonyls such as $Os_3(CO)_{12}$, $Ru_3(CO)_{12}$, and $Ir_4(CO)_{12}$ are effective homogeneous catalysts. The mechanism in Scheme 5.32 was suggested by Masters.[170]

Scheme 5.32

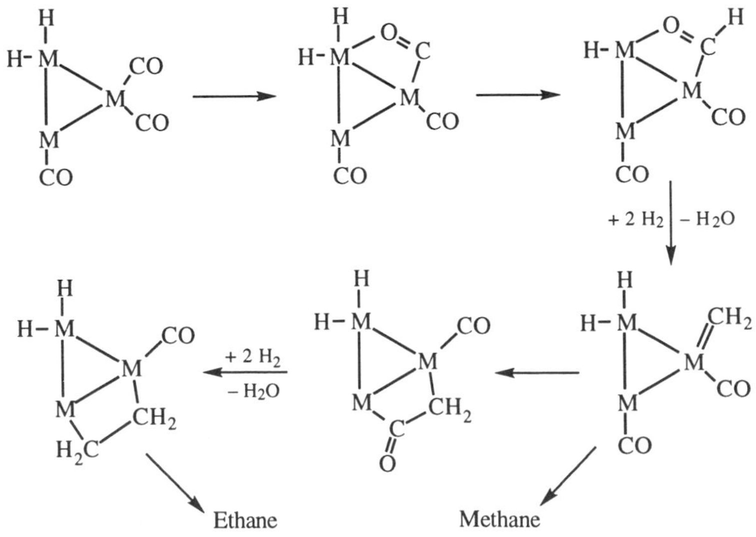

It involves formation of a metal carbene ($M=CH_2$) as an important step in the process. The apparent requirement for at least two metal centers in the catalyst has stimulated much work on dimetal species and metal cluster complexes. It has been suggested that a key intermediate is an η^2 CO in which the η^2 bonding serves to weaken the CO triple bond and make it susceptible to reduction.

5.6.d Ethylene–Butadiene Codimerization

The purpose of the ethylene–butadiene codimerization process is to produce *trans*-1,4-hexadiene from ethylene and butadiene as in reaction (5.49). The product is an important monomer in synthetic rubber production.

$$H_2C=CH_2 + H_2C=CHCH=CH_2 \longrightarrow H_2C=CHCH_2CH=CHCH_3 \quad (5.49)$$

The reaction is catalyzed in the presence of $RhCl_3$ in aqueous HCl and has been studied kinetically by Cramer.[171] It is proposed that under the reducing influence of the ethylene, an unspecified rhodium hydride species ($Rh(Cl)_2H$?) forms as the catalyst and reacts with the ethylene to give an η^1-ethyl complex that may be converted to the η^2-ethyl complex by complexing with butadiene. This is followed by H transfer to give a π-crotyl complex and finally the desired product, as shown in Scheme 5.33.

Scheme 5.33

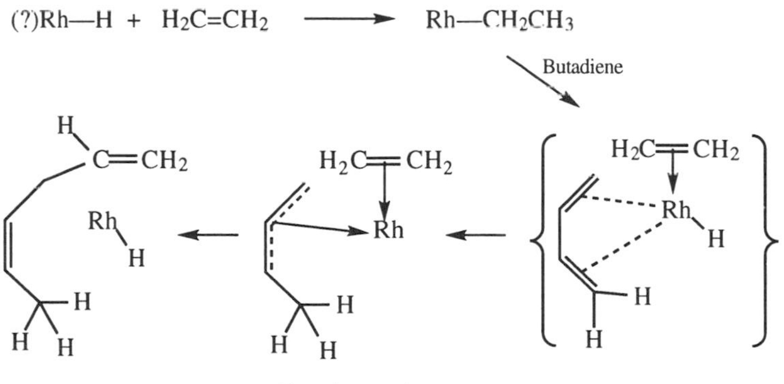

π-Crotyl complex

Tolman[172] found that $Ni(P(OEt)_3)_4H$ also is a good catalyst for this reaction, and showed that it reacts by dissociation of a phosphite ligand to form a π-crotyl complex with butadiene. Then, another phosphite is lost and the π-ethylene + π-crotyl complex forms and rearranges similarly to the Rh system in Scheme 5.33.

References

1. Darensbourg, D. J. *Adv. Organomet. Chem.* **1982**, *21*, 113.
2. Basolo, F. *Coord. Chem. Rev.* **1982**, *49*, 7.
3. Howell, J. A. S.; Burkinshaw, P. M. *Chem. Rev.* **1983**, *83*, 557.
4. Basolo, F. *J. Organomet. Chem.* **1990**, *383*, 579.
5. Shi, Q.-Z.; Richmond, T. G.; Trogler, W. C.; Basolo, F. *J. Am. Chem. Soc.* **1984**, *106*, 71.
6. Day, J. P.; Basolo, F.; Pearson, R. G. *J. Am. Chem. Soc.* **1968**, *90*, 6927.
7. Noack, K.; Ruch, M. *J. Organomet. Chem.* **1969**, *17*, 309.
8. Shen, J.-K.; Gao, Y.-C.; Shi, Q.-Z.; Basolo, F. *Inorg. Chem.* **1989**, *28*, 4304.
9. Huq, R.; Pöe, A. J.; Chawla, S. *Inorg. Chim. Acta* **1980**, *38*, 121.
10. Poliakoff, M.; Turner, J. J. *J. Chem Soc., Faraday Trans. 2* **1974**, *70*, 93.
11. Li, J.; Schreckenbach, G.; Ziegler, T. *J. Am. Chem. Soc.* **1995**, *117*, 486.
12. Ehlers, A. W.; Frenking, G. *Organometallics*, **1995**, *14*, 423.
13. King, R. B. *J. Inorg. Nucl. Chem.* **1969**, *5*, 906.
14. Brower, K. R.; Chen, T.-S. *Inorg. Chem.* **1973**, *12*, 2198.
15. Wojcicki, A.; Basolo, F. *J. Am. Chem. Soc.* **1961**, *83*, 527.
16. Johnson, B. F. G.; Lewis, J.; Meiser, J. R.; Robinson, B. H.; Robinson, P. W.; Wojcicki, A. *J. Chem. Soc. A* **1968**, 522; Kaesz, H. D.; Bau, R.; Hendrikson, D.; Smith, M. *J. Am. Chem. Soc.* **1967**, *89*, 2844; Robinson, P. W.; Cohen, M. A.; Wojcicki, A. *Inorg. Chem.* **1971**, *10*, 2081; Berry, A.; Brown, T. L. *Inorg. Chem.* **1972**, *11*, 1165.
17. Atwood, J. D.; Brown, T. L. *J. Am. Chem. Soc.* **1975**, *97*, 3380.
18. Chen, L.; Poë, A. J. *Inorg. Chem.* **1989**, *28*, 3641.
19. Lorsbach, B. A.; Bennett, D. M.; Prock, A.; Giering, W. P. *Organometallics* **1995**, *14*, 869; Wilson, M. R.; Liu, H.; Prock, A.; Giering, W. P. *Organometallics* **1993**, *12*, 2044, and references therein.
20. Bartik, T.; Himmler, T.; Schulte, H.-G.; Seevogel, K. *J. Organomet. Chem.* **1984**, *272*, 29.
21. Zhang, S.; Shen, J. K.; Basolo, F.; Ju, T. D.; Lang, R. F.; Kiss, G.; Hoff, C. D. *Organometallics* **1994**, *13*, 3692.
22. Schuster-Woldan, H. G.; Basolo, F. *J. Am. Chem. Soc.* **1966**, *88*, 1657.
23. Cramer, R.; Seiwell, L. P. *J. Organomet. Chem.* **1975**, *92*, 245; Bleeke, J. R.; Peng, W. J. *Organometallics* **1986**, *5*, 635.
24. Ji, L.-N.; Rerek, M. E.; Basolo, F. *Organometallics* **1984**, *3*, 740.
25. Habib, A.; Tanke, R. S.; Holt, E. M.; Crabtree, R. H. *Organometallics* **1989**, *8*, 1225.
26. Bang, H.; Lynch, T. J.; Basolo, F. *Organometallics* **1992**, *11*, 40; Casey, C. P.; O'Connor, J. M. *Organometallics* **1985**, *4*, 384.
27. Ji, L.-N.; Rerek, M. E.; Basolo, F. *Organometallics* **1984**, *3*, 740.
28. Jones, D. J.; Mawby, R. J. *Inorg. Chim. Acta* **1972**, *6*, 157.
29. Gamasa, M. P.; Gimeno, J.; Gonzalez-Bernardo, C.; Martin-Vaca, B. M.; Monti, D.; Bassetti, M. *Organometallics* **1996**, *15*, 302.
30. Pevear, K. A.; Banaszak Holl, M. M.; Carpenter, G. B.; Rieger, A. L.; Sweigart, D. A. *Organometallics* **1995**, *14*, 512.

31. Bonifaci, C.; Ceccon, A.; Santi, S.; Mealli, C.; Zoellner, R. W. *Inorg. Chim. Acta* **1995**, *240*, 541.

32. Kowaleski, R. M.; Basolo, F.; Trogler, W. C.; Ernst, R. D. *J. Am. Chem. Soc.* **1986**, *108*, 6046; Kowaleski, R. M.; Basolo, F.; Trogler, W. C.; Gedridge, R. W.; Newbound, T. D.; Ernst, R. D. *J. Am. Chem. Soc.* **1986**, *108*, 6046.

33. Morris, D. E.; Basolo, F. *J. Am. Chem. Soc.* **1968**, *90*, 2531.

34. Darensbourg, D. J.; Darensbourg, M. Y.; Walker, N. *Inorg. Chem.* **1981**, *20*, 1918; Cotton, F. A.; Darensbourg, D. J.; Kolthammer, B. W. S.; Kudaroski, R. *Inorg. Chem.* **1982**, *21*, 1656.

35. Darensbourg, D. J.; Klausmeyer, K. K.; Reibenspies, J. H. *Inorg. Chem.* **1995**, *34*, 4933.

36. Darensbourg, D.J.; Klausmcycr, K. K.; Reibenspies, J. H. *Inorg. Chem.* **1996**, *35*, 1529; Ibid. **1996**, *35*, 1535.

37. Rossi, A. G.; Hoffmann, R. *Inorg. Chem.* **1975**, *14*, 365.

38. Lichtenberger, D. L.; Brown, T. L. *J. Am. Chem. Soc.* **1978**, *100*, 366.

39. Poliakoff, M. *Inorg. Chem.* **1976**, *15*, 2892.

40. Cotton, F. A.; Darensbourg, D. J.; Klein, S.; Kolthammer, B. W. S. *Inorg. Chem.* **1982**, *21*, 294.

41. Ewen, J. E.; Darensbourg, D. J. *J. Am. Chem. Soc.* **1975**, *97*, 6874.

42. Kubas, G. J. *Acc. Chem. Res.* **1988**, *21*, 120.

43. Gonzalez, A. A.; Zhang, K.; Hoff, C. D. *Inorg. Chem.* **1989**, *28*, 4285.

44. Graham, J. R.; Angelici, R. J. *Inorg. Chem.* **1967**, *6*, 992.

45. Memering, M. N.; Dobson, G. R. *Inorg. Chem.* **1973**, *12*, 2490.

46. Halverson, D. E.; Reisner, G. M.; Dobson, G. R.; Bernal, I.; Mulcahy, T. L. *Inorg. Chem.* **1982**, *21*, 4285.

47. Macholdt, H.-T.; van Eldik, R.; Dobson, G. R. *Inorg. Chem.* **1986**, *25*, 1914.

48. Cao, S.; Reddy, K. B.; Eyring, E. M.; van Eldik, R. *Organometallics*, **1994**, *13*, 91.

49. Wawersik, H.; Basolo, F. *Inorg. Chim. Acta* **1969**, *3*, 113.

50. Fawcett, J. P.; Poë, A.; Sharma, K. R. *J. Am. Chem. Soc.* **1976**, *98*, 1401; Fawcett, J. P.; Poë, A. *J. Chem. Soc., Dalton Trans.* **1977**, 1302.

51. Wegman, R. W.; Olsen, R. J.; Gard, D. R.; Faulkner, L. R.; Brown, T. L. *J. Am. Chem. Soc.* **1981**, *103*, 6089.

52. Sonnenberger, D.; Atwood, J. D. *J. Am. Chem. Soc.* **1980**, *102*, 3484.

53. Coville, N. J.; Stolzenberg, A. M.; Muetterties, E. L. *J. Am. Chem. Soc.* **1983**, *105*, 2499.

54. Poë, A. *Inorg. Chem.* **1981**, *20*, 4029, 4032.

55. Atwood, J. D. *Inorg. Chem.* **1981**, *20*, 4031.

56. Darensbourg, D. J.; Zalewski, D. J.; Delford, T. *Organometallics* **1984**, *3*, 1210.

57. Kennedy, J. R.; Basolo, F.; Trogler, W. C. *Inorg. Chim. Acta* **1988**, *146*, 75.

58. Sonnenberger, D. C.; Atwood, J. D. *J. Am. Chem. Soc.* **1982**, *104*, 2113; Darensbourg, D. J.; Baldwin-Zuschke, B. J. *J. Am. Chem. Soc.* **1982**, *104*, 3906.

59. Chen, L.; Poë, A. J. *Can. J. Chem.* **1989**, *67*, 1924.
60. Poë, A. J.; Sekhar, V. C. *Inorg. Chem.* **1985**, *24*, 4376.
61. Brodie, M. J.; Poë, A. J. *Can. J. Chem.* **1995**, *73*, 1187.
62. Shen, J.-K.; Basolo, F. *Organometallics* **1993**, *12*, 2942.
63. Neubrand, A.; Poë, A. J.; van Eldik, R. *Organometallics* **1995**, *14*, 3249.
64. Poë, A. J.; Farrar, D. H.; Zheng, Y. *J. Am. Chem. Soc.* **1992**, *114*, 5146.
65. Farrar, D. H.; Poë, A. J.; Zheng, Y. *J. Am. Chem. Soc.* **1994**, *116*, 6252.
66. Shen, J.-K.; Basolo, F.; Nombel, P.; Lugan, N.; Lavigne, G. *Inorg. Chem.* **1996**, *35*, 755.
67. Absi-Halabi, M.; Brown, T. L. *J. Am. Chem. Soc.* **1977**, *99*, 2983.
68. Byers, B. H.; Brown, T. L. *J. Am. Chem. Soc.* **1977**, *99*, 2527.
69. Byers, B. H.; Brown, T. L. *J. Organomet. Chem.* **1977**, *127*, 181.
70. Sweany, R. L.; Halpern, J. *J. Am. Chem. Soc.* **1977**, *99*, 8335.
71. Bullock, R. M.; Samsel, E. G. *J. Am. Chem. Soc.* **1991**, *112*, 6886.
72. Herrinton, T. R.; Brown, T. L. *J. Am. Chem. Soc.* **1985**, *107*, 5700.
73. McCullen, S. B.; Walker, H. W.; Brown, T. L. *J. Am. Chem. Soc.* **1982**, *104*, 4007.
74. Fox, A.; Malito, J.; Poë, A. *J. Chem. Soc., Chem. Commun.* **1981**, 1052.
75. Therien, M. J.; Ni, C.-L.; Anson, F. C.; Osteryoung, J. G.; Trogler, W. C. *J. Am. Chem. Soc.* **1986**, *108*, 4037.
76. Therien, M. J.; Trogler, W. C. *J. Am. Chem. Soc.* **1988**, *110*, 4942.
77. Noack, K.; Calderazzo, F. *J. Organomet. Chem.* **1967**, *10*, 101.
78. Flood, T. C.; Jensen, J. E.; Statler, J. A. *J. Am. Chem. Soc.* **1981**, *103*, 4410.
79. Wright, S. C.; Baird, M. C. *J. Am. Chem. Soc.* **1985**, *107*, 6899.
80. Jablonski, C.; Bellachioma, G.; Cardacci, G.; Reichenbach, G. *J. Am. Chem. Soc.* **1990**, *112*, 1632.
81. Flood, T. C.; Campbell, K. D. *J. Am. Chem. Soc.* **1984**, *106*, 2853.
82. Brunner, H.; Hammer, B.; Bernal, I.; Draux, M. *Organometallics* **1983**, *2*, 1595.
83. van Leeuwen, P. W. N. M.; Roobeek, C. F.; van der Heijden, H. *J. Am. Chem. Soc.* **1994**, *116*, 12117.
84. Mawby, R. J.; Basolo, F.; Pearson, R. G. *J. Am. Chem. Soc.* **1964**, *86*, 3994.
85. Mawby, R. J.; Basolo, F.; Pearson, R. G. *J. Am. Chem. Soc.* **1964**, *86*, 5043.
86. Boese, W. T.; Ford, P. C. *J. Am. Chem. Soc.* **1995**, *117*, 8381.
87. Butler, I. S.; Basolo, F.; Pearson, R. G. *Inorg. Chem.* **1967**, *6*, 2074.
88. Wax, M. J.; Bergman, R. G. *J. Am. Chem. Soc.* **1981**, *103*, 7028.
89. Collman, J. P.; Finke, R. G.; Cawse, J. N.; Brauman, J. I. *J. Am. Chem. Soc.* **1978**, *100*, 4766.
90. Wojcicki, A. *Adv. Organomet. Chem.* **1974**, *12*, 31.
91. Whitesides, G. M.; Boschetto, D. J. *J. Am. Chem. Soc.* **1971**, *93*, 1529.
92. Darensbourg, D. J.; Bauch, C. G.; Reibenspies, J. H.; Rheingold, A. L. *Inorg. Chem.* **1988**, *27*, 4203.
93. Leitner, W. *Angew. Chem., Int. Ed.* **1995**, *34*, 2207.

94. Braunstein, P.; Matt, D.; Nobel, D. *Chem. Rev.* **1988**, *88*, 747; Culter, A. R.; Hanna, P. K.; Vites, J. C. *Chem. Rev.* **1988**, *88*, 1363.
95. Sakaki, S.; Musashi, Y. *Inorg. Chem.* **1995**, *34*, 1914.
96. Darensbourg, D. J.; Hanckel, R. K.; Bauch, C. G.; Pala, M.; Simmons, D.; White, J. N. *J. Am. Chem. Soc.* **1985**, *107*, 7463; Darensbourg, D. J.; Grötsch, G. *J. Am. Chem. Soc.* **1985**, *107*, 7473.
97. Darensbourg, D. J.; Sanchez, K. M.; Reibenspies, J. H.; Rheingold, A. L. *J. Am. Chem. Soc.* **1989**, *111*, 7094.
98. Simpson, R. D.; Bergman, R. G. *Organometallics* **1992**, *11*, 4306.
99. Abu-Hasanayan, F.; Goldman, A. S.; Krogh-Jespersen, K. *Inorg. Chem.* **1994**, *33*, 5122.
100. Burk, M. J.; McGrath, M. P.; Wheeler, R.; Crabtree, R. H. *J. Am. Chem. Soc.* **1988**, *110*, 5034.
101. Kubas, G. J.; Unkefer, C. J.; Swanson, B. I.; Fukushima, E. *J. Am. Chem. Soc.* **1986**, *108*, 7000, and references therein.
102. Heinekey, D. M.; Olham, W. J. *Chem. Rev.* **1993**, *93*, 913.
103. Hay, P. J. *J. Am. Chem. Soc.* **1987**, *109*, 705.
104. Saillard, J.-Y.; Hoffmann, R. *J. Am. Chem. Soc.* **1984**, *106*, 2006.
105. Choi, H. W.; Muetterties E. L. *J. Am. Chem. Soc.* **1982**, *104*, 153.
106. Van-Catledge, F. A.; Ittel, S. D.; Jesson, J. P. *Organometallics* **1985**, *4*, 18.
107. Heinekey, D. M.; Hinkle, A. S.; Close, J. D. *J. Am. Chem. Soc.* **1996**, *118*, 5353.
108. Demachy, I.; Esteruelas, M. A.; Jean, Y.; Lledós, A.; Maseras, F.; Oro, L. A.; Valero, C.; Volatron, F. *J. Am. Chem. Soc.* **1996**, *118*, 8388.
109. Eckert, J.; Jensen, C. M.; Koetzle, T. F.; Husebo, T. L.; Nicol, J.; Wu, P. *J. Am. Chem. Soc.* **1995**, *117*, 7271.
110. Griffin, T. R.; Cook, D. B.; Haynes, A.; Pearson, J. M.; Monti, D.; Morris, G. E. *J. Am. Chem. Soc.* **1996**, *118*, 3029.
111. Shaw, B. L.; Stainbank, R. E. *J. Chem. Soc., Dalton Trans.* **1972**, 223; Miller, E. M.; Shaw, B. L. *J. Chem. Soc., Dalton Trans.* **1974**, 480.
112. Labinger, J. A.; Osborn, J. A. *Inorg. Chem.* **1980**, *19*, 3230; Labinger, J. A.; Osborn, J. A.; Coville, N. J. *Inorg. Chem.* **1980**, *19*, 3236.
113. Williams, G. M.; Schwartz, J. *J. Am. Chem. Soc.* **1982**, *104*, 1122.
114. Basil, J. D.; Murray, H. H.; Fackler, J. P., Jr.; Tocher, J.; Mazany, A. M.; Trzcinska-Bancroft, B.; Knachel, H.; Dudis, D.; Delord, T. J.; Marler, D. O. *J. Am. Chem. Soc.* **1985**, *107*, 6908.
115. Kramer, A. V.; Labinger, J. A.; Bradley, J. S.; Osborn, J. A. *J. Am. Chem. Soc.* **1974**, *96*, 7145; Kramer, A. V.; Osborn, J. A. *J. Am. Chem. Soc.* **1974**, *96*, 7832.
116. Stoutland, P. O.; Bergman, R. G. *J. Am. Chem. Soc.* **1988**, *110*, 5732.
117. Baker, M. V.; Field, L. D. *J. Am. Chem. Soc.* **1986**, *108*, 7436.
118. Ghosh, C. K.; Hoyano, J. K.; Krentz, R.; Graham, W. A. G. *J. Am. Chem. Soc.* **1989**, *111*, 5480.
119. McNamara, B.; Becher, D. M.; Towns, M. H.; Grant, E. R. *J. Phys. Chem.* **1994**, *98*, 4622.

120. Becker, P. N.; Bergman, R. G. *J. Am. Chem. Soc.* **1983**, *105*, 2985.
121. Osborn, J. A.; Jardine, F. H.; Young, J. F.; Wilkinson, G. *J. Chem. Soc. A* **1966**, 1711; Jardine, F. H.; Osborn, J. A.; Wilkinson, G. *J. Chem. Soc. A* **1967**, 1574; Montelatici, S.; van der Ent, A.; Osborn, J. A.; Wilkinson, G. *J. Chem. Soc. A* **1968**, 1054.
122. Arai, H.; Halpern, J. *J. Chem. Soc., Chem. Commun.* **1971**, 1571.
123. Meakin, P.; Jesson, J. P.; Tolman, C. A. *J. Am. Chem. Soc.* **1972**, *94*, 3240.
124. Halpern, J.; Wong, S. W. *J. Chem. Soc., Chem. Commun.* **1973**, 629.
125. Tolman, C. A.; Meakin, P. Z.; Lindner, D. L.; Jesson, J. P. *J. Am. Chem. Soc.* **1974**, *96*, 2762.
126. Halpern, J.; Okamoto, T.; Zakhariev, A. *J. Mol. Catal.* **1977**, *2*, 65.
127. Halpern, J.; Riley, D. P.; Chan, A. S. C.; Pluth, J. J. *J. Am. Chem. Soc.* **1977**, *99*, 8055.
128. Chan, A. S. C.; Pluth, J. J.; Halpern, J. *J. Am. Chem. Soc.* **1980**, *102*, 5952.
129. Landis, C. R.; Halpern, J. *J. Am. Chem. Soc.* **1987**, *109*, 1746.
130. Sun, Y.; Landau, R. N.; Wang, J.; LeBlond, C.; Blackmond, D. G. *J. Am. Chem. Soc.* **1996**, *118*, 1348.
131. Brunner, H. *Top. Stereochem.* **1988**, *18*, 129; Noyori, R. *Chem. Soc. Rev.* **1989**, *18*, 187.
132. Noyori, R. *Asymmetric Catalysis in Organic Synthesis*; Wiley Interscience: New York, 1994; Chapter 2.
133. Takahashi, H.; Sakuraba, S.; Takeda, H.; Achiwa, K. *J. Am. Chem. Soc.* **1990**, *112*, 5876.
134. Chan, Y. N. C.; Osborn, J. A. *J. Am. Chem. Soc.* **1990**, *112*, 9400.
135. Garland, M.; Blaser, H.-U. *J. Am. Chem. Soc.* **1990**, *112*, 7048.
136. Janowicz, A. H.; Bergman, R. G. *J. Am. Chem. Soc.* **1982**, *104*, 352.
137. Hoyano, J. K.; Graham, W. A. G. *J. Am. Chem. Soc.* **1982**, *104*, 3723.
138. Brookhart, M.; Green, M. L. H.; Wong, L.-L. *Prog. Inorg. Chem.* **1988**, *36*, 1.
139. Crabtree, R. H. *Chem. Rev.* **1985**, *85*, 245.
140. Rothwell, I. P. *Acc. Chem. Res.* **1988**, *21*, 153.
141. Thompson, M. E.; Bercaw, J. E. *Pure Appl. Chem.* **1984**, *56*, 1.
142. Perutz, R. N. *Chem. Soc. Rev.* **1993**, 361.
143. Arndtsen, B. A.; Bergman, R. G.; Mobley, T. A.; Peterson, T. H. *Acc. Chem. Res.* **1995**, *28*, 154.
144. Ghosh, C. K.; Graham, W. A. G. *J. Am. Chem. Soc.* **1987**, *109*, 4726.
145. Ghosh, C. K.; Graham, W. A. G. *J. Am. Chem. Soc.* **1989**, *111*, 375.
146. Sakakura, T.; Sodeyama, T.; Sasaki, K.; Wada, K.; Tanaka, M. *J. Am. Chem. Soc.* **1990**, *112*, 7221.
147. Margl, P.; Ziegler, T.; Blöchl, P. E. *J. Am. Chem. Soc.* **1996**, *118*, 5412.
148. Nolan, S. P.; Hoff, C. D.; Stoutland, P. D.; Newman, L. J.; Buchanan, J. M.; Bergman, R. G.; Yang, G. K.; Peters, K. S. *J. Am. Chem. Soc.* **1987**, *109*, 3143.
149. Marx, D. E.; Lees, A. J. *Inorg. Chem.* **1988**, *27*, 1121.

150. Drolet, D. P.; Lees, A. J. *J. Am. Chem. Soc.* **1990**, *112*, 5878.
151. Schulz, R. H.; Bengali, A. A.; Tauber, M. J.; Wasserman, E. P.; Weiller, B. H.; Kyle, K. R.; Moore, C. B.; Bergman, R. G. *J. Am. Chem. Soc.* **1994**, *116*, 7369; Weiller, B. H.; Wasserman, E. P.; Bergman, R. G.; Moore, C. B.; Pimentel, G. C. *J. Am. Chem. Soc.* **1989**, *111*, 8288.
152. Bengali, A. A.; Schultz, R. H.; Moore, C. B.; Bergman, R. G. *J. Am. Chem. Soc.* **1994**, *116*, 9585.
153. Wasserman, E. P.; Morse, C. B.; Bergman, R. G. *Science* **1992**, *255*, 315.
154. Bengali, A. A.; Bergman, R. G.; Moore, C. B. *J. Am. Chem. Soc.* **1995**, *117*, 3879.
155. Musaev, D. G.; Morokuma, K. *J. Am. Chem. Soc.* **1995**, *117*, 799; Song, J.; Hall, M. B. *Organometallics* **1993**, *12*, 3118; Ziegler, T.; Tschinke, V.; Fan, L.; Becke, A. D. *J. Am. Chem. Soc.* **1989**, *111*, 9177.
156. Siegbahn, P. E. M. *J. Am. Chem. Soc.* **1996**, *118*, 1487.
157. Whittlesey, M. K.; Mawby, R. J.; Osman, R.; Perutz, R. N.; Field, L. D.; Wilkinson, M. P.; George, M. W. *J. Am. Chem. Soc.* **1993**, *115*, 8627; Cronin, L.; Nicasio, C.; Perutz, R. N.; Peters, R. G.; Roddick, D. M.; Whittlesey, M. K. *J. Am. Chem. Soc.* **1995**, *117*, 10047.
158. Klassen, J. K.; Selke, M.; Sorensen, A. A.; Yang, G. K. *J. Am. Chem. Soc.* **1990**, *112*, 1267.
159. Belt, S. T.; Grevels, F.-W.; Klotzbücher, W. E.; McCamley, A.; Perutz, R. N. *J. Am. Chem. Soc.* **1989**, *111*, 8373.
160. Bullock, R. M.; Headford, C. E. L.; Hennessy, K. M.; Kegley, S. E.; Norton, J. R. *J. Am. Chem. Soc.* **1989**, *111*, 3897.
161. Buchanan, J. M.; Stryker, J. M.; Bergman, R. G. *J. Am. Chem. Soc.* **1986**, *108*, 1537.
162. Luinstra, G. A.; Wang, L.; Stahl, S. S.; Labinger, J. A.; Bercaw, J. E. *Organometallics* **1994**, *13*, 755; Stahl, S. S.; Labinger, J. A.; Bercaw, J. E. *J. Am. Chem. Soc.* **1996**, *118*, 5961.
163. Siegbahn, P. E. M.; Crabtree, R. H. *J. Am. Chem. Soc.* **1996**, *118*, 4442.
164. Pruett, R. L. *Adv. Organomet. Chem.* **1979**, *17*, 1.
165. Heck, R. F.; Breslow, D. S. *J. Am. Chem. Soc.* **1961**, *83*, 4023.
166. Yagupsky, G.; Brown, C. K.; Wilkinson, G. *J. Chem. Soc. A* **1970**, 1392; Brown, C. K.; Wilkinson, G. *J. Chem. Soc. A* **1970**, 2753.
167. Kang, H.; Mauldin, C. H.; Cole, T.; Slegeir, W.; Cann, K.; Pettit, R. *J. Am. Chem. Soc.* **1977**, *99*, 8323.
168. Thomas, M. G.; Beier, B. F.; Muetterties, E. L. *J. Am. Chem. Soc.* **1976**, *98*, 1296; Demitras, G. C.; Muetterties, E. L. *J. Am. Chem. Soc.* **1977**, *99*, 2796.
169. Laine, R. M.; Rinker, R. G.; Ford, P. C. *J. Am. Chem. Soc.* **1977**, *99*, 252.
170. Masters, C. *Adv. Organomet. Chem.* **1979**, *17*, 61.
171. Cramer, R. *Acc. Chem. Res.* **1968**, *1*, 186.
172. Tolman, C. A. *Chem. Rev.* **1977**, *77*, 313.

6

Oxidation–Reduction Reactions

For most purposes, inorganic reactions can be classified as either substitution reactions or oxidation–reduction reactions. The latter involve the transfer of at least one electron from the reducing agent to the oxidizing agent. Such reactions are widely used in analytical procedures and are important in many biological processes. Oxidation–reduction reactions have been classified in two general ways; the first historically is by stoichiometry and the second is by mechanism.

6.1 CLASSIFICATION OF REACTIONS

6.1.a Stoichiometric Classification

The stoichiometric classification only requires a knowledge of the reaction stoichiometry but has limited kinetic applicability.

6.1.a.i Complementary Reactions

The change in oxidation state of the reducing agent is the same as the change in oxidation state of the oxidizing agent. Some examples are

$$Cr^{2+} + Ag^+ \longrightarrow Ag^0 + Cr^{3+}$$
$$Zn^0 + Cu^{2+} \longrightarrow Cu^0 + Zn^{2+}$$

(6.1)

6.1.a.ii Noncomplementary Reactions

The oxidizing agent and the reducing agent undergo different net changes in oxidation state. Some examples are

$$2\,Cr^{2+} + Tl^{3+} \longrightarrow 2\,Cr^{3+} + Tl^+$$
$$Zn^0 + 2\,Fe^{3+} \longrightarrow Zn^{2+} + 2\,Fe^{2+}$$

(6.2)

This classification has no direct mechanistic implications. However, it is a qualitative observation that complementary reactions are faster than noncomplementary reactions. This "rule" can be useful in designing analytical and preparative procedures, but it is by no means universal. The mechanistic rationale for this qualitative kinetic observation is based on the assumption that the reactions occur in bimolecular steps. Therefore, noncomplementary reactions must normally proceed through an unstable oxidation state of one of the reactants, as in the following examples:

$$Cr^{2+} + Tl^{3+} \longrightarrow Cr^{3+} + \{Tl^{2+}\}$$
$$Cr^{2+} + \{Tl^{2+}\} \longrightarrow Cr^{3+} + Tl^{+}$$

$$(6.3)$$

or

$$Cr^{2+} + Tl^{3+} \longrightarrow \{Cr^{4+}\} + Tl^{+}$$
$$\{Cr^{4+}\} + Cr^{2+} \longrightarrow 2\,Cr^{3+}$$

$$(6.4)$$

For the reaction sequence in (6.3), Tl^{2+} is an unstable oxidation state of thallium and in (6.4), Cr^{4+} is an unstable state for chromium.

6.1.b Mechanistic Classification

The mechanistic classification obviously requires a knowledge of the reaction mechanism. A major theme of this area is the elucidation of the kinetic features that allow one to determine the type of mechanism.

6.1.b.i Inner-Sphere Electron Transfer

At least one ligand is shared in the first coordination sphere of the oxidant and reductant in the transition state for the electron-transfer process:

$$(L)_5M^{III}X + M^{II}(Y)_6 \longrightarrow \{(L)_5M^{III}\!-\!X\!-\!M^{II}(Y)_5\} \longrightarrow \text{Products} \ (6.5)$$
$$+ \ Y$$

The formation of the transition state involves a substitution reaction on one of the metal centers in order to form the intermediate or transition state with the bridging ligand —X— in reaction (6.5). Then, electron transfer occurs, possibly via the bridging ligand, and the final products are determined by the substitution lability of the metal centers.

This mechanism was first demonstrated by Taube and co-workers[1] in a system that exploited the ideas of substitution lability and inertness that Taube was developing at the same time. Complexes of cobalt(III) (low-spin d^6) are inert, of chromium(II) (high-spin d^4) are labile, of cobalt(II) (high-spin d^7) are labile and of chromium(III) (d^3) are inert. This combination of conditions allowed the reaction products to establish definitively an inner-sphere mechanism for the reaction in Scheme 6.1. The metal ions retain their original oxidation states in the precursor complex. The relative labilities of Co(II) and Cr(III) cause the successor complex to decompose to the Cr(III) product in which the Cl$^-$ ligand originally on cobalt(III) has been transferred to chromium(III).

Scheme 6.1

$$(H_3N)_5CoCl^{2+} + Cr(OH_2)_6^{2+} \xrightarrow[5 \text{ H}_2\text{O}]{5 \text{ H}^+} Co(OH_2)_6^{2+} + 5 \text{ NH}_4^+ + (H_2O)_5CrCl^{2+}$$

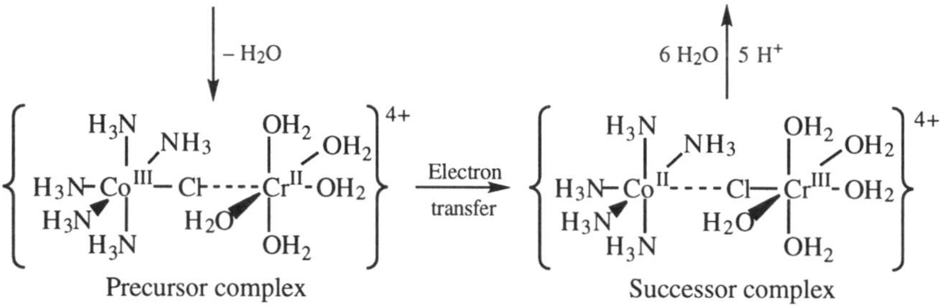

Precursor complex   Successor complex

The example in Scheme 6.1 was crucial to establishing that inner-sphere or bridged electron transfer can occur, but only rarely does one have the proper combination of substitution lability patterns so that the products uniquely define the mechanism. More commonly, this mechanism is inferred if the electron transfer is unusually fast and sensitive to the chemical nature of the bridging group. A wide range of bridging ligands is known, such as the halide ions, pseudohalide ions, hydroxide ion, and carboxylate ions. Ammonia cannot act as a bridging ligand because it does not have a second unshared pair of electrons, which is necessary for simultaneous bonding to the oxidizing and reducing centers. The effect of the bridging group is discussed later in the chapter.

6.1.b.ii Outer-Sphere Electron Transfer

The coordination spheres of the oxidant and reductant remain intact during the electron transfer, as shown in the following reaction:

$$M^{III}(L)_6 + M^{II}(Y)_6 \rightleftharpoons \{(L)_5M^{III}L\}\{YM^{II}(Y)_5\} \rightarrow M^{II}(L)_6 + M^{III}(Y)_6 \tag{6.6}$$

The reactants are considered as hard charged spheres, and an electrostatic approach can be used to anticipate the rates of these reactions. To date, this is one of the few areas in which theory has provided useful guidelines for mechanistic studies. A mechanistic decision between the inner- and outer-sphere possibilities is often based on whether or not the reaction rate corresponds reasonably to the predictions of the outer-sphere theory.

6.2 OUTER-SPHERE ELECTRON-TRANSFER THEORY

The outer-sphere theory has been developed using an electrostatic approach to calculate the energy necessary to bring reactants together, to reorganize the solvent around the transition state and to prepare the metal centers for electron transfer.

In North America the theory is associated with the name of Marcus and referred to as the Marcus theory.[2] However, Hush[3] in Australia and Levitch[4] and Dogonadze[5] in the U.S.S.R. have made original contributions. Marcus started from the transition-state theory for ionic reactions, Hush from solid-state electron-transfer theory and Levitch from a consideration of reactions at electrodes. All arrived at essentially the same result, although using different terminologies.

The *Franck–Condon principle* is fundamental to the theory. This principle states that electron movement is much faster than nuclear motion; thus, internuclear distances do not change during the instant of electron transfer. Therefore, it is assumed that on approaching the transition state, the bond lengths of the reactants will adjust to approach those of the products.

The electron transfer is assumed to be an *adiabatic process in the Ehrenfest sense* so that the transmission coefficient, ρ, in the transition-state theory expression, Eq. (6.7), is equal to one.

$$k = \rho \frac{k\,T}{h} \exp\left(-\frac{\Delta H^*}{R\,T} + \frac{\Delta S^*}{R} \right) \qquad (6.7)$$

This implies that there is enough interaction between reactants in the transition state to make the probability of electron transfer equal to one. There are numerous occasions when this condition is thought not to be entirely true, and the effect of "nonadiabaticity" on the reaction rate is sometimes used as a rationale for differences between observed and predicted rate constants.

6.2.a Marcus Cross Relationship from Thermodynamics

One of the most important results to evolve from the theoretical treatment of Marcus is now referred to as the Marcus cross relationship.

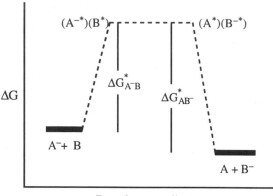

Figure 6.1. Reaction coordinate diagram for an outer-sphere electron-transfer reaction.

This important relationship was developed later by Ratner and Levine[6] from a thermodynamic perspective, and this formulation provides a simple basis for understanding some of the concepts and assumptions in the more microscopic molecular theory of Marcus.

The electron-transfer reaction between a reductant A^- and an oxidant B is given by the following net reaction:

$$A^- + B \longrightarrow A + B^- \qquad (6.8)$$

It is assumed that the reactants come together to form a precursor complex $(A^{-*})(B^*)$; then, electron transfer occurs to give the successor complex $(A^*)(B^{-*})$, which decomposes to products, as shown in the reaction coordinate diagram in Figure 6.1.

Ratner and Levine assumed that in the precursor and successor complexes, one can define thermodynamic properties for the individual partners, (A^{-*}), (B^*), (A^*), and (B^{-*}). This amounts to assuming that there is no significant bonding between the partners in the precursor and successor complexes. For such a condition, it was shown earlier by Levine[7] that detailed balance requires that the free energies of the precursor and successor complexes must be equal, so that

$$G^0(A^{-*}) + G^0(B^*) = G^0(A^*) + G^0(B^{-*}) \qquad (6.9)$$

and this is shown by the horizontal dashed line between the activated complexes in Figure 6.1. The net free energy change for the reaction is

$$\Delta G^0_{A^--B} = G^0(A) + G^0(B^-) - G^0(A^-) - G^0(B) \qquad (6.10)$$

The cross relationship involves the relationship between the rate constants for the following reactions, where (6.11) and (6.12) are called self-exchange reactions and (6.13) is called the cross reaction:

$$A^- + A \xrightarrow{k_{AA}} A + A^- \tag{6.11}$$

$$B + B^- \xrightarrow{k_{BB}} B^- + B \tag{6.12}$$

$$A^- + B \xrightarrow{k_{AB}} A + B^- \tag{6.13}$$

It is assumed that the free energies of the individual partners in the self-exchange reactions are the same as in the cross reaction; therefore

$$\Delta G^*_{AA} = G^o(A^{-*}) + G^o(A^*) - G^o(A^-) - G^o(A) \tag{6.14}$$

$$\Delta G^*_{BB} = G^o(B^{-*}) + G^o(B^*) - G^o(B^-) - G^o(B) \tag{6.15}$$

$$\Delta G^*_{AB} = G^o(A^{-*}) + G^o(B^*) - G^o(A^-) - G^o(B) \tag{6.16}$$

If Eq. (6.16) is multiplied by 2 and substitution is made from Eq. (6.9), one obtains

$$2\,\Delta G^*_{AB} = G^o(A^{-*}) + G^o(B^*) + G^o(A^*) + G^o(B^{-*}) \\ - 2\,G^o(A^-) - 2\,G^o(B) \tag{6.17}$$

Then, substitution from Eq. (6.10), rearrangement, and substitution from Eqs. (6.14) and (6.15) gives

$$\Delta G^*_{AB} = 1/2 \left(\Delta G^*_{AA} + \Delta G^*_{BB} + \Delta G^o_{A-B} \right) \tag{6.18}$$

This is the Marcus cross relationship in terms of free energies.

The thermodynamic development of the cross relationship depends on the *assumptions* that:

1. The activation process for each reactant is independent of the other reactant.
2. The activated species are the same for the self-exchange and the cross reaction.

Clearly, these assumptions will not be valid for an inner-sphere mechanism. They will also be invalid if there are special attractive or repulsive forces between A and B that are not present between A and A^- or B and B^-.

From the transition-state theory, the free energy of activation and the rate constant are related by

$$k_{ij} = Z_{ij} \exp\left(- \frac{\Delta G_{ij}^*}{RT} \right) \tag{6.19}$$

where Z_{ij} is the collision frequency. If this expression and the thermodynamic relationship $\Delta G_{A-B}^0 = - RT(\ln K_{AB})$ are substituted into Eq. (6.18), then one obtains

$$k_{AB} = \left(k_{AA} k_{BB} K_{AB} \frac{Z_{AB}^2}{Z_{AA} Z_{BB}} \right)^{1/2}$$

$$= \left(k_{AA} k_{BB} K_{AB} F_{AB} \right)^{1/2} \tag{6.20}$$

This is the cross relationship in terms of rate constants. There is often reason to believe (or need to assume) that $F_{AB} \approx 1$, and then Eq. (6.20) reduces to what is often called the simplified Marcus cross relationship. This is particularly useful because a knowledge of any three of the values, k_{AB}, k_{AA}, k_{BB}, or K_{AB} allows one to predict the fourth.

The assumption that $F_{AB} \approx 1$ means that $Z_{AB}^2 \approx Z_{AA} Z_{BB}$. For ionic reactants, this would seem quite reasonable if the reaction has charge symmetry (e.g., $A^{2+} + B^{3+} \rightarrow A^{3+} + B^{2+}$) since the charges of the species in the self-exchange reactions are the same as those in the net reaction. The assumption is somewhat less valid if the species are of the same charge type but the reaction lacks charge symmetry (e.g., $A^{2+} + B^{3+} \rightarrow A^{1+} + B^{4+}$). In this example, Z_{AA} for $A^{2+} + A^{1+}$ will be larger than Z_{BB} for $B^{3+} + B^{4+}$, whereas Z_{AB} for $A^{2+} + B^{3+}$ may be intermediate between the two, and it may still be true that, within a factor of 10, $Z_{AB}^2 \approx Z_{AA} Z_{BB}$. However, if the reactants are of opposite charge type, such as $A^{2-} + B^{3+} \rightarrow A^{3-} + B^{2+}$, then the attractive force between the ions of opposite charge will make Z_{AB} larger than either Z_{AA} or Z_{BB} and F_{AB} will be much larger than 1.

The preceding development illustrates the assumptions that are necessary to develop the cross relationship. The more detailed theory that follows provides further understanding in terms of the molecular properties of the reactants and solvent.

6.2.b Marcus Theory Details

The details of the Marcus theory have been described in several reviews[8-10] and in books by Reynolds and Lumry[11] and Cannon.[12] The following discussion will simply outline the features of the theory and give the physical factors that are predicted to be important in determining the rates of outer-sphere electron-transfer reactions.

The reactants are considered to be two hard spheres of charge z_1 and z_2 and radii a_1 and a_2. Work will be required to bring the reactants together to a separation of $r = a_1 + a_2$, which is considered to be the reactant separation in the transition state. From simple electrostatics, this coulombic work contribution to the free energy of activation is given by

$$\Delta G^*_{coul} = \left(\frac{N z_1 z_2 e^2}{4 \pi \varepsilon_0} \right) \left(\frac{1}{\varepsilon_s r} \right) \qquad (6.21)$$

where e is the charge on the electron and ε_s is the bulk dielectric constant of the solvent. In more recent treatments, the formation of the precursor complex has been considered as a diffusion-controlled equilibrium reaction, with the equilibrium constant, K_{os}, equal to the ratio of the forward and reverse rate constants and given by

$$K_{os} = \frac{4 \pi N r^3}{3000} \exp\left(\frac{U}{1 + B r \sqrt{\mu}} \right) \qquad (6.22)$$

where

$$U = \frac{z_1 z_2 e^2}{4 \pi \varepsilon_0 \varepsilon_s r k T}$$

and the terms in U have been defined previously in Eq. (1.79). The B is the Debye–Hückel factor discussed in Eq. (1.78) and μ is the ionic strength. The latter terms obviously are introduced in an attempt to correct for ionic strength variations. Alternative approaches to the ionic strength effect have been described in Chapter 1 and by Tembe et al.[13] In the transition state theory equation, either ΔG^*_{coul} is a contribution to the overall ΔG^* or K_{os} is a pre-exponential factor.

When the reactants come together, they are considered to form a spherical transition state of diameter r. The solvent molecules will reorganize around the transition state, and this solvent or outer-sphere reorganization contribution to the overall ΔG^* is given by

$$\Delta G^*_{solv} = (\Delta q)^2 \left(\frac{N e^2}{16 \pi \varepsilon_0} \right) \left(\frac{1}{2 a_1} + \frac{1}{2 a_2} - \frac{1}{r} \right) \left(\frac{1}{n_s^2} - \frac{1}{\varepsilon_s} \right) \qquad (6.23)$$

where Δq is the number of electrons transferred (one for the majority of reactions) and n_s is the refractive index of the solvent (n_s^2 is the high-frequency dielectric constant of the solvent).

In order to satisfy the Franck–Condon principle, the ligands around the metal ions will adjust their bond lengths in the transition state toward the lengths they will have in the products. This factor is the inner-sphere reorganization energy contribution to the overall ΔG^* and is given by

$$\Delta G_{in}^* = N\left(\frac{n\,f_1}{2}\left(d_1^o - d_1^*\right)^2 + \frac{n\,f_2}{2}\left(d_2^o - d_2^*\right)^2\right) \qquad (6.24)$$

where n is the number of ligands, f_1 and f_2 are the force constants for the symmetrical breathing vibrational mode, which is assumed to generate the appropriate bond lengthening or shortening, and d_i^o and d_i^* are the ground-state and transition-state metal–ligand bond lengths, respectively. The ΔG_{in}^* is the most difficult term to calculate because d_i^* is unknown and the force constants require an assignment of the vibrational spectrum in the difficult region of ~200 to 800 cm^{-1}. One approach is to let $d^* = d_1^* = d_2^*$ and then solve for d^* when ΔG_{in}^* is a minimum (i.e., when $d(\Delta G_{in}^*)/d(d^*) = 0$). Then

$$d^* = \frac{f_1\,d_1^o + f_2\,d_2^o}{f_1 + f_2} \qquad (6.25)$$

and

$$\Delta G_{in}^* = N\left(\frac{n\,f_1\,f_2}{2\,(f_1 + f_2)}\right)\left(d_2^o - d_1^o\right)^2 \qquad (6.26)$$

To determine the force constants, f_i, the appropriate potential energy function, U, should be used to calculate $f = d^2(U)/d^2(d)$. A simple harmonic diatomic oscillator is often assumed, so that $f_i = (2\pi\nu_i c)^2\,m_r$, where ν_i is the vibrational frequency in cm^{-1}, c is the speed of light in cm s^{-1}, and m_r is the reduced mass of the ligand in kg.

Nuclear tunneling has also been included as a pre-exponential factor, Γ_n, in the rate constant expression. It is given by $\Gamma_n = 3 \times 10^{10}\nu_n$, where ν_n is the nuclear vibrational frequency of the reactants and

$$\nu_n^{\,2} = \frac{\nu_{solv}^{\,2}\,\Delta G_{solv}^* + \nu_{in}^{\,2}\,\Delta G_{in}^*}{\Delta G_{solv}^* + \Delta G_{in}^*} \approx \frac{\nu_{in}^{\,2}\,\Delta G_{in}^*}{\Delta G_{solv}^* + \Delta G_{in}^*} \qquad (6.27)$$

where it is assumed that the frequency for the outer-sphere solvent molecules, ν_{solv} ~30 cm^{-1}, is more than 10 times smaller than the inner-sphere frequency, ν_{in} ~400 cm^{-1}, and $\Delta G_{in}^* \approx \Delta G_{solv}^*$. The value of ν_{in} is taken as the average of the breathing-mode frequencies of the two reactants as calculated from

$$\nu_{in}^{\,2} = \frac{2\,\nu_1^{\,2}\,\nu_2^{\,2}}{\nu_1^{\,2} + \nu_2^{\,2}} \qquad (6.28)$$

The final feature to be considered as a pre-exponential factor is the electronic transmission coefficient, κ_{el}, which is equal to one if the electron-transfer step is adiabatic and is smaller otherwise.

All of the pre-exponential factors and contributions to the overall ΔG^* are incorporated into the transition state theory equation to obtain the calculated outer-sphere rate constant as

$$k_{calc} = \kappa_{el}\, \Gamma_n \, K_{os}\, exp\left(\frac{-\left(\Delta G^*_{solv} + \Delta G^*_{in} \right)}{RT} \right) \qquad (6.29)$$

Clearly, the next step is to compare calculated rate constants from Eq. (6.29) to some experimental results. Because of their overall symmetry, it is easier to do the calculation for electron-exchange reactions such as (6.30) where M and L are the same and only the oxidation state is different in the two reactants:

$$M(L)_n{}^{z+} + *M(L)_n{}^{(z-1)+} \longrightarrow M(L)_n{}^{(z-1)+} + *M(L)_n{}^{z+} \qquad (6.30)$$

Results for some such systems are given in Table 6.1, and they show that the main feature controlling the relative rates for different metal ions is the differences in ΔG^*_{in}, which are in turn controlled by the differences in bond lengths of the two oxidation states. The agreement of the calculated and observed values is quite good for the Fe and Co systems but poor for $Ru(OH_2)_6{}^{2/3+}$. Clearly, some approximations have gone into these calculations in addition to those involved in the theory. The most critical factor is the bond length or size difference between the two oxidation states because ΔG^*_{in} depends on the square of this difference. For example, the observed and calculated values for $Ru(OH_2)_6{}^{2/3+}$ agree if the size difference is 0.125 Å instead of 0.09 Å and a change from 0.04 Å to 0.095 Å gives agreement for $Ru(NH_3)_6{}^{2/3+}$. The force constants used in the calculations might also be questioned because of the simplicity of the model, although the errors will be somewhat self-compensating, as can be seen from the form of Eq. (6.26). Bernhard and Ludi[14] have done a normal coordinate analysis on the $Ru(OH_2)_6{}^{2/3+}$ system and obtained values of 191 and 298 N m^{-1} for the 2+ and 3+ oxidation states, respectively, and the simple model based on a harmonic oscillator for the symmetric breathing mode gives 190 and 300 N m^{-1}, respectively. This factor does not seem to be a source of concern. The term most commonly blamed for lack of agreement between k_{calc} and k_{obsd} is the electronic transmission coefficient, κ_{el}, because calculated values are usually higher than observed, so that nonadiabaticity is invoked with κ_{el} in the range of 0.1 to 10^{-3} to bring the values into agreement.

There has been much discussion of the solvent dependence of the self-exchange rates.[15] The theory is based on a solvent dielectric continuum model and predicts a variation in rate constant with the solvent properties due to the $(n_s{}^{-2} - \varepsilon_s{}^{-1})$ dependence of ΔG^*_{solv}.

Table 6.1. Comparison of Relevant Parameters and Calculated and Observed Self-Exchange Rate Constants for $M(L)_n^{2/3+}$ Systems

Parameter	$Fe(OH_2)_6^{2/3+}$	$Ru(OH_2)_6^{2/3+}$	$Ru(NH_3)_6^{2/3+}$	$Co(NH_3)_6^{2/3+}$
a_1, a_2 (Å)	3.33, 3.19	3.32, 3.23	3.37, 3.33	3.41, 3.19
v_1, v_2 (cm^{-1})	390, 490	424, 532	442, 500	357, 394
v_n (cm^{-1})	319	286	151	347
$10^{-12} \times \Gamma_n$ (s^{-1})	9.6	8.6	4.5	10.4
μ (M)	0.55	5.0	0.10	2.5
K_{os} (M^{-1})	0.055	0.23	0.017	0.093
ΔG^*_{solv} (kJ mol^{-1})	29.3	29.2	28.5	29.0
ΔG^*_{in} (kJ mol^{-1})	35.0	17.1	3.2	73.4
k_{calc} (M^{-1} s^{-1}, 25°C)	2.8	1.5×10^5	2.2×10^5	1.1×10^{-6}
k_{obsd} (M^{-1} s^{-1}, 25°C)	4.2 [a]	20 [b]	6.7×10^3 [c]	8×10^{-6} [d]

[a] Brunschwig, B. S.; Creutz, C.; Macartney, D. H.; Sham, T.-K.; Sutin, N. *Disc. Faraday Soc.* **1982**, *74*, 113; Jolley, W. H.; Stranks, D. R.; Swaddle, T. W. *Inorg. Chem.* **1990**, *29*, 1948.

[b] Bernhard, P.; Helm, L.; Ludi, A.; Merbach, A. E. *J. Am. Chem. Soc.* **1985**, *107*, 312.

[c] Extrapolated with $\Delta H^* = 5$ kcal mol^{-1} from measurements at 4°C, Smolenaers, P. J.; Beattie, J. K. *Inorg. Chem.* **1986**, *25*, 2259.

[d] At 40°C with trifluoromethanesulfonate counterion, Hammershoi, A.; Geselowitz, D.; Taube, H. *Inorg. Chem.* **1984**, *23*, 979.

Chan and Wahl[16] found that the continuum prediction was followed for the tris(hexafluoroacetylacetonato)ruthenium (II)/(III) system. Later studies[17] on the $Fe(C_5H_5)_2^{0/+}$ system also showed the expected general trend, although the correlation is not perfect. More recent studies[18] on other metallocenes led to the suggestion of a correlation with "solvent friction", as measured by solvent NMR relaxation times. This correlation has been criticized by Drago and Ferris,[19] who analyze the results in terms of the S' parameter of the "united solvation model" and the E_B and C_B values of the solvents. Abbott and Rusling[20] found a correlation with the Kamlet–Taft solvent parameters. Lay et al.[21] studied the effect of solvent on the reduction potentials of several cobalt(III) complexes and found a correlation with hydrogen-bonding basicities that indicates the limitations of the conventional continuum model for solvent effects. A similar conclusion might be drawn from Swaddle's detailed analysis[22] of the solvent effects on the volumes of activation for electron-exchange reactions. More recently, Curtis and co-workers[23] have studied several ruthenium(II/III) ammine systems and find a correlation of the self-exchange rates with the Gutmann donor number of the solvent. They suggest that this results from hydrogen bonding between the ammine complex and the solvent.

Table 6.2. Reduction Potentials and Self-Exchange Rate Constants (25°C) for Selected Reagents

Redox Couple	$E°$ (V)[a]	k_{AA} (M^{-1} s^{-1})[b]
$V(OH_2)_6^{2/3+}$	-0.255	1×10^{-2}
$Cr(OH_2)_6^{2/3+}$	-0.40	$\sim 1 \times 10^{-5}$
$Fe(OH_2)_6^{2/3+}$	$+0.74$	4.0
$Co(OH_2)_6^{2/3+}$	$+1.96$	~ 5
$Co(en)_3^{2/3+}$	-0.13	2.0×10^{-5}
$Co(phen)_3^{2/3+}$	$+0.42$	6.7 [c]
$Co(sep)^{2/3+}$	-0.26	5.1
$Co(NH_3)_6^{2/3+}$	$+0.058$	8×10^{-6} [d, e]
$Ru(NH_3)_6^{2/3+}$	$+0.051$	6.7×10^3 [d]
$Ru(en)_3^{2/3+}$		1.7×10^4 [f]
$Ru(bpy)_3^{2/3+}$	$+1.26$	2.0×10^9

[a] Creaser, I. I.; Sargeson, A. M.; Zanella, A. W. *Inorg. Chem.* **1983**, *22*, 4022.
[b] See Chou, M.; Creutz, C.; Sutin, N. *J. Am. Chem. Soc.* **1977**, *99*, 5615 for original references, unless otherwise indicated.
[c] Grace, M. R.; Swaddle, T. W. *Inorg. Chem.* **1993**, *32*, 5597, in 0.11 M Cl$^-$.
[d] See Table 6.1.
[e] At 40°C.
[f] Beattie, J. K.; Smolenaers, P. J. *J. Phys. Chem.* **1986**, *90*, 3684.

The self-exchange rates and reduction potentials for several metal ion and ligand systems are given in Table 6.2. For the cobalt(II)/(III) complexes, the k_{AA} values are quite sensitive to the ligand environment and are quite small compared to the Ru(II)/(III) analogues. Initially, it was thought that the cobalt systems were slow because of a large inner-sphere reorganization energy caused by the greater bond-length differences in the $M(NH_3)_6^{2/3+}$ complexes of cobalt (Co^{2+} 2.11 Å, Co^{3+} 1.94 Å) as compared to ruthenium (Ru^{2+} 2.14 Å, Ru^{3+} 2.10 Å). This results largely because $Co(NH_3)_6^{2+}$ is in the high-spin state. Electronic factors also have been invoked, based on the idea that Co must transfer an e_g electron [Co(III) (t_{2g}^6), Co(II) $(t_{2g}^5 e_g^2)$], whereas ruthenium transfers a t_{2g} electron [Ru(III) (t_{2g}^5), Ru(II) (t_{2g}^6)]. A further possibility for the cobalt system has been suggested in which cobalt(II) may need to undergo a spin-state change[24] before electron transfer can occur, as in Co(II) $(t_{2g}^5 e_g^2) \rightarrow$ Co(II) $(t_{2g}^6 e_g^1)$. These arguments seem to be qualitatively consistent with the fact that $Co(phen)_3^{2+}$ is close to the low-spin state and has a larger k_{AA} than the hexaammine system.

The situation has been clarified by recent studies of Sargeson and co-workers on macrocyclic cobalt and ruthenium complexes. The self-exchange rate at 25°C[25] for Co(sepulchrate)$^{2/3+}$ is 5.1 M^{-1} s^{-1} in 0.2 M NaCl and 4.8 M^{-1} s^{-1} in 0.2 M LiClO$_4$, although the spin state and bond lengths (Co^{II}—N, 2.16 Å; Co^{III}—N, 1.96 Å) are essentially the same as in the Co(NH$_3$)$_6$$^{2/3+}$ system (sepulchrate is a cage hexadentate N-donor ligand). The much larger rate constant (~10^6 times) has been attributed to strain in the cage ligand because the Co(III) is a bit too small to fit into the cage and the Co(II) is a bit too large.[26] As a result, bond length distortion toward the transition state is particularly favorable. Similar factors may apply to Co(tacn)$_2$$^{2/3+}$ (an N$_3$ macrocycle) with k = 0.18 M^{-1} s^{-1}, compared to Ru(tacn)$_2$$^{2/3+}$ with k = 5 x 10^4 M^{-1} s^{-1}.[27] For Co(ttacn)$_2$$^{2/3+}$ (an S-donor chelate) and Co(azacapten)$^{2/3+}$ (an S$_3$N$_3$ cage), the bond lengths are almost the same in the two oxidation states because the Co(II) is low spin and the rate constants are 1.3 x 10^4 and 4.5 x 10^3 M^{-1} s^{-1}, respectively.[28,29] In summary, it appears that bond-length differences are a major influence on the self-exchange rates for the cobalt(II)/(III) systems.

6.2.c Applications of the Marcus Cross Relationship

From the detailed theory, Marcus recognized certain simplifications that led to a cross relationship of the same form as that developed by Ratner and Levine. In Marcus' terms, this relationship is given by

$$k_{AB} = (k_{AA} k_{BB} K_{AB} f_{AB})^{1/2} \qquad (6.31)$$

where

$$\log f_{AB} = \frac{(\log K_{AB})^2}{4 \log\left(\dfrac{k_{AA} k_{BB}}{Z^2} \right)}$$

and Z is the collision number for the ions in solution ($\approx 10^{11}$ M^{-1} s^{-1}). The f$_{AB}$ factor is analogous to the F$_{AB}$ in the Ratner and Levine development and f$_{AB} \approx 1$ unless K$_{AB}$ is large. If f$_{AB} \approx 1$, then Eq. (6.31) reduces to Eq. (6.32), called the simplified Marcus equation:

$$k_{AB} = (k_{AA} k_{BB} K_{AB})^{1/2} \qquad (6.32)$$

For a one-electron transfer at 25°C, the logarithmic form of this equation is given by

$$\log(k_{AB}) = 0.5 [\log(k_{AA}) + \log(k_{BB})] + 0.5 (16.9) \Delta E° \quad (6.33)$$

where $\Delta E°$ is the reduction potential in volts. Sutin and co-workers[30] tested this equation using a series of Fe(phenX)$_3$$^{2/3+}$ complexes, for which k_{BB} is expected to be constant, reacting with aqueous iron(II) and cerium(IV). They found that plots of $\log(k_{AB})$ versus $\Delta E°$ have close to the expected slope, but the intercepts are smaller than predicted.

This equation can be applied to calculate one of the self-exchange rate constants, k_{AA} or k_{BB}, or to calculate k_{AB}, in order to compare it to an experimental value. These applications have been reviewed by Sutin and co-workers,[31] who conclude that k_{AB} can be predicted to within a factor of about 25 if the reaction is outer sphere. Since the self-exchange rate constant depends on the square of k_{AB}, one could expect to predict k_{AA} or k_{BB} to within 25^2. Some typical results from the cross relationship are given in Table 6.3. The first three examples are typical of the type of agreement that is generally viewed as acceptable. However, the calculations involving the Fe(II)/Fe(III) couple, with $k_{AA} = 4$ M^{-1} s^{-1}, consistently give predicted values that are larger than the experimental values and seem more consistent with a value of k_{AA} that is $\sim 10^3$ smaller than the measured value. This effect will be discussed further with regard to the results in Table 6.4.

Bernhard and Sargeson[32] have applied Eq. (6.31) to determine the self-exchange rates of some encapsulated ruthenium, manganese, iron, and nickel complexes. They used five reagents with known self-exchange rates and studied 19 cross reactions, in order to determine five new self-exchange rates by using a least-squares fit of the data to Eq. (6.31). Some representative results are given in Table 6.4. The authors allowed Z to be a fitting variable and obtained a best-fit value five times smaller than that of 10^{11} M^{-1} s^{-1} commonly assumed, but this is not a major influence on the fit because f_{AB} is near unity for most systems. Bernhard and Sargeson find that the measured self-exchange rate for the $Fe(OH_2)_6^{2/3+}$ system of 4 M^{-1} s^{-1} does not fit the results and treated this as a variable to obtain a self-consistent value of 6.2×10^{-3} M^{-1} s^{-1}. Deviations such as this were noted in the survey by Sutin et al. and have been examined more extensively by Hupp and Weaver,[33] who also suggest that a value of $\sim 10^{-3}$ M^{-1} s^{-1} is more consistent for a number

Table 6.3. Comparison of Some Observed Rate Constants (M^{-1} s^{-1}, 25°C) with Those Calculated from the Marcus Cross Relationship

Reactants	log K_{AB}	$k_{AB\ obsd}$	$k_{AB\ calc}$
$Ru(NH_3)_6^{2+} + Co(phen)_3^{3+}$	6.25	1.5×10^4	3.5×10^5
$V(OH_2)_6^{2+} + Co(en)_3^{3+}$	2.12	5.8×10^{-4}	3.1×10^{-3}
$V(OH_2)_6^{2+} + Ru(NH_3)_6^{3+}$	5.19	1.3×10^3	2.2×10^3
$V(OH_2)_6^{2+} + Fe(OH_2)_6^{3+}$	16.9	1.8×10^4	1.6×10^6
$Ru(NH_3)_6^{2+} + Fe(OH_2)_6^{3+}$	11.7	3.4×10^5	1.2×10^7
$Fe(OH_2)_6^{2+} + Ru(bpy)_3^{3+}$	8.81	7.2×10^5	3.6×10^8

Table 6.4. Self-Exchange Rate Constants (M^{-1} s^{-1}, 25°C) Obtained by Fitting to the Marcus Cross Relationship

Reagent		E_A (V)	E_B (V)	$k_{AA}^{a, b}$	$k_{BB}^{a, c}$	10^{-4} x k^a	
A	B					obsd	calc
$Ru(sar)^{2+}$	$Ru(NH_3)_5(py)^{3+}$	0.29	0.302	1.2 x 10^5	1.1 x 10^5	10.5	14
$Ru(sar)^{2+}$	$Ru(NH_3)_5(nic)^{3+}$	0.29	0.362	1.2 x 10^5	1.1 x 10^5	28	44
$Ru(sar)^{2+}$	$Ru(NH_3)_5(isn)^{3+}$	0.29	0.384	1.2 x 10^5	1.1 x 10^5	52	66
$Ru(sar)^{2+}$	$Ru(tacn)^{3+}$	0.29	0.366	1.2 x 10^5	5.4 x 10^4	73	34
$Ru(sar)^{2+}$	$Mn(sar)^{3+}$	0.29	0.519	1.2 x 10^5	1.7 x 10^1	17	9
$Ru(sar)^{2+}$	$Fe(OH_2)_6^{3+}$	0.29	0.74	1.2 x 10^5	6.2 x 10^{-3} ^{b}	7.2	6.7
$Mn(sar)^{3+}$	$Ru(NH_3)_5(py)^{2+}$	0.519	0.302	1.7 x 10^1	1.1 x 10^5	3.7	7.2
$Mn(sar)^{3+}$	$Ru(NH_3)_5(isn)^{2+}$	0.519	0.384	1.7 x 10^1	1.1 x 10^5	1.4	1.7
$Mn(sar)^{3+}$	$Ru(tacn)^{2+}$	0.519	0.366	1.7 x 10^1	1.2 x 10^5	2.9	1.6
$Mn(sar)^{3+}$	$Fe(OH_2)_6^{3+}$	0.519	0.74	1.7 x 10^1	6.2 x 10^{-3} ^{b}	1.2 ^{d}	2.0 ^{d}

^{a} $\mu = 0.1$ M, selected from Ref. 27, where structures of the ligands are given.
^{b} Determined from a least-squares fit to Eq. (6.31) with $Z = 1.9$ x 10^{10} M^{-1} s^{-1}.
^{c} Fixed values known independently unless otherwise indicated.
^{d} Values of 10^{-1} x k.

of reactions. The standard explanation for this deviation is that the $Fe(OH_2)_6^{2/3+}$ exchange is inner sphere with a bridging water molecule. It has been suggested[34] that the self-exchange rate of 20 M^{-1} s^{-1} for the system $Ru(OH_2)_6^{2/3+}$ also implies a much smaller value for $Fe(OH_2)_6^{2/3+}$ because the inner-sphere rearrangement is greater for Fe than for Ru (see Table 6.1). In a similar vein, it should be noted that the difference in ionic radii of V(II) and V(III) is very similar to that of Fe(II) and Fe(III), so that the $Fe(OH_2)_6^{2/3+}$ exchange rate might be expected to be ~0.01 M^{-1} s^{-1}. On the other hand, for the $Fe(OH_2)_6^{2/3+}$ exchange[35], the ΔV^* of -11 cm^3 mol^{-1} has been taken as evidence for an outer-sphere mechanism, especially by comparison to the value of 0.8 cm^3 mol^{-1} for $Fe(OH_2)_5OH^{2+}/Fe(OH_2)_6^{2+}$ which is believed to be an inner-sphere reaction. It is somewhat ironic that the mechanism for one of the most studied and analyzed reactions remains controversial.

The cross relationship can also be used to estimate self-exchange rates when these rates cannot be measured directly. If the least-squares analysis of Bernhard and Sargeson is not used, then the calculation is somewhat cyclical because k_{BB} also appears in f_{AB}, but f_{AB} is often ~1 and not strongly dependent on k_{BB}. Macartney and Sutin[36] applied this method to various ascorbate radicals and their parents and calculated the following self-exchange rate constants (M^{-1} s^{-1}, 25°C): 2 x 10^3 for $H_2A/H_2A^{\bullet+}$; 1 x 10^5 for $HA/HA^{\bullet}$; and ~2 x 10^5 for $A^{2-}/A^{\bullet-}$.

Confidence in such applications is tempered by attempts to calculate the self-exchange rate constant for the dioxygen/superoxide, O_2/O_2^-, couple. Taube and co-workers[37] studied the oxidation of three Ru(II) ammine complexes and analyzed the results, using Eq. (6.31) to obtain a fairly self-consistent self-exchange rate constant of 1×10^3 M^{-1} s^{-1} for O_2/O_2^-. Espenson and co-workers[38] expanded the earlier study with more reducing agents and refined the analysis by including so-called work terms that attempt to correct for asymmetries in the charge and size of the species involved. The work term corrections are given by the following equations[39]:

$$k_{AB} = \left(k_{AA} \, k_{BB} \, K_{AB} \, f_{AB} \right)^{1/2} W_{AB} \tag{6.34}$$

where f_{AB} and W_{AB} are given by

$$\ln f_{AB} = \frac{\left(\ln K_{AB} + \left(w_{AB} - w_{BA} \right)/RT \right)^2}{4 \ln\left(k_{AA} \, k_{BB} / Z^2 \right) + \left(w_{AA} + w_{BB} \right)/RT}$$

$$W_{AB} = \exp\left(\frac{-\left(w_{AB} + w_{BA} - w_{AA} - w_{BB} \right)}{2RT} \right)$$

The individual work terms are calculated from

$$w_{ij} = \frac{4.225 \times 10^3 \, z_A \, z_B}{r \left(1 + 0.329 \, r \sqrt{\mu} \right)} \tag{6.35}$$

where r is the sum of the radii of the reaction partners in Å, μ is the ionic strength, the numerical constants are for water at 25°C and the w_{ij} are in cal mol^{-1}. Some of these results are given in Table 6.5. The O_2 oxidations give values of k_{BB} that are consistent within the accepted limits, but the O_2^- reductions give quite different and divergent values. Finally, the self-exchange rate constant has been measured[40] by oxygen isotope exchange and found to be $(4.5 \pm 1.6) \times 10^2$ M^{-1} s^{-1} (in 0.3 M 2-propanol, 0.02 M NaOH). Several rationalizations for the discrepancy between the direct and Marcus relationship values may be offered. The direct O_2/O_2^- reaction may not be truly outer sphere in that there might be some bonding interaction during the encounter of the reactants. In that case, the Marcus relationship might be giving the true outer-sphere rate constant by an argument analogous to that used for the $Fe(OH_2)_6^{2/3+}$ system. The small values obtained for the reductions by O_2^- present the unanswered problem of whether they represent the true outer-sphere rate constant or are due to some unexpected chemical problem. However, it may be that the Marcus formulation is not entirely justified for small and possibly strongly solvated and hydrogen-bonded species such as aqueous O_2^-.

Table 6.5. O_2/O_2^- Self-Exchange Rate Constants $(25°C)^a$ as Calculated from the Marcus Cross Relationship

Reactants	k_{AB} (M^{-1} s^{-1})	$a_{ML}{}^b$ (Å)	W_{AB}	f_{AB}	$k_{BB}{}^c$ (M^{-1} s^{-1})
$Cr(bpy)_3{}^{2+} + O_2$	6×10^5	6.8	4.0	0.78	0.6
$Cr(phen)_3{}^{2+} + O_2$	1.5×10^6	6.8	4.0	0.72	1.9
$Ru(NH_3)_6{}^{2+} + O_2$	6.3×10^1	3.4	29.0	0.84	3.1
$Ru(en)_3{}^{2+} + O_2$	3.6×10^1	4.0	15.9	0.44	2.7×10^2
$Co(sep)^{2+} + O_2$	4.3×10^1	4.5	11.3	0.82	7.0×10^{-2}
$Fe(CN)_6{}^{3-} + O_2^-$	3×10^2	4.5	3.65	0.090	1.5×10^{-8}
$Fe(C_5H_5)_2{}^+ + O_2^-$	8.6×10^6	5.0	1.4	0.016	5.7×10^{-4}

^{a} Further data and original references are given by Zahir, K.; Espenson, J. H.; Bakac, A. *J. Am. Chem. Soc.* **1988**, *110*, 5059.

^{b} The radii of O_2 and O_2^- used are 1.21 and 1.33 Å, respectively, and the oxidized form of the metal complex was assumed to be 0.05 Å smaller than the reduced form, except for equal values for Ru complexes.

^{c} $Z = 1 \times 10^{11}$ M^{-1} s^{-1} is used for all reactions.

The results in Table 6.5 show that the corrections due to W_{AB} can be substantial because k_{BB} depends on $W_{AB}{}^2$. The small f_{AB} values for the O_2^- reactions result mainly from the large equilibrium constants for these reactions, which produce small values of f_{AB} because the denominator in the ln f_{AB} equation is negative.

This cross relationship is often applied to metalloenzyme systems to determine their self-exchange rates, because techniques are seldom available to measure the values directly. These applications have variable success, the difficulties usually are attributed to varying points of attack on the enzyme and induced conformational changes in the enzyme.

Volumes of activation have been studied for a number of outer-sphere reactions. Swaddle and co-workers[41] have shown that the ΔV analogue of Eq. (6.18) should apply for adiabatic reactions of modest driving force and found that the agreement between experimental and predicted values of ΔV^* is good for $Fe(OH_2)_6{}^{3+} + Co([9]aneS_3)_2{}^{2+}$ but poor for $Fe(OH_2)_6{}^{3+} + Co(sep)^{2+}$. They suggest that the larger driving force for the latter causes the cross relationship to fail. Tregolan and co-workers[42] used cyclic voltammetry to measure the pressure dependence of the reduction potential for a number of $M^{III/II}$ systems and attempted to separate the intrinsic or inner-sphere and electrostriction components of the volume change. The latter is in the +19 to +28 cm^3 M^{-1} range for ammonia, en, bipy and phen complexes of Fe, Cr, Co, and Ru, but is only +3 to +7 for the hexaaqua complexes of Fe and Ru. The authors suggest that this is due to specific hydrogen bonding effects with the water solvent that give more structure to the solvation layer.

6.3 DIFFERENTIATION OF INNER-SPHERE AND OUTER-SPHERE MECHANISMS

Several criteria can be used to differentiate the two mechanisms:

1. The best method is to identify a product to which the bridging group has been transferred, as in the classic study by Taube and co-workers discussed earlier. Unfortunately, the appropriate combination of labile and inert metals is seldom available. Aside from Cr(II), other reagents, such as $Co(CN)_5^{3-}$ and bis(dimethylglyoxime)cobalt(II), can be used. Sometimes, both metal centers in the product are inert and the dimeric product can be identified as in the following examples[43,44]:

$$Fe(CN)_6^3 + Co(CN)_5^{3-} \longrightarrow (NC)_5FeCNCo(CN)_5^{6-} \quad (6.36)$$

$$Ir(Cl)_6^{2-} + Cr(OH_2)_6^{2+} \longrightarrow (Cl)_5IrClCr(OH_2)_5 + H_2O \quad (6.37)$$

2. If the rate constant for the oxidation–reduction reaction is larger than the rate of ligand substitution on either metal, then an outer-sphere mechanism is required. For example, $V(OH_2)_6^{2+}$ has a water exchange rate of 90 s^{-1} and substitution is by an I_a mechanism, so that the substitution rate constants should be <100 $M^{-1} s^{-1}$. For the following reactions, the rate constant is much larger than this.

$$(H_3N)_5Co(OPO_3) + V(OH_2)_6^{2+} \xrightarrow{\; k = 1.4 \times 10^7 \; M^{-1} \; s^{-1}\;} \quad (6.38)$$

$$Fe(OH_2)_6^{3+} + V(OH_2)_6^{2+} \xrightarrow{\; k = 1.8 \times 10^4 \; M^{-1} \; s^{-1}\;} \quad (6.39)$$

In the last example, substitution might occur on Fe, but it has a water exchange rate of ~150 s^{-1} and the rate of the oxidation–reduction is too fast for this pathway. Therefore, both reactions must be using an outer-sphere mechanism. Clearly, if both reactants are quite inert to substitution, an outer-sphere mechanism is almost certain.

3. If all the ligands on both reactants have no unshared electron pairs, it will not be possible to form a bridged intermediate. For the chelates, one must be certain that ring opening does not precede electron transfer for this criterion to be valid.

4. If the reaction(s) obey the predictions of the Marcus theory, then an outer-sphere mechanism is often assumed. This is a dangerous criterion, because it has been observed that inner-sphere reaction rates[45,46] also show a correlation with the overall ΔG° of the reaction that is predicted by the Marcus theory for outer-sphere reactions. Murdoch[47] has shown that such linear free-energy correlations may be more general than might originally have been expected.

6.4 BRIDGING LIGAND EFFECTS IN INNER-SPHERE REACTIONS

An inner-sphere mechanism consists of two processes, precursor complex formation followed by electron transfer:

$$(L)_5MX \; + \; N(Y)_6 \; \overset{K}{\rightleftharpoons} \; (L)_5M{-}X{-}N(Y)_5 \; \overset{k_e}{\longrightarrow} \; \text{Products} \quad (6.40)$$
$$+ \; Y$$

so that the experimental rate constant is given by

$$k_{exp} \; = \; K \, k_e \qquad (6.41)$$

Therefore, the variation of k_{exp} with the nature of the bridging ligand may reflect changes in either or both K and k_e. Rate constants for some inner-sphere reactions with different bridging ligands are given in Table 6.6.

Table 6.6. Rate Constants (25°C) for Reduction of $Co^{III}(NH_3)_5X$ Complexes by Cr^{2+} at $\mu = 1.0$ M

X	k $(M^{-1} \, s^{-1})$	ΔH^* (kcal mol^{-1})	ΔS^* (cal mol^{-1} K^{-1})
F$^-$ [a]	2.5×10^5		
Cl$^-$ [a]	6×10^5		
Br$^-$ [b]	1.4×10^6		
HCO$_2^-$ [c]	7.2	8.3	−27
H$_3$CCO$_2^-$ [c]	3.5×10^{-1}	8.2	−33
Cl$_2$HCO$_2^-$ [c]	7.5×10^{-2}	8.1	−36
F$_3$CCO$_2^-$ [c]	1.7×10^{-2}	9.3	−35
(CH$_3$)$_3$CCO$_2^-$ [c]	7.0×10^{-3}	11.1	−31
HO$_2$CCO$_2^-$ [d]	1.0×10^2		
$^-$O$_2$CCO$_2^-$ [d]	4.6×10^4	2.3	−20
H$_3$CC(=O)CO$_2^-$ [d]	1.1×10^4	5.8	−21
H$_3$CC(OH)$_2$CO$_2^-$ [e]	2.6×10^1		

[a] Candlin, J. P.; Halpern, J. *Inorg. Chem.* **1965**, *4*, 766.

[b] Moore, M. C.; Keller, R. N. *Inorg. Chem.* **1971**, *10*, 747, at $\mu = 0.1$ M.

[c] Barrett, M. B.; Swinehart, J. H.; Taube, H. *Inorg. Chem.* **1971**, *10*, 1983.

[d] Price, H. J.; Taube, H. *Inorg. Chem.* **1968**, *7*, 1.

[e] Sisley, M. J.; Jordan, R. B. *Inorg. Chem.* **1989**, *28*, 2714.

Several kinetic trends have been interpreted as largely reflecting changes in K. When $Co(NH_3)_5X^{2+}$ complexes with $X = F^-$, Cl^-, and Br^- are reduced by Cr^{2+}, the order of reactivity is $Br^- > Cl^- > F^-$, but the opposite order is observed when Eu^{2+} is the reducing agent. This is rationalized from the knowledge that Eu^{2+} is a harder acid and forms stronger complexes with F^-, whereas Cr^{2+} forms stronger complexes with Br^-. If X is a carboxylate anion, then the order of reactivity of $Co(NH_3)_5X^{2+}$ with Cr^{2+} is $HCO_2^- > CH_3CO_2^- > CHCl_2CO_2^- > CF_3CO_2^- > (CH_3)_3CCO_2^-$. This is interpreted as a combination of steric and electron-withdrawing effects, causing K to decrease for the bridged intermediate. The unusually high reactivity of the formate complex has been attributed[48] to its ability to form the more sterically accessible conformer on the right in (6.42), therefore giving an even larger K than expected.

$$
\begin{bmatrix}
\text{NH}_3 \\
\text{H}_3\text{N}-\text{Co}-\text{O} \\
\text{H}_3\text{N} \quad \text{H}_3\text{N} \quad \text{O}
\end{bmatrix}^{2+}
\rightleftharpoons
\begin{bmatrix}
\text{NH}_3 \\
\text{H}_3\text{N}-\text{Co}-\text{O} \\
\text{H}_3\text{N} \quad \text{H}_3\text{N} \quad \text{H}
\end{bmatrix}^{2+}
\tag{6.42}
$$

The much higher rates for the oxalato and keto form of pyruvato complexes[49] are attributed to stabilization of the bridged intermediate by chelation. The same effect may be operating, but less effectively, for the pyruvate hydrate complex.[50]

The kinetic product[51] shown in reaction (6.43) indicates that the simple carboxylate ions probably use the β-oxygen in bridging.

$$
\begin{bmatrix}
(\text{H}_3\text{N})_5\text{Co}-\text{S} \quad \text{O} \\
\text{C} \\
\text{H}-\text{N}-\text{CH}_3
\end{bmatrix}^{2+}
\xrightarrow[11\ \text{H}_2\text{O},\ 5\ \text{H}^+]{\text{Cr}^{2+}}
\begin{bmatrix}
\text{S} \quad \text{O}-\text{Cr(OH}_2)_5 \\
\text{C} \\
\text{H}-\text{N}-\text{CH}_3
\end{bmatrix}^{2+}
\tag{6.43}
$$

$$+ \ \text{Co(OH}_2)_6^{2+} + 5\ \text{NH}_4^+$$

The O-bonded Cr(III) complex isomerizes to the stable S-bonded form. It was also shown that the O-bonded Co(III) complex produces the S-bonded Cr(III) product.

With Cr(II) as the reducing agent, the Cr(III) product provides evidence of an inner-sphere mechanism, but in a more general sense it would be useful to know the necessary properties for a bridging ligand. The minimum requirement is two electron pairs, one to bond to the oxidizing agent and the other to the reducing agent. Jordan and Balahura[52] have suggested that in ligands more complex than the

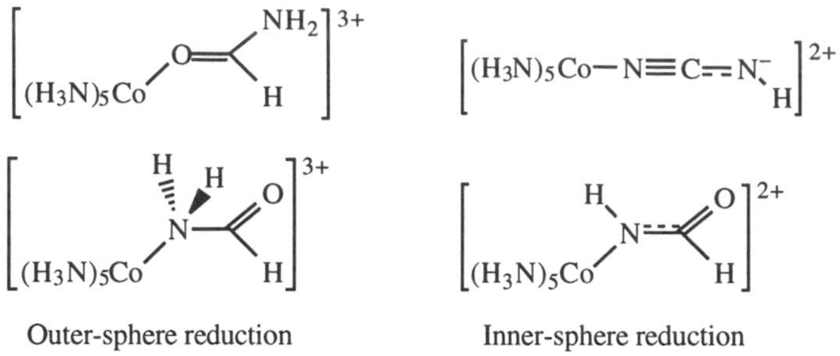

Outer-sphere reduction   Inner-sphere reduction

Figure 6.2. Cobalt(III) complexes whose chromium(II) reduction products indicate the reaction mechanism and the properties needed for a bridging ligand.

halides and hydroxide, two metal centers are unlikely to bond to the same atom and must be bonded to atoms that are part of a conjugated system. Then, the electron may transfer through the π or π^* orbital of the conjugated system. The evidence for this was based on observations of the Cr(III) products from the Cr(II) reductions of the Co(III) complexes in Figure 6.2.

The observation that $Co(NH_3)_5(OH_2)^{3+}$ is reduced very slowly by Cr(II)[53] indicates that water is not an effective bridging ligand. However, OH^- is an effective bridge in the same system.

Inner-sphere attack of Cr(II) on the adjacent atom is observed with $(H_3N)_5Co—SCN^{2+}$ to yield $(H_2O)_5Cr—SCN^{2+}$ ($k = 0.8 \times 10^5$ M^{-1} s^{-1}) as well as the remote attack product, $(H_2O)_5Cr—NCS^{2+}$ ($k = 1.9 \times 10^5$ M^{-1} s^{-1}).[54] The linkage isomer, $(H_3N)_5Co—NCS^{2+}$, gives only one product, $(H_2O)_5Cr—SCN^{2+}$. In the Co—NCS case, the adjacent N atom does not have a lone pair available to bond to the reducing agent, whereas the S atom does have a lone pair in the former.

In contrast to the outer-sphere case, theory has been of limited help in analyzing inner-sphere reactivity. Schwarz and Endicott[55] studied the electron exchange between $Co^{III}(N_4)(OH_2)X^{2+}$ and $Co^{II}(N_4)(OH_2)^{2+}$, where (N_4) is an N-donor macrocycle and X is Cl$^-$, Br$^-$, or N_3^-. The authors provide a detailed analysis framed in the general terminology of the Marcus theory and conclude, among other things, that the inner-sphere reorganization energy is about half the outer-sphere energy.

The mechanistic classifications are not always as straightforward as the preceding examples might imply. Reduction of the N-bonded glycine complex[56] leads to transfer of glycine to Cr(III), although there is no conjugation between the $-NH_2$ and $-CO_2^-$ groups. This can be rationalized as due to a *bridged outer-sphere mechanism*, illustrated in Scheme 6.2. It is proposed that glycine serves to hold the oxidizing and reducing centers close together but that electron transfer proceeds by an outer-sphere process rather than through the glycine.

Scheme 6.2

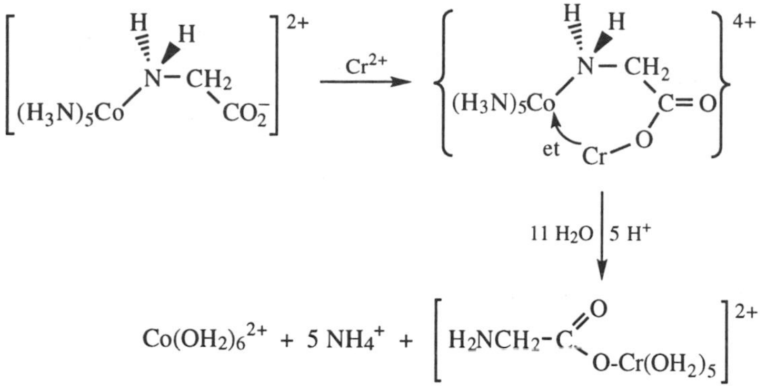

$$\left[(H_3N)_5Co\overset{H}{\underset{}{\overset{\vdots}{N}}}\overset{H}{\underset{CO_2^-}{\overset{}{-CH_2}}}\right]^{2+} \xrightarrow{Cr^{2+}} \left\{(H_3N)_5Co\overset{H\;\;H}{\underset{et\;\;Cr\overset{}{\underset{}{-O}}}{\overset{\vdots}{N}}}\overset{}{\underset{C=O}{\overset{-CH_2}{}}}\right\}^{4+}$$

$$\Big\downarrow 11\ H_2O\ |\ 5\ H^+$$

$$Co(OH_2)_6{}^{2+}\ +\ 5\ NH_4^+\ +\ \left[H_2NCH_2-C\overset{\displaystyle O}{\underset{O-Cr(OH_2)_5}{\big\langle}}\right]^{2+}$$

The systems listed in Table 6.7 are examples of what is called *remote attack*, in which the Cr(II) is attached at a ligand atom considerably removed from the Co(III). In such systems, the rate constant is sensitive to the nature of the remote substituent through its effect on K. The rate constants also correlate generally with the ease of reduction of the bridging ligand, which seems to reflect an influence on k_e.

Table 6.7. Rate Constants (25°C) for Reduction by Remote Attack of Cr^{2+}

X:Co(NH$_3$)$_5$	k (M^{-1} s^{-1})
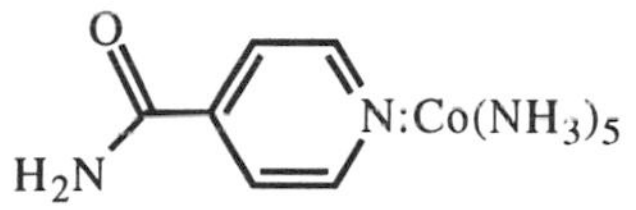 (H$_2$N–CO–C$_6$H$_4$–)N:Co(NH$_3$)$_5$	17.4
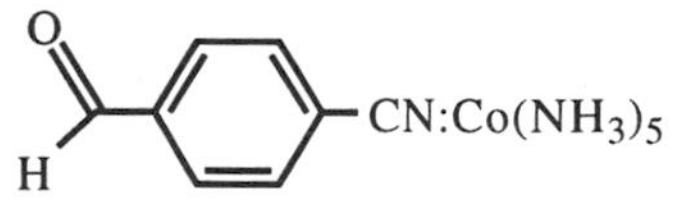 (H–CO–C$_6$H$_4$–)CN:Co(NH$_3$)$_5$	2.5 x 10^5
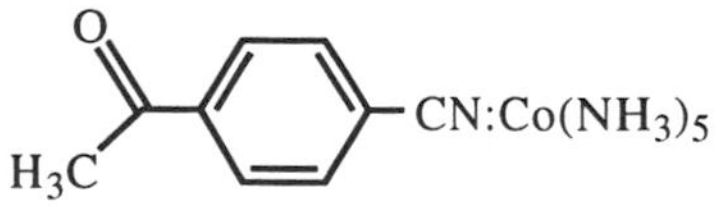 (H$_3$C–CO–C$_6$H$_4$–)CN:Co(NH$_3$)$_5$	6 x 10^3
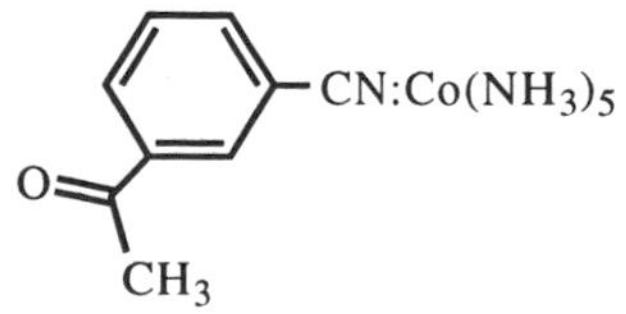 (O=... CH$_3$ meta-substituted C$_6$H$_4$–)CN:Co(NH$_3$)$_5$	0.28

The isonicotinamide system[57] gives an initial product with Cr(III) bonded to the oxygen of the amide group and provides clear evidence for a remote attack mechanism. The meta isomer, nicotinamide, reacts about 500 times slower and gives about 70 percent Cr(III)–amide complex. The smaller rate of the *m*-acetylcyanobenzene complex compared to the *p*-acetyl isomer[58] also reflects the importance of conjugation between the remote group and the lead-in group at cobalt. These types of observations have led to the suggestion that such systems may proceed by a *chemical mechanism* in which the electron is actually transferred to the bridging group to form a radical intermediate.

In the case of the *p*-formylbenzoato complex, the rate law has an H^+-dependent path that is consistent with the mechanism in Scheme 6.3.[59]

Scheme 6.3

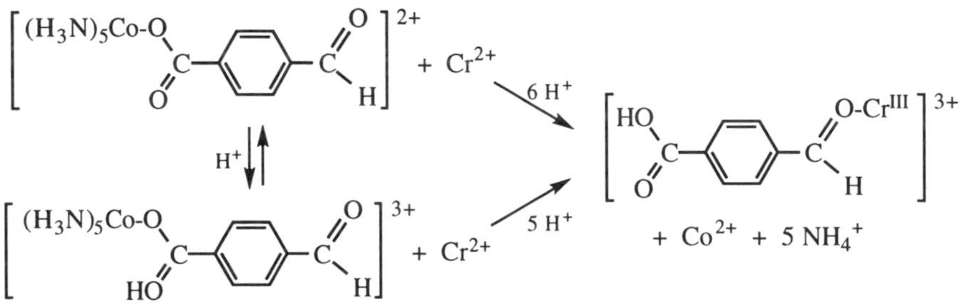

$$\text{Rate} = (5.3 + 380\,[H^+])\,[Co(III)]\,[Cr(II)]$$

Protonation of the adjacent carboxylate group may increase the rate by improving the reducibility of the bridging ligand and/or improving the conjugation between the metal centers in the bridged intermediate. In analogous systems with simple carboxylate ligands, such as acetate and benzoate, protonation inhibits reduction,[60] presumably because the adjacent group cannot accommodate both H^+ and Cr^{2+}.

Tsukahara and Wilkins[61] studied the product of the reaction of $Co(NH_3)_5(mbpy)^{4+}$, where mbpy is 1-methyl-4,4'-bipyridinium, with $CO_2^{\bullet -}$ at pH 7.2 and 25°C. The $CO_2^{\bullet -}$ was formed by pulse radiolysis. The initial product was assigned as the radical complex $Co(NH_3)_5(mbpy\bullet)^{3+}$, which then undergoes intramolecular electron transfer ($k_e = 8.7 \times 10^2$ s^{-1}) and bimolecular electron transfer to $Co(NH_3)_5(mbpy)^{4+}$ ($k_2 = 5.4 \times 10^7$ M^{-1} s^{-1}). They also observed that the reaction of $Ru(NH_3)_5(mbpy)^{4+}$ and $CO_2^{\bullet -}$ produces no detectable radical intermediate and attributed this to fast intramolecular electron transfer ($k_e > 10^6$ s^{-1}). The difference in reactivity of the Co and Ru systems was suggested to be due to the fact that both the acceptor and donor orbitals are of π symmetry in the Ru system while the acceptor orbital is of σ symmetry in the Co system.

The radical mechanism has been characterized for a series of nitrobenzoate-type ligands attached to Co(III) in which moderately persistent radicals can be generated by pulse radiolysis.[62] The coordinated radical is produced by electron transfer from a reactive radical, R•, such as e^-, $CO_2^{•-}$, and $(H_3C)_2(C•)OH$ produced in the radiolysis pulse, and the reaction proceeds as shown in Scheme 6.4.

Scheme 6.4

$$(H_3N)_5Co-O_2C-X-C_6H_4NO_2^{2+} + R•$$

Fast $\downarrow$

$$\{(H_3N)_5Co-O_2C-X-C_6H_4\overset{•}{N}O_2^+\} + R^+$$

$k_e \downarrow$

$$^-O_2C-X-C_6H_4NO_2 + Co^{2+} + 5\,NH_3$$

The rate of the intramolecular electron transfer, k_e, can then be determined from the disappearance of the nitrobenzoate radical intermediate and some results are given in Table 6.8.

It is notable that these intramolecular reactions are not exceptionally fast. The ortho isomers are much more reactive because of effective conjugation and possibly some "outer-sphere" transfer from the ($•NO_2$) group, which is near the Co(III). This is consistent with the lower rate when a CH=CH group is introduced. The meta isomers are least reactive because of poor conjugation. The moderate reactivity of the saturated $CH_2CH_2CH_2$ derivative is ascribed to the "outer-sphere" path allowed by the flexibility of the $CH_2CH_2CH_2$. The OC(NH)CH$_2$ derivative is less reactive because of the rigidity imposed by the planar OC(NH) group.

Table 6.8. Rate Constants (25°C) for the Reduction of Some Nitrobenzoate Radicals Coordinated to $Co^{III}(NH_3)_5$

X (isomer)	k_e (s^{-1})	X (isomer)	k_e (s^{-1})
- (o)	4.0×10^5	CH=CH (o)	1.7×10^3
- (m)	1.5×10^2	CH=CH (m)	3.1
- (p)	2.6×10^3	CH=CH (p)	4.8×10^2
CH$_2$ (o)	3.5×10^4	CH$_2$CH$_2$CH$_2$ (p)	1.5×10^2
CH$_2$ (m)	1.0×10^2	OC(NH)CH$_2$ (p)	5.8
CH$_2$ (p)	3.9×10^2		

Another potential method of separating the K and k_e effects on electron-transfer rates is actually to prepare the bridged complex using inert oxidizing and reducing centers. For example, Taube and co-workers[63] studied the following system:

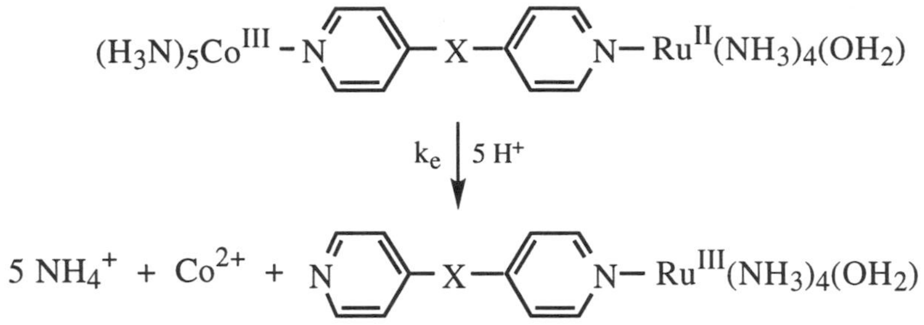

$$(H_3N)_5Co^{III} - N \bigcirc - X - \bigcirc N - Ru^{II}(NH_3)_4(OH_2)$$

$$k_e \downarrow 5\,H^+ \tag{6.44}$$

$$5\,NH_4^+ + Co^{2+} + N \bigcirc - X - \bigcirc N - Ru^{III}(NH_3)_4(OH_2)$$

The values of k_e are rather insensitive to X (CH_2–CH_2, CH=CH) and small (0.1×10^{-2} to 4×10^{-2} s^{-1}, 25°C), especially compared to the rate constant of 3 M^{-1} s^{-1} for the reaction of $Co(NH_3)_5(OH_2)^{3+}$ with $Ru(NH_3)_6^{2+}$. It was suggested that the reactions are slow because of inner-sphere reorganization analogous to that proposed for outer-sphere reactions of cobalt(III) complexes.

A similar study, by Haim and co-workers,[64] has been done on the analogous $(H_3N)_5Co$—L—$Fe(CN)_5$ system. The values of k_e are all in the range of 1.5×10^{-3} to 5×10^{-3} s^{-1}, except for much smaller values when X = CH_2 or CO. The rate constants are similar to the values for the outer-sphere reactions of $Co(NH_3)_5(L')^{3+} + Fe(CN)_5L^{3-}$, where L and L' are pyridine derivatives. These reactions proceed through a strong ion pair ($K_{ip} \approx 900$ M^{-1}) so that k_e in the ion pair is measured.[65] The implication is that the binuclear systems with more flexible X, such as $(CH_2)_2$ and $(CH_2)_3$, use a bridged outer-sphere mechanism.

6.5 INTERVALENCE ELECTRON TRANSFER

There has been intense interest in the process referred to as intervalence electron transfer in mixed-valence (oxidation state) species ever since the initial preparation[66] of the Taube–Creutz compound:

$$(H_3N)_5Ru^{II} - N \bigcirc N - Ru^{III}(NH_3)_5$$

The intervalence electron transfer involves electron exchange between the two metal centers that are in different oxidation states. This process seems to be typified by transitions in the near-IR region of the electronic spectra of such species. For the preceding ion in water, this band occurs at 1560 nm (6,400 cm^{-1}). There is some debate as to whether the preceding description (localized or trapped valence) of this ion is correct or whether a delocalized picture is more appropriate. The trapped-valence model seems correct[67] for the complex in which the bridging ligand is 4,4'-bipyridine, since the two pyridine rings are not planar and therefore are not in full conjugation.

The unique electronic absorbance observed in these mixed-valence systems is usually assigned as a metal–metal charge transfer band, MMCT. Meyer and Hupp[68] have noted that it should not be a single band because of the lower than O_h symmetry and spin-orbit coupling in Ru. The t_{2g} level will split into three nondegenerate levels so that three closely spaced bands may actually be observed. Since the MMCT process is equivalent to electron transfer from one metal to another, the interest in these systems has centered on the energetics of the MMCT process. Hush[69] suggested that the energy of the MMCT band would be the sum of the inner- and outer-sphere reorganization energies, $\Delta G_{in}^* + \Delta G_{solv}^*$, and it therefore presents a way of studying these features. Calculations by Creutz[67] indicate that ΔG_{in}^* is ~1400 cm^{-1} for Ru(II)/(III), so that ΔG_{solv}^* would seem to be the dominant factor, given that the MMCT energies are in the 6000 to 12000 cm^{-1} range.

Hupp and Meyer measured the MMCT energy, E_{OP}, as a function of solvent for the 4,4'-bipyridine dimer, in order to test the prediction of Hush that E_{OP} should be given by

$$E_{OP} = \Delta G_{in}^* + \frac{N e^2}{16 \pi \varepsilon_0} \left(\frac{1}{a} - \frac{1}{r} \right) \left(\frac{1}{n_s^2} - \frac{1}{\varepsilon_s} \right) \qquad (6.45)$$

where it is assumed that $a_1 = a_2 = a$ in Eq. (6.23). The variation of E_{OP} with $(n_s^{-2} - \varepsilon_s^{-1})$ is linear, but the slope of 78,100 cm^{-1} is much larger than the one predicted, 22,500 cm^{-1} (a = 3.5 Å, r = 11.3 Å), and the intercept of 4820 cm^{-1} is rather far from the one expected, 1400 cm^{-1}. Meyer and Hupp have offered several explanations for these discrepancies, including the band multiplicity problem mentioned earlier.

More recently, Meyer and co-workers[70] have correlated the solvent dependence of the reduction potentials and electronic spectra with the Gutmann donor number for $(bpy)_2(Cl)Os(L)Ru(NH_3)_5^{4+}$ systems, where L = 4,4'-bipyridine or pyrazine. They find that the two metal centers respond differently to the solvent so that the oxidation states are Os^{III}–Ru^{II} for DN < 14 and Os^{II}–Ru^{III} for DN > 15. They also conclude that intramolecular electron transfer involves coupled electronic and nuclear motions and has no simple relationship to thermal electron transfer.

The variation of E_{OP} with the structure of the bridging ligand has been the subject of several studies. It is hoped that this will provide some information on long-range electron-transfer reactions in biological systems. Some results are given in Table 6.9 from studies by Sutton and Taube[71] and Spangler and co-workers (fifth entry)[72] on the pyridine derivatives and by Stein et al.[73] on the thiospiranes. The pyridine systems have been extended in recent work by Spangler and co-workers.[74] These examples fall into two categories. In the pyridine systems, the metal centers show significant coupling and the molar absorption coefficients, ε, are in the 10^2 to 10^3 M^{-1} cm^{-1} range; in the thiospiranes, there is very weak coupling between the metals and the ε values are <50 M^{-1} cm^{-1}. This difference means that electron transfer may be considered as being adiabatic for the pyridines, but probably is nonadiabatic for the thiospiranes.

Table 6.9. Intervalence Absorption Energies for Some Complexes of the Type $(H_3N)_5Ru^{III}$—L—$Ru^{II}(NH_3)_5^{5+}$

L	d (Å)	E_{OP} (cm^{-1})	ε (M^{-1} cm^{-1})
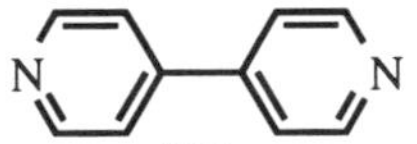	11.3	9,710	920
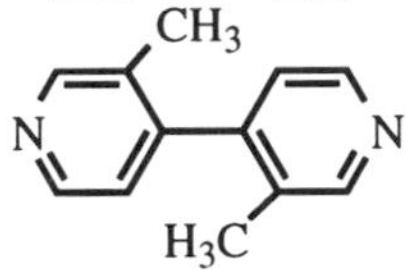	11.3	11,240	165
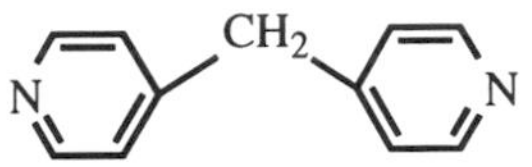	10.5	12,350	30
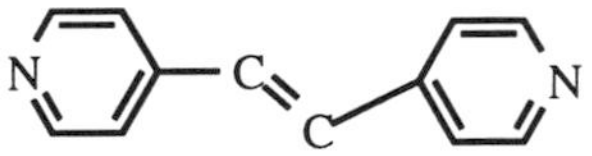	13.8	10,420	760
	15.8	9,615	1430
	11.3	10,990	43
	14.4	12,300	9
	17.6	14,500	2.3

The substituents in the bridge clearly affect the coupling between the metal centers in a rational way that is reflected in the ε values. The two CH_3 substituents twist the pyridine rings further out of conjugation and reduce the coupling and the ε value, as does the saturated —CH_2— bridge. In the thiospiranes, the ε decreases as the Ru—Ru distance increases, indicating reduced coupling. The E_{OP} increases as the distance increases in the thiospiranes. The latter trends are expected for coupling through space or through σ bonds.

The conjugated —CH_2=CH_2— bridge allows conjugation between rings and increases ε. It should be noted that the earlier results of Spangler and co-workers[72] were reanalyzed by Reimers and Hush,[75] who have corrected the energies for band overlap and these corrections have been incorporated in later work. Recently, it has been found that ferrocene linked by —CH_2=CH_2— bridges behaves similarly to the $Ru(NH_3)_5$ analogues.[76]

The nature of the relationship between intervalence electronic bands and electron-transfer processes remains an open question. There appears to be a relationship between E_{OP} and ΔG^*_{solv}, but the quantitative interpretation in terms of the Hush equation is less than satisfactory, as noted by Meyer and Hupp.[68] A correlation between the ΔG^* for electron transfer and the metal–metal separation for $(H_3N)_5Co$—L—$Fe(CN)_5$ and $(H_3N)_5Co$—L—$Ru(NH_3)_4(OH_2)$ systems has been noted by Haim.[77] In $(H_3N)_5Ru^{II}$—L—$M^{III}(NH_3)_5$ systems, Geselowitz[78] found that E_{OP} for M = Ru correlates with ΔG^* for electron transfer for M = Co. He concluded that such systems have adiabatic electron transfer and the correlation works because E_{OP} is related to ΔG^*_{solv} when L gives reasonable coupling between metal centers. The suggestion of a complex relationship by Meyer et al. is noted above.

The standard interpretation for weakly coupled systems has assumed that the inner- and outer-sphere rearrangement energies are not dependent on the separation between the metal centers and that the distance dependence of E_{OP} and ΔG^* for electron transfer is due to the decrease in electronic coupling between the centers with increasing distance. This electronic factor will affect the probability of electron transfer and therefore is lumped together with ΔS^* in transition-state theory interpretations. The dependence of the electronic factor, κ_e, on distance is taken to be related to that of the exchange integral, H_{AB}, between the metal centers, and quantum mechanics predicts that

$$\kappa_e = \kappa_{eo} \exp[-\beta(r-r_0)] \qquad (6.46)$$

where $\beta \approx 1$ Å^{-1} (values of 0.9 to 1.2 are often used), r is the metal–metal separation in the system under study, and r_0 is the separation at which the transfer will be adiabatic (when $\kappa_e = \kappa_{eo}$). The value of r_0 is not known and is sometimes taken to be 0, although 3 to 4 Å might be more reasonable. The electron-transfer rate should have an exponential dependence on the metal–metal separation if other factors are constant.

Sutin and co-workers[79] have questioned the assumption that the nuclear factors, $\Delta G_{in}^* + \Delta G_{solv}^*$, are independent of the separation. They find a correlation between separation and E_{OP} for Ru^{II}—L—Ru^{III} systems and the same correlation between the separation and ΔH^* for electron transfer in Os^{II}—iso$(Pro)_n$—Ru^{III} systems, where iso$(Pro)_n$ is a polyproline bridge: with $n = 1$, $r = 12.2$ Å; $n = 2$, $r = 14.8$ Å; $n = 3$, $r = 18.1$ Å. This implies that the ΔH^* is dependent on distance and that it is not valid to assume that separation affects only the probability of electron transfer through the electronic factor.

6.6 ELECTRON TRANSFER IN METALLOPROTEINS

The metalloproteins consist of a metal complex imbedded in and bonded (through the ligand) to a protein net of covalently bonded amino acids. The most commonly studied systems are the myoglobins and cytochromes, which contain an iron (II or III) porphyrin complex or the copper blue proteins, which have Cu(II) or Cu(I) complexed most often by histidine nitrogens and cysteine and methionine sulfurs from the protein. The metalloproteins can be oxidized or reduced by standard transition-metal complex reagents, and the latter usually are chosen to ensure outer-sphere electron transfer. This area has been the subject of several reviews.[80–85]

Since the metal center is surrounded by the protein system, these electron transfers occur at much longer distances (10–20 Å) than normal for small molecules and therefore have an analogy to the weakly coupled bimetal systems discussed in the previous section. The distance dependence of the rate of electron transfer has been investigated using modified proteins in which, typically, $Ru^{III}(NH_3)_5$ is attached at a specific site; then, it is reduced to Ru(II) and the rate of electron transfer from the latter to the metal in the protein is measured.[86]

The interpretation of the kinetic results on these systems has revolved around the distance dependence, as discussed in the previous section. Dutton and co-workers[87] first noted a correlation with the distance, r Å, between donor and acceptor, of the following form:

$$k\,(s^{-1}) = 10^{13} \exp[-1.4\,(r-3.6)] \qquad (6.47)$$

But a study[88] of Ru-modified myoglobins gave the following smaller dependence on distance:

$$k\,(s^{-1}) = 7.8 \times 10^8 \exp[-0.91\,(r-3)] \qquad (6.48)$$

Theoretical work indicates that such interpretations may be overly simplistic. The theory proposed by Hopfield and co-workers[89] suggests that the electron transfers through the σ bonds in the protein by a tunneling mechanism. The rate is then related to the most favorable electron-transfer path that can be found from a reagent at a particular site on the metalloprotein to its metal center. The theory has been applied[90] and found to be consistent with the relative rate constants observed for Ru-modified cytochrome c, but the simple-distance theory is also consistent with these observations. Beratan and Onuchic and co-workers[91] have surveyed several ruthenated proteins and devised methods for predicting the most favorable electron-transfer pathway. Ulstrup and co-workers[92] have given an extended Hückel theory analysis of the electron-transfer routes along different amino acid sequences in platocyanins.

Recently, Jay-Gerin and co-workers[93] have surveyed many such systems and proposed a model based on through-space coupling of donor and acceptor. The coupling depends on the energy of the protein's orbitals relative to those of the donor and acceptor, the associated band width, and the density of nonhydrogen sites in the protein. They suggest that the distance dependence is strongly dependent on the energy of the protein's orbitals and can be variable for different proteins.

References

1. Taube, H.; Myers, H.; Rich, R. L. *J. Am. Chem. Soc.* **1953**, *75*, 4118; Taube, H.; Myers, H. *J. Am. Chem. Soc.* **1954**, *76*, 2103.
2. Marcus, R. A. *Annu. Rev. Phys. Chem.* **1964**, *15*, 155; *J. Chem. Phys.* **1965**, *43*, 679.
3. Hush, N. S. *Trans. Farad. Soc.* **1961**, *57*, 557; *Electrochim. Acta* **1968**, *13*, 1005; *Prog. Inorg. Chem.* **1967**, *8*, 391.
4. Levich, V. G. *Adv. Electrochem. Eng.* **1966**, *4*, 249.
5. Dogonadze, R. R. In *Reactions of Molecules at Electrodes*; Hush, N. S., Ed.; Wiley-Interscience: New York, 1971, Ch. 3.
6. Ratner, M. A.; Levine, R. D. *J. Am. Chem. Soc.* **1980**, *102*, 4898.
7. Levine, R. D. *J. Phys. Chem.* **1979**, *83*, 159.
8. Sutin, N. *Acc. Chem. Res.* **1968**, *1*, 225.
9. Newton, T. W. *J. Chem. Educ.* **1968**, *45*, 571.
10. Sutin, N. *Prog. Inorg. Chem.* **1983**, *30*, 441.
11. Reynolds, W. L.; Lumry, R. W. *Mechanisms of Electron Transfer*; Ronald Press: New York, 1966.
12. Cannon, R. D. *Electron Transfer Reactions*; Butterworths: London, 1980.
13. Tembe, B. L.; Friedman, H. L.; Newton, M. J. *J. Chem. Phys.* **1982**, *76*, 1490.
14. Bernhard, P.; Ludi, A. *Inorg. Chem.* **1984**, *23*, 870.
15. Wherland, S. *Coord. Chem. Rev.* **1993**, *123*, 169.
16. Chan, M.-S.; Wahl, A. C. *J. Chem. Phys.* **1982**, *86*, 126.

17. Nielson, R. M.; McManis, G. E.; Safford, L. K.; Weaver, M. J. *J. Chem. Phys.* **1989**, *93*, 2152.
18. Weaver, M. J. *Chem. Rev.* **1992**, *92*, 463.
19. Drago, R. S.; Ferris, D. C. *J. Phys. Chem.* **1995**, *99*, 6563.
20. Abbott, A. P.; Rusling, J. F. *J. Phys. Chem.* **1990**, *94*, 8910.
21. Lay, P. A.; McAlpine, N. S.; Hupp, J. T.; Weaver, M. J.; Sargeson, A. M. *Inorg. Chem.* **1990**, *29*, 4322.
22. Swaddle, T. W. *Inorg. Chem.* **1990**, *29*, 5017.
23. Mao, W.; Qian, Z.; Yen, H.-J.; Curtis, J. C. *J. Am. Chem. Soc.* **1996**, *118*, 3247.
24. Binstead, R. A.; Beattie, J. K.; Dewey, T. G.; Turner, D. H. *J. Am. Chem. Soc.* **1980**, *102*, 6442.
25. Creaser, I. I.; Geue, R. J.; Harrowfield, J. McB.; Herlt, A. J.; Sargeson, A. M.; Snow, M. R.; Springborg, J. *J. Am. Chem. Soc.* **1982**, *104*, 6016.
26. Geselowitz, D. *Inorg. Chem.* **1981**, *20*, 4457.
27. Bernhard, P.; Sargeson, A. M. *Inorg. Chem.* **1988**, *27*, 2582.
28. Küppers, H.-J.; Neves, A.; Pomp, C.; Ventur, D.; Wieghardt, K.; Nuber, B.; Weiss, J. *Inorg. Chem.* **1986**, *25*, 2400.
29. Dubs, R. V.; Gahan, L. R.; Sargeson, A. M. *Inorg. Chem.* **1983**, *22*, 2523.
30. Dulz, G.; Sutin, N. *Inorg. Chem.* **1963**, *2*, 917; Ford-Smith, M. H.; Sutin, N. *J. Am. Chem. Soc.* **1961**, *83*, 1830.
31. Chou, M.; Creutz, C.; Sutin, N. *J. Am. Chem. Soc.* **1977**, *99*, 5615.
32. Bernhard, P.; Sargeson, A. M. *Inorg. Chem.* **1987**, *26*, 4122.
33. Hupp, J. T.; Weaver, M. J. *Inorg. Chem.* **1983**, *22*, 2557.
34. Bernhard, P.; Helm, L.; Ludi, A.; Merbach, A. E. *J. Am. Chem. Soc.* **1985**, *107*, 312.
35. Jolley, W. H.; Stranks, D. R.; Swaddle, T. W. *Inorg. Chem.* **1990**, *29*, 1948.
36. Macartney, D. H.; Sutin, N. *Inorg. Chim. Acta* **1983**, *74*, 221.
37. Stanbury, D. M.; Haas, O.; Taube, H. *Inorg. Chem.* **1980**, *19*, 518.
38. Zahir, K.; Espenson, J. H.; Bakac, A. *J. Am. Chem. Soc.* **1988**, *110*, 5059.
39. Sutin, N. *Acc. Chem. Res.* **1982**, *15*, 275.
40. Lind, J.; Shen, X.; Merényi, G.; Jonsson, B. O. *J. Am. Chem. Soc.* **1989**, *111*, 7655.
41. Grace, M. R.; Takagi, H.; Swaddle, T. W. *Inorg. Chem.* **1994**, *33*, 1915.
42. Sachinidis, J. I.; Shalders, R. D.; Tregolan, P. A. *Inorg. Chem.* **1996**, *35*, 2497.
43. Haim, A.; Wilmarth, W. K. *J. Am. Chem. Soc.* **1961**, *83*, 509.
44. Sykes, A. G.; Thorneley, R. N. F. *J. Chem. Soc. A* **1970**, 232.
45. Hua, L. H.-C.; Balahura, R. J.; Fanchiang, Y.-T.; Gould, E. S. *Inorg. Chem.* **1978**, *17*, 3692.
46. Linck, R. G. *Inorg. React. Methods* **1986**, *15*, 68.
47. Murdoch, J. R. *J. Am. Chem. Soc.* **1972**, *94*, 4410.
48. Balahura, R. J.; Jordan, R. B. *Inorg. Chem.* **1973**, *12*, 1438.
49. Price, H. J.; Taube, H. *Inorg. Chem.* **1968**, *7*, 1.
50. Sisley, M. J.; Jordan, R. B. *Inorg. Chem.* **1989**, *28*, 2714.

51. Balahura, R. J.; Johnson, M. D.; Black, T. *Inorg. Chem.* **1989**, *28*, 3933.
52. Jordan, R. B.; Balahura, R. J. *J. Am. Chem. Soc.* **1971**, *93*, 625.
53. Toppen, D. L.; Linck, R. G. *Inorg. Chem.* **1971**, *10*, 2635.
54. Shea, C.; Haim, A. *J. Am. Chem. Soc.* **1971**, *93*, 3055.
55. Schwarz, C. I.; Endicott, J. F. *Inorg. Chem.* **1995**, *34*, 4572.
56. Kupferschmidt, W. C.; Jordan, R. B. *Inorg. Chem.* **1981**, *20*, 3469.
57. Nordmeyer, F.; Taube, H. *J. Am. Chem. Soc.* **1968**, *90*, 1162.
58. Balahura, R. J.; Purcell, W. L. *J. Am. Chem. Soc.* **1976**, *98*, 4457.
59. Zannella, A.; Taube, H. *J. Am. Chem. Soc.* **1972**, *94*, 6403.
60. Barrett, M. B.; Swinehart, J. H.; Taube, H. *Inorg. Chem.* **1971**, *10*, 1983.
61. Tsukahara, K.; Wilkins, R. G. *Inorg. Chem.* **1989**, *28*, 1605.
62. Whitburn, K. D.; Hoffman, M. Z.; Simic, M. G.; Brezniak, N. V. *Inorg. Chem.* **1980**, *19*, 3180; Whitburn, K. D.; Hoffman, M. Z.; Brezniak, N. V.; Simic, M. G. *Inorg. Chem.* **1986**, *25*, 3037.
63. Fischer, H.; Tom, G. M.; Taube, H. *J. Am. Chem. Soc.* **1976**, *98*, 5512.
64. Jwo, J.-J.; Gaus, P. L.; Haim, A. *J. Am. Chem. Soc.* **1979**, *101*, 6189.
65. Gaus, P. L.; Villanueva, J. L. *J. Am. Chem. Soc.* **1980**, *102*, 1934.
66. Creutz, C.; Taube, H. *J. Am. Chem. Soc.* **1969**, *91*, 3988; Ibid. **1973**, *95*, 1086.
67. Creutz, C. *Inorg. Chem.* **1978**, *17*, 3723.
68. Hupp, J. T.; Meyer, T. J. *Inorg. Chem.* **1987**, *26*, 2332.
69. Hush, N. S. *Inorg. Chem.* **1967**, *8*, 391.
70. Neyhart, G. A.; Hupp, J. T.; Curtis, J. C.; Timpson, C. J.; Meyer, T. J. *J. Am. Chem. Soc.* **1996**, *118*, 3724; Neyhart, G. A.; Timpson, C. J.; Bates, W. D.; Meyer, T. J. *J. Am. Chem. Soc.* **1996**, *118*, 3730.
71. Sutton, J. E.; Taube, H. *Inorg. Chem.* **1981**, *20*, 3125.
72. Woitellier, S.; Launay, J. P.; Spangler, C. W. *Inorg. Chem.* **1989**, *28*, 758.
73. Stein, C. A.; Lewis, N. A.; Seitz, G. *J. Am. Chem. Soc.* **1982**, *104*, 2596.
74. Ribou, A.-C.; Launay, J.-P.; Takahashi, K.; Nihira, T.; Tarutani, S.; Spangler, C. W. *Inorg. Chem.* **1994**, *33*, 1325.
75. Reimers, J. R.; Hush, N. S. *Inorg. Chem.* **1990**, *29*, 4510.
76. Ribou, A.-C.; Launay, J.-P.; Sachtleben, M. L.; Li, H.; Spangler, C. W. *Inorg. Chem.* **1996**, *35*, 3735.
77. Haim, A. *Pure Appl. Chem.* **1983**, *55*, 89.
78. Geselowitz, D. *Inorg. Chem.* **1987**, *26*, 4135.
79. Isied, S. S.; Vassilian, A.; Wishart, J. F.; Creutz, C.; Schwarz, H. A.; Sutin, N. *J. Am. Chem. Soc.* **1988**, *110*, 635.
80. Isied, S. S. *Prog. Inorg. Chem.* **1984**, *32*, 443.
81. Sykes, A. G. *Chem. Soc. Rev.* **1985**, *14*, 283.
82. McLendon,G.; Guarr, T.; McGuire, M.; Simolo, K.; Strauch, S.; Taylor, K. *Coord. Chem. Rev.* **1985**, *64*, 113.
83. Marcus, R. A.; Sutin, N. *Biochim. Biophys. Acta* **1985**, *811*, 265.
84. Gray, H. B. *Chem. Soc. Rev.* **1986**, *15*, 17.
85. McLendon, G. *Acc. Chem. Res.* **1988**, *21*, 160.
86. Scott, R. A.; Mauk, A. G.; Gray, H. B. *J. Chem. Educ.* **1985**, *62*, 932; Karas, J. L.; Lieber, C. M.; Gray, H. B. *J. Am. Chem. Soc.* **1988**, *110*, 599.

87. Moser, C. C.; Keske, J. M.; Warncke, K.; Faird, R. S.; Dutton, P. L. *Nature* **1992**, *355*, 796.
88. Axup, A. W.; Albin, M.; Mayo, S. L.; Crutchley, R. J.; Gray, H. B. *J. Am. Chem. Soc.* **1988**, *110*, 435.
89. Beratan, D. N.; Onuchic, J. N.; Hopfield, J. J. *J. Chem. Phys.* **1987**, *86*, 4488; Cowan, J. A.; Upmacis, R. K.; Beratan, D. N.; Onuchic, J. N.; Gray, H. B. *Ann. N. Y. Acad. Sci.* **1989**, *550*, 68.
90. Bowler, B. E.; Meade, T. J.; Mayo, S. L.; Richards, J. H.; Gray, H. B. *J. Am. Chem. Soc.* **1989**, *111*, 8757.
91. Beratan, D. N.; Onuchic, J. N.; Betts, J. N.; Bowler, B. E.; Gray, H. B. *J. Am. Chem. Soc.* **1990**, *112*, 7915; Betts, J. N.; Beratan, D. N.; Onuchic, J. N. *J. Am. Chem. Soc.* **1992**, *114*, 4043; Skourtis, S. S.; Regan, J. J.; Onuchic, N. J. *J. Phys. Chem.* **1994**, *98*, 3379.
92. Christensen, H. E. M.; Conrad, L. S.; Mikkelsen, K. V.; Nielsen, M. K.; Ulstrup, J. *Inorg. Chem.* **1990**, *29*, 2808.
93. Lopez-Castillo, J.-M.; Filali-Mouhim, A.; Van Binh-Otten, E. N.; Jay-Gerin, J.-P. *J. Am. Chem. Soc.* **1997**, *119*, 1978; Lopez-Castillo, J.-M.; Filali-Mouhim, A.; Plante, I. L.; Jay-Gerin, J.-P. *J. Phys. Chem.* **1995**, *99*, 6864.

7

Inorganic Photochemistry

Electromagnetic radiation in the form of UV and visible light has long been used as a reactant in inorganic reactions. The energy of light in the 200- to 800-nm region varies between 143 and 36 kcal mol^{-1}, so it is not surprising that chemical bonds can be affected when a system absorbs light in this readily accessible region. Systematic mechanistic studies in this area have benefited greatly from the development of lasers that provided intense monochromatic light sources and from improvements in actinometers to measure the light intensity. Prior to the laser era, it was necessary to use filters to limit the energy of the light used to a moderately narrow region or to just cut off light below a certain wavelength. Pulsed-laser systems also allow much faster monitoring of the early stages of the reaction and the detection of primary photolysis intermediates.

The systems discussed in this chapter have been chosen because of their relationship to substitution reaction systems discussed previously. For a broader assessment of this area, various books[1-5] and review articles[6-12] should be consulted.

7.1 BASIC TERMINOLOGY

Mechanistic photochemistry incorporates features of both electron transfer and substitution reactions, but the field has some of its own terminology,[13] which is summarized as follows:

Quantum Yield
The quantum yield Φ is the number of defined events, in terms of reactant or product, that occur per photon absorbed by the system. An einstein E is defined as a mole of photons, and if n is the moles of reactant consumed or product formed, then $\Phi = n/E$. For simple reactions $\Phi \leq 1$ but can be >1 for chain reactions.

Actinometer
An actinometer is a device used to measure the number of einsteins emitted at a particular wavelength by a particular light source. Photon-

counting devices are now available and secondary chemical actinometers have been developed, such as that based on the Reineckate ion,[14] $Cr(NH_3)_2(NCS)_4^-$, as well as the traditional iron(III)–oxalate and uranyl–oxalate actinometers. An early problem in this field was the lack of an actinometer covering the 450- to 600-nm range and the Reineckate actinometer solved this problem.

Reactant Photolyzed
The number of moles, n, of reactant photolyzed is determined by appropriate analytical techniques. A combination of spectrophotometry and chromatography is commonly used. This is not a trivial problem because photochemical studies typically follow only the first 5 to 15 percent of the reaction, so that the change in reactant concentration is small, as is the amount of product formed.

Internal Filtration
If the reaction products absorb light at the wavelength being used, then the quantum yield will decrease as the reaction proceeds because reactants are not absorbing all the light. To minimize this problem, only the initial stages of the photochemical process are studied.

Secondary Photolysis
Secondary photolysis refers to the photolysis of the initial products to give secondary products. Again, only the initial part of the primary reaction is followed to minimize this problem.

Stern–Volmer Plots
Stern–Volmer plots are used to test the dependence of the quantum yield on the concentration of reactants. The form of the plot depends on the photochemical mechanism proposed.

Fluorescence
Fluorescence refers to the emission of light when an electronic excited state decays to another state of the same spin multiplicity. The emission is usually very fast.

Phosphorescence
Phosphorescence refers to the emission of light when an electronic excited state decays to another state of different spin multiplicity. This process is usually slower than fluorescence and is typically on the millisecond to microsecond time scale for transition-metal complexes.

Sensitizer
A sensitizer is a substance that makes a reaction more sensitive to photolysis. Sensitizers absorb light more strongly than the reactant and then transfer the absorbed energy to the reactant. The sensitizer must have an electronic excited state that is sufficiently long-lived to allow for

the energy transfer to the reactant, and the energy of this state must be similar to that of the acceptor state of the reactant.

Quencher
A quencher is a substance that reduces the quantum yield of a process by accepting energy from the photoexcited state(s) of the reactant. The energy of the excited state of the quencher must be similar to that of the reactant.

Photostationary State
A photostationary state can occur in a kinetically labile system, initially at equilibrium, in which only the forward reaction is promoted by light. Under photochemical conditions, more reactant will be converted to product and a new equilibrium condition will be established in which the forward and reverse rates are the same; this is referred to as a photostationary state. If the light source is removed, the system will return to the thermodynamic equilibrium position.

Intersystem Crossing
Intersystem crossing refers to the process whereby an electronic excited state may be converted to another excited state of similar or lower energy. Back intersystem crossing refers to the reverse process.

7.2 KINETIC FACTORS AFFECTING QUANTUM YIELDS

Most photochemical systems have some common features that affect the lifetime of the photoexcited states and thereby the quantum yields. Figure 7.1 describes a general system with a ground state, G, which absorbs a photon at a rate of dE/dt to produce an excited state, I. Excited vibrational levels have been omitted for simplicity.

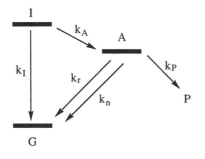

k_I Nonradiative deactivation

k_A Conversion of I to A

k_r Radiative deactivation of A

k_n Nonradiative deactivation of A

k_P Production of product P

$$\Phi = \frac{dP/dt}{dE/dt} = \frac{dP}{dE}$$

Figure 7.1. A general photochemical system with a ground state G, an initially activated state I, and a photoactive state A.

The excited state can undergo intersystem crossing to produce the photoactive state A, which decays to products P or back to the ground state. It is normal to assume a steady state for the excited states, A and I, and their steady-state concentrations are given by

$$[A] = \frac{k_A\,[I]}{(k_P + k_r + k_n)} \tag{7.1}$$

$$[I] = \frac{1}{k_A + k_I}\left(\frac{d\,E}{d\,t}\right) \tag{7.2}$$

The substitution for [I] from (7.2) into (7.1) gives

$$[A] = \frac{k_A}{(k_A + k_I)(k_P + k_r + k_n)}\left(\frac{d\,E}{d\,t}\right) \tag{7.3}$$

The rate of formation of product is given by

$$\frac{d\,[P]}{d\,t} = k_P\,[A] \tag{7.4}$$

and substitution for [A] from (7.3) gives

$$\frac{d\,[P]}{d\,t} = \frac{k_P\,k_A}{(k_A + k_I)(k_P + k_r + k_n)}\left(\frac{d\,E}{d\,t}\right) \tag{7.5}$$

From its definition, and substitution from (7.5), the quantum yield is

$$\Phi = \frac{d\,[P]}{d\,t}\left(\frac{d\,E}{d\,t}\right)^{-1} = \frac{k_P\,k_A}{(k_A + k_I)(k_P + k_r + k_n)} \tag{7.6}$$

The point of this development is to show that several factors in addition to k_P can affect the quantum yield. It is not uncommon in this area to study the effect of changing ligand substituents and solvents on Φ and to use the results to infer the mechanism of the k_P step. But such changes in conditions may affect k_A, k_I, k_r, and/or k_n and thereby make any mechanistic conclusions very tenuous. Similar ambiguities can arise when the solvent is changed.

The system in Figure 7.1 could be expanded to include formation of products from the initially populated state I or from other photoactive states produced from I or A. Furthermore, there could be back intersystem crossing from A to I. It is also possible to have the simpler case in which I is also the photoactive state.

7.3 PHOTOCHEMISTRY OF COBALT(III) COMPLEXES

Cobalt(III) forms a wide range of substitution-inert (low-spin d^6) complexes whose thermal aquation and anation reactions have been thoroughly studied. These provide useful comparisons for photochemical work. In addition, the substitution inertness of the products is an advantage for product studies.

7.3.a Co^{III}(L)$_6$ Complexes

The electronic spectroscopy of Co^{III}(L)$_6$ systems is well understood and described by ligand field theory. The electronic states that involve the $3d$ orbitals are shown in Figure 7.2. The electronic spectra generally show two absorptions in the visible region, $^1A_{1g} \rightarrow {}^1T_{1g}$ (~500 nm) and $^1A_{1g} \rightarrow {}^1T_{2g}$ (~350 nm). The spin-forbidden transitions to the $^3T_{1g}$ and $^5T_{2g}$ states are normally too weak to be observed. In addition, there often is a ligand-to-metal charge-transfer band in the UV region.

The diagram at the right in Figure 7.2 shows the potential energy surfaces for four of these levels in Co(NH$_3$)$_6{}^{3+}$, as suggested by the low-temperature spectroscopic study of Wilson and Solomon.[15] The Co—N bond lengths change by ΔÅ from the ground-state values. The photoaquation of Co(NH$_3$)$_6{}^{3+}$ has a very low quantum yield of 3.1×10^{-4} mol einstein^{-1} for wavelengths in the visible region. The photoactive state is usually assumed to be the $^3T_{1g}$, which is populated by intersystem crossing from the $^1T_{1g}$ and $^1T_{2g}$ states. The low quantum yield could be associated with radiationless deactivation of these states.

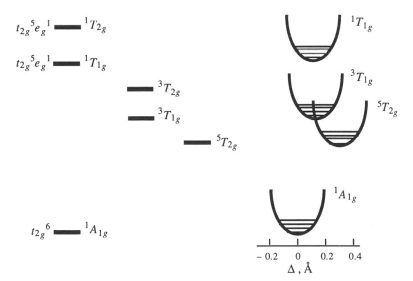

Figure 7.2. The ligand field electronic state energies and potential energy surfaces for octahedral cobalt(III).

Wilson and Solomon suggest that the $^3T_{1g}$ may decay to the $^5T_{2g}$ state, which then decays efficiently to the ground state because of the overlap of the potential energy surfaces.

According to Scandola et al.[16] and Nishazawa and Ford,[17] the photoaquation of $Co(CN)_6^{3-}$ is much more efficient ($\Phi = 0.31$) for wavelengths in the visible region. Wilson and Solomon have suggested that the $^5T_{2g}$ state is at higher energy than the $^3T_{1g}$ state in this system because of the larger Dq of CN^- compared to NH_3; therefore, deactivation through the quintet state is not effective in $Co(CN)_6^{3-}$.

Scandola et al. studied the solvent dependence of the quantum yield in water–glycerol solutions and found that Φ decreases from 0.31 to 0.1 with increasing amounts of glycerol. They interpreted this as a viscosity effect on a cage intermediate, shown in Scheme 7.1.

Scheme 7.1

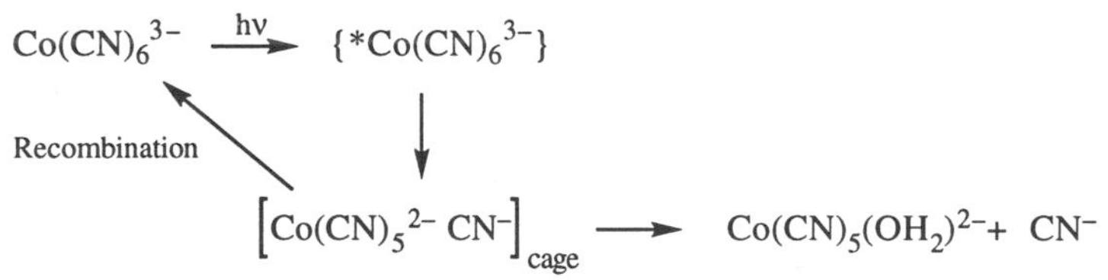

It was suggested that increasing viscosity inhibited CN^- release from the cage and therefore favored recombination to give a lower quantum yield. This type of explanation is not uncommon in photochemical studies, but it is now recognized that solvent changes can also affect the lifetimes of the photoactive states by changing the efficiency of the various decay mechanisms, and meaningful interpretations require information about excited-state lifetimes in the solvent mixtures. In the case of $Co(CN)_6^{3-}$, Wong and Kirk[18] found that $Co(CN)_5(glycerol)^{2-}$ is actually formed.

7.3.b   Co^{III}(L)$_5$X Complexes

The Co^{III}(CN)$_5$X complexes photoaquate cleanly to Co^{III}(CN)$_5$(OH$_2$)$^{2-}$ with quantum yields in the range of 0.05 to 0.3, depending on X. Kirk and Kneeland[19] studied the competition of SCN^- with H_2O to compare the thermal and photochemical reactions of Co^{III}(CN)$_5$X^{3-} complexes. The ratio of the two linkage isomers Co^{III}(CN)$_5$(SCN)$^{3-}$ and Co^{III}(CN)$_5$(NCS)$^{3-}$ is independent of the leaving group X, but the ratio is different for the thermal and photochemical reactions, as shown in Table 7.1. The conclusion is that the transition states for the two processes are different because they give different isomer ratios, but both appear to be dissociative because the ratio is reasonably independent of the leaving group. Kirk and Kneeland suggested that the

Table 7.1. Comparison of the S- and N- Linkage Isomers from Thermal and Photochemical Aquation of $Co(CN)_5X^{3-}$ in the Presence of SCN^-

Ligand X	Thermal S/N Ratio	Photochemical[a] S/N Ratio
Cl^-	4.3	8.5
Br^-	4.5	8.8
I^-	4.7	9.3
N_3^-	4	8.7
CN^-		8.1
OH^-		8.7

[a] Photolysis at 313 nm with 0.01 M $Co(CN)_5X^{3-}$ and 2 M NaSCN.

photoactive state, possibly $^3T_{1g}$, has a more diffuse electronic distribution and is therefore "softer", favoring bonding to the "softer" S end of thiocyanate. More recent work by the same authors[20] has given revised photochemical S/N ratios of ~12 and has shown that this ratio and the total quantum yield vary little with the cation, but the amount of thiocyanate capture decreases in the order $Li^+ > Na^+ > K^+ > NH_4^+$. This is attributed to ion pairing favoring anation of the photochemical intermediate.

The photolysis of $Co^{III}(NH_3)_5X$ complexes can be complicated by a photoredox process in which Co(III) is reduced to Co(II) and a ligand is oxidized. This is the dominant process when absorption is into the ligand-to-metal charge-transfer band (≤ 350 nm), and it has quantum yields in the 0.1 to 0.5 range, depending on X. The general reactions and intermediates suggested to rationalize the behavior of these systems are summarized in Scheme 7.2, where R is H or CH_3.

Scheme 7.2

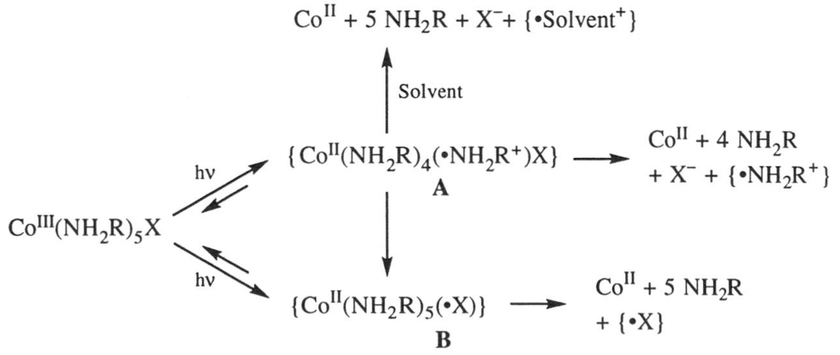

This process was studied extensively by Endicott and co-workers[21] for the $Co^{III}(NH_3)_5X$ complexes, and Weit and Kutal[22] have examined the $Co^{III}(NH_2CH_3)_5X$ systems with X = Cl$^-$ and Br$^-$. The NH_3 and NH_2CH_3 systems are more different than might be expected from the rather simple addition of a methyl group. The electronic spectral properties indicate that species **A** will be favored by higher-energy irradiation compared to species **B**. The NH_3 systems appear to undergo internal conversion to **B** or react with solvent, based on the solvent sensitivity of the reaction, and do not form any amine-type radicals. The NH_2R systems are different in that they show a wavelength dependence of the quantum yield, a different solvent variation in glycerol–water, and an increase in quantum yield in the presence of O_2. Weit and Kutal interpret these differences as due to shielding of the complex from the solvent by the CH_3 group so that reaction from **A** is observed. The O_2 effect is ascribed to scavenging of the •NH_2R radicals by O_2. Following earlier observations of Hennig and co-workers,[23] Wang and Kutal[24] found that these photoredox reactions are sensitized by ion pairing with BPh_4^- in CH_3OH and $CH_3OH:CH_2Cl_2$, with significant quantum yields even at 436 nm.

The photolysis of $Co(NH_3)_4(OCO_2)^+$ at 254 nm yields a mixture of $Co(II)$ and $Co(NH_3)_4(OH_2)(OCO_2H)^{2+}$.[25] A recent flash photolysis study[26] suggests that the initial species formed is $Co^{II}(NH_3)_4(OCO_2•)^+$ and that it undergoes chelate ring opening followed by competitive loss of NH_3 and $OCO_2•$ and back electron transfer to give the various products.

Poznyak et al.[27] have shown that aminocarboxylate complexes of cobalt(III) undergo photolysis by UV light to produce products with stable Co—C bonds. The observations can be understood by the reaction sequence in Scheme 7.3. Studies by Kawaguchi et al.[28] with various isomers have shown that the reaction does not always go with retention, probably due to fluxionality in the Co(II) intermediate(s).

Scheme 7.3

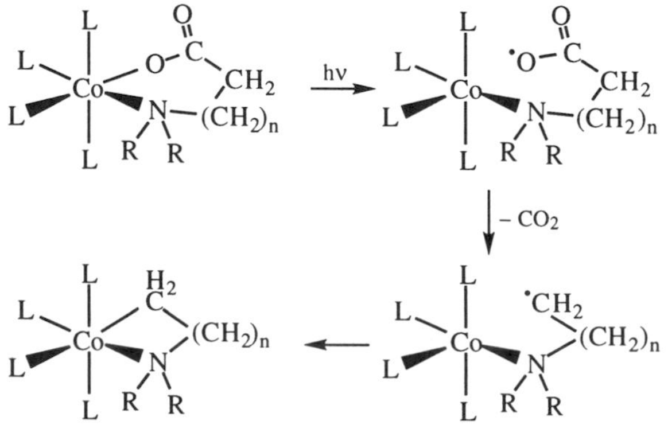

Table 7.2. Quantum Yields (25°C) for the Loss of X and NH_3 for Complexes of the Type $Co^{III}(NH_3)_5X$

X	Φ_X	Φ_{NH_3}
F^- [a]	5.5×10^{-4}	2.0×10^{-3}
Cl^- [a]	1.7×10^{-3}	5.1×10^{-3}
Br^- [b]	2.0×10^{-3}	5.1×10^{-4}

[a] Zribush, R. A.; Poon, C. K.; Bruce, C. M.; Adamson, A. W. *J. Am. Chem. Soc.* **1974**, *96*, 3027, at pH 2, 488 nm; Langford, C. H.; Malkhasian, Y. S. *J. Am. Chem. Soc.* **1987**, *109*, 2682 give λ dependence with X = Cl^-.
[b] Zanella, A. W.; Ford, K. H.; Ford, P. C. *Inorg. Chem.* **1978**, *17*, 1051, at pH 3.4, 546 nm.

If a $Co^{III}(NH_3)_5X$ complex is photolyzed by irradiation into the *d–d* bands (wavelength ≥450 nm), then photoaquation is observed with low quantum yields in the range of 1×10^{-2} to 1×10^{-3}. These reactions show a significant *antithermal pathway* in which NH_3 is released to yield $Co(NH_3)_4(OH_2)X^{2+}$ mainly as the trans isomer. Some examples of quantum yields for the two paths are given in Table 7.2.

7.4 PHOTOCHEMISTRY OF RHODIUM(III) COMPLEXES

Rhodium(III) systems are formally analogous to cobalt(III) in that both are d^6 systems. The larger Dq values for second-row transition metals makes all the Rh(III) complexes more analogous to the cyano complexes of Co(III). Studies with rhodium(III) have the advantage that the rate of decay of the photoexcited states can be measured. The photoactive state appears to be the $^3T_{1g}$ (^3E for Rh(L)$_5$X symmetry), and the rate constants for phosphorescent decay from this state in the solids at 77 K are given in Table 7.3.[29]

Table 7.3. Rate Constants (77 K) for Decay of the Triplet State for Some Rhodium(III) Complexes

Complex	k_n (s^{-1})
$Rh(NH_3)_5(OH_2)^{3+}$	2.9×10^5
$Rh(NH_3)_5Cl^{2+}$	8.2×10^4
$Rh(NH_3)_5(NH_3)^{3+}$	5.1×10^4
$Rh(ND_3)_5(ND_3)^{3+}$	0.8×10^4

The dramatic effect of changing hydrogen for deuterium in the hexaammine complex indicates that the energy of the excited state is dissipated to ligand vibrational modes. This is probably a common feature for many complexes.

Solvent effects have been investigated[30] for the photolysis of $Rh(NH_3)_5Cl^{2+}$. The reaction proceeds with loss of either NH_3 or Cl^- in proportions that vary with the solvent, as shown by the quantum yields given in Table 7.4. The quantum yields do not correlate with the lifetime of the excited state, and the chloride loss process is especially sensitive to the solvent. A complicating feature may be the increasing recombination with Cl^- in the solvent cage when solvation of the chloride ion is less favorable.

The volumes of activation for the photoaquation of $Rh(NH_3)_5Cl^{2+}$ have been determined[31] as -8.6 and 9.3 cm^3 mol^{-1} for the production of $Rh(NH_3)_5(OH_2)^{3+}$ and $trans$-$Rh(NH_3)_4(OH_2)Cl^{2+}$, respectively. The large difference in these values implies that leaving group solvation is an important factor, but the analysis is complicated by the uncertain volumes of the electronic excited state(s). The photoaquation of cis-$Rh(bpy)_2(Cl)_2^+$ to cis-$Rh(bpy)_2(OH_2)Cl^{2+}$ has $\Delta V^* = -9.7$ cm^3 mol^{-1}. These two negative activation volumes for Cl^- release pose a problem for rationalizations[32] in terms of a dissociative mechanism.

Morrison and co-workers[33] have studied the photochemically activated reaction of cis-$Rh(phen)_2(Cl)_2^+$ with deoxyadenosine, deoxyguanosine, and uric acid. Based on the variation of products in the presence or absence of dioxygen, they suggest that the photoactivated cis-$Rh(phen)_2(Cl)_2^+$ is reduced by deoxyguanosine to $Rh^{II}(phen)_2Cl^+$ and by uric acid to $Rh^I(phen)_2^+$. In the absence of dioxygen, the latter reacts with the oxidized form of uric acid to give a Rh—uric acid adduct.

Irradiation of the ligand-to-metal charge-transfer bands of $Rh(NH_3)_5I^{2+}$ gives $trans$-$Rh(NH_3)_4(OH_2)I^{2+}$ with a quantum yield about half of that for irradiation of the lower energy d–d bands.[34] This indicates that deactivation of the charge-transfer states is very

Table 7.4. Photolysis (25°C) of $Rh(NH_3)_5Cl^{2+}$ in Various Solvents

Solvent	Φ_{Cl^-}	Φ_{NH_3}	k_n (s^{-1})
Water	0.18	0.02	7.0×10^7
Formamide	0.057	<0.011	4.5×10^7
Dimethylsulfoxide	<0.006	0.029	2.8×10^7
Methanol	0.008	0.11	$<5 \times 10^7$
Dimethylformamide	0.004	0.070	3.1×10^7

competitive with intersystem crossing to the lower-energy states. Flash photolysis studies in the presence of traces of I^- reveal the presence of transient $\cdot I_2^-$ and the redox mechanism which was proposed is shown in Scheme 7.4.

Scheme 7.4

$$Rh(NH_3)_5I^{2+} \xrightarrow[\text{Charge transfer}]{h\nu} Rh(NH_3)_4^{2+} + NH_3 + \cdot I$$

$$I^- + \cdot I \longrightarrow \cdot I_2^-$$

$$Rh(NH_3)_4^{2+} + \cdot I_2^- + H_2O \longrightarrow \textit{trans-}Rh(NH_3)_4(OH_2)I^{2+} + I^-$$

This behavior is quite different from that of the cobalt complexes in the preceding section, possibly because Rh(II) remains in the low-spin state, while Co(II) goes to the labile high-spin state.

7.5 PHOTOCHEMISTRY OF CHROMIUM(III) COMPLEXES

Chromium(III) systems have long been the prototype for inorganic photochemical studies. They have the same chemical advantages as cobalt(III) systems, but the quantum yields are much larger, the bands in the electronic spectra are well separated, and there is no photoredox process.

Some general observations can be made:
1. Quantum yields are typically 0.1 to 0.8.
2. The quantum yield is independent of the incident light energy.
3. The reactions often appear to be antithermal.

For example, photolysis[35] of $Cr(NH_3)_5(NCS)^{2+}$ gives $\Phi_{NH_3} = 0.46$ and $\Phi_{NCS} = 0.03$ at 373 nm, and $\Phi_{NH_3} = 0.47$ and $\Phi_{NCS} = 0.021$ at 492 nm. Ammonia loss is the dominant photochemical process, but it is not observed thermally.

7.5.a $Cr^{III}(L)_6$ Complexes

The electronic states involving the valence d orbitals in $Cr^{III}(L)_6$ complexes are shown in Figure 7.3. Electronic transitions to the $^4T_{2g}$ and $^4T_{1g}$ states are observed at 550 to 600 nm and 350 to 400 nm, respectively. Weak spin-forbidden transitions to the $^2T_{1g}$ and ^2E_g states can sometimes be observed in the 600-nm region.

There has long been a controversy about the photoactive state(s) in these complexes, primarily concerning the $^4T_{2g}$ and doublet states ($^2T_{1g}$, ^2E_g). Some evidence favoring the quartet state was consistent with the idea that population of the e_g orbital would promote a dissociative mechanism. In the doublet states, the metal ligand bonds would not be

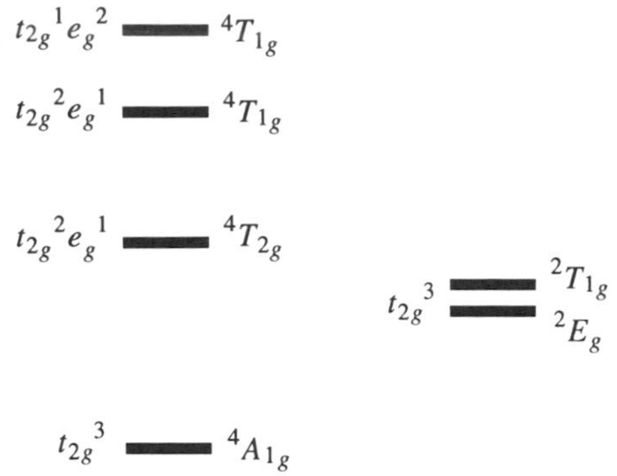

$$t_{2g}^{\ 1}e_g^{\ 2} \quad \rule{2em}{1pt} \quad {}^4T_{1g}$$

$$t_{2g}^{\ 2}e_g^{\ 1} \quad \rule{2em}{1pt} \quad {}^4T_{1g}$$

$$t_{2g}^{\ 2}e_g^{\ 1} \quad \rule{2em}{1pt} \quad {}^4T_{2g}$$

$$t_{2g}^{\ 3} \quad \rule{2em}{1pt} \quad {}^2T_{1g}$$
$$\quad \rule{2em}{1pt} \quad {}^2E_g$$

$$t_{2g}^{\ 3} \quad \rule{2em}{1pt} \quad {}^4A_{1g}$$

Figure 7.3. The d-orbital electronic states for $Cr^{III}(L)_6$ complexes.

weakened much compared to the ground state, but the empty t_{2g} orbital could favor an associative substitution mechanism.

Waltz and Lillie[36] have used pulsed-laser techniques to measure the phosphorescent lifetimes of $Cr(NH_3)_6^{3+}$ and conductivity to monitor the following photoaquation reaction:

$$Cr(NH_3)_6^{3+} \xrightarrow[\text{H}_2\text{O, H}^+]{h\nu} Cr(NH_3)_5(OH_2)^{3+} + NH_4^+ \tag{7.7}$$

They observed an initial fast decrease in conductivity ($t_{1/2} \approx 1 \times 10^{-6}$ s) followed by a slower change ($t_{1/2} \approx 10 \times 10^{-6}$ s). The major process (>67 percent) is the slower one, and it has the same rate as the phosphorescent decay, which is from a doublet state. They concluded that the fast conductivity change is due to photoaquation from the quartet state and the slower and dominant decay is from the doublet state. The general implication of these observations is that both doublet and quartet states can be photoactive, with their relative amounts depending on the efficiency of the intersystem crossing between these states, which in turn will depend on the ligands on the chromium(III).

More recently, Waltz and co-workers[37] have studied the solvent and pressure dependence of the emission lifetimes and quantum yields of $Cr(NH_3)_6^{3+}$. The lifetimes increase with increasing donor number of the solvent, with a ΔV^* of ~ 4 cm^{-3} mol^{-1}. However, the quantum yield for solvolysis is relatively unaffected by the change in solvent and has a ΔV^* of about -4 cm^{-3} mol^{-1}. The authors suggest that the dominant decay route is via nonradiative back intersystem crossing from the ^2E_g state to the $^4T_{2g}$ state.

For $Cr(CN)_6^{3-}$, Wasgestian[38] found that the emission lifetime of the doublet state(s) is very solvent dependent in N,N-dimethylformamide–water mixtures, but the quantum yield for aquation (0.11) is unaffected by the solvent composition. This shows that reactivity is not from the doublet state(s) and suggests that the quartet state is photoactive. From the activation volume[39] for photoaquation of 2.7 cm^{-3} mol^{-1} (15°C, 364.5 nm), an I_d mechanism was suggested. In the related complex, $Cr(CN)_5(NH_3)^{2-}$, quartet state activity also is indicated,[40] since $Co(sep)^{3+}$ quenches the phosphorescence in DMSO but does not change the quantum yields.

7.5.b Cr(III) Amine Complexes

There has been a great deal of work on the photoaquation of complexes of the general type $Cr(A)_5X$, $Cr(A)_4(X)_2$, $Cr(A)_3(X)_3$, and their isomers, where A is an amine ligand. The reactions are often, but not always, antithermal, showing loss of the amine ligand. The first attempt to explain the variation in products and quantum yields was made by Adamson.[41] These led to what are termed *Adamson's rules*, which can be summarized as follows:

1. The axis with the lowest average 10 Dq will be the most labilized.
2. If two different ligands are on that axis, the one with the larger Dq will photoaquate.
3. The quantum yield will be about the same as that for the $Cr(L)_6$ complex, where ligands of type L are on the lower Dq axis.

Some complexes that obey Adamson's rules are given by the first four examples in Table 7.5.

The fluoride complexes do not obey the "rules" because the Dq of F$^-$ is smaller than that of the amine nitrogen. This led to considerable soul searching and the rationalization that what is really required is a measure of the Cr—X bond strength. With F$^-$, it was suggested that the

Table 7.5. Quantum Yields for Photoaquation of NH$_3$ (Φ_N) and X (Φ_X) from Some Chromium(III) Amine Complexes[a]

Complex	Φ_N	Φ_X
$Cr(NH_3)_5Cl^{2+}$	0.36	0.005
trans-$Cr(en)_2(Cl)_2^+$	<0.001	0.32
trans-$Cr(NH_3)_4(Cl)_2^+$	0.003	0.44
trans-$Cr(en)_2(NH_3)Cl^{2+}$	0.34	<0.01
trans-$Cr(en)_2(F)_2^{+\ b}$	0.20	0.02
trans-$Cr(en)_2(NH_3)F^{2+}$	0.27	0.14

[a] Unless otherwise indicated, original references are given by Kirk, A. D. *Coord. Chem. Rev.* **1981**, *39*, 225.

[b] Manfrin, M. F.; Sandrini, A.; Juris, A.; Gandolfi, M. T. *Inorg. Chem.* **1978**, *17*, 90.

Dq was anomalously low because of π-bonding effects, so that Dq does not reflect the σ-bond strength. This resulted in ligand field theory rationalizations such as those proposed by Zinck[42] and Vanquickenborne and Cuelemans.[43] The observation[44] that *trans*-$Cr(H_2N(CH_2)_3NH_2)_2(F)_2^+$ gives mainly F^- loss ($\Phi_F = 0.34$, $\Phi_N = 0.18$) and is much different from the ethylenediamine analogue shows that subtle effects (steric or ring strain) can influence the course of the photochemical process.

The quantum yields for competitive water exchange and NH_3 aquation[45] also are not consistent with the theoretical approaches. Some of these results are given in Table 7.6. The first entry is consistent with Adamson's and other predictions, but the second should predominantly give NH_3 loss, and the third and fourth should mainly give water exchange; the last entry conforms to the "rules".

The antithermal nature of the photochemical reactions has been reassessed.[46] The thermal reaction of $Cr(NH_3)_5(OH_2)^{3+}$ at 50°C is mainly water exchange ($k = 1.37 \times 10^{-3}$ s^{-1}, $\Delta H^* = 99.1$ kJ mol^{-1}), but a minor pathway produces *cis*-$Cr(NH_3)_4(OH_2)_2^{3+}$ ($k = 4.03 \times 10^{-6}$ s^{-1}, $\Delta H^* = 110.5$ kJ mol^{-1}). If one happened to work at a high enough temperature, then the ammonia loss could be the dominant thermal reaction. It was suggested that the photochemical process provides access, in the electronic ground state, to a pentagonal bipyramid transition state that is about 10 kJ mol^{-1} higher in energy than the transition state for thermal water exchange.

Evidence for stereomobility of the photochemical transition state has been obtained[47] with *trans*-$Cr(cyclam)(Cl)_2^+$, a rigid macrocyclic system that has a very low quantum yield for Cl^- aquation of 3×10^{-4}. This has been attributed to the macrocycle preventing rearrangements that normally lead to less energetic configurations in the photoactive states that lead to aquation. The *trans*-$Cr(cyclam)(NH_3)_2^{3+}$ is similar to the dichloro, but *cis*-$Cr(cyclam)(NH_3)_2^{3+}$ has $\Phi_{NH_3} = 0.2$,[48] although the trans isomer has a much longer-lived doublet state ($\tau = 55 \times 10^{-6}$ s, versus 2×10^{-6} s at 293 K). The authors propose that the photoactive

Table 7.6. Quantum Yields for the Water Exchange (Φ_{exch}) and NH_3 Loss (Φ_{NH_3}) from Some Chromium(III) Complexes

Complex	Φ_{exch}	Φ_{NH_3}
$Cr(NH_3)_5(OH_2)^{3+}$	0.078	0.195
cis-$Cr(NH_3)_4(OH_2)_2^{3+}$	0.057	0.058
trans-$Cr(NH_3)_4(OH_2)_2^{3+}$	0.001	0.025
fac-$Cr(NH_3)_3(OH_2)_3^{3+}$	0.040	0.053
trans-$Cr(NH_3)_2(OH_2)_4^{3+}$	0.072	0.004

Table 7.7. Volumes of Activation for Photochemical and Thermal Aquation of Some Chromium(III) Complexes

Complex	ΔV^* (cm^3 mol^{-1})		
	Φ_{NH_3}	Φ_{X^-}	Thermal
$Cr(NH_3)_5Cl^{2+}$	-9.4	-13.0	-10.8
$Cr(NH_3)_5Br^{2+}$	-10.2	-12.2	-10.2
$Cr(NH_3)_5(NCS)^{2+}$	-11.4	-9.8	-8.6
$Cr(NH_3)_6^{3+}$	-12.6		

state is the quartet state, which may be reached by back intersystem crossing. The latter process is less effective with the trans isomer because the energy separation is 22.2 kcal mol^{-1} compared to 19.0 kcal mol^{-1} for the cis isomer.

Volumes of activation have been measured by Angermann et al.[49] for some of the reactions, and the results are given in Table 7.7. The authors suggest that the negative values of ΔV^* are consistent with associative activation for the photochemical process and the thermal reaction. Solvent electrostriction effects should make a larger negative contribution when X$^-$ is the leaving group compared to NH$_3$. Since this difference is not reflected in the ΔV^* values, Angermann et al. proposed that the ammonia release process is "more associative" than the X$^-$ release.

7.6 RU(II)POLYPYRIDINE COMPLEXES

Much of the work in the Ru(II)polypyridine area has concentrated on Ru(bpy)$_3{}^{2+}$ and its derivatives. The electronic spectra are dominated by an intense charge-transfer band in the visible region ($\lambda_{max} \approx 450$ nm, $\varepsilon_{max} \approx 10^4$ M^{-1} cm^{-1}); Ru(bpy)$_3{}^{2+}$ shows modest photosubstitution activity[50] with quantum yields <0.1. Photochemical methods have proved useful in the preparation of several derivatives,[51] including the unusual *trans*-Ru(bpy)$_2$(OH$_2$)$_2{}^{2+}$[52] and Ru(bpy)$_2$(dmbpy)(NCCH$_3$)$^{2+}$, where dmbpy is monodentate 3,3'-dimethyl-2,2'-bipyridine.[53]

There is interest in Ru(bpy)$_3{}^{2+}$ because of its use in photochemical energy transfer, and its photophysics and applications have been the subject of several recent reviews.[54–58] Parris and Brandt[59] first observed that this complex has a relatively long-lived emission ($\tau \approx 600$ ns in water). This, together with its electronic spectral properties, allows the system to trap visible light energy long enough for the photoexcited state to undergo further chemical reactions. The long-lived excited state is a metal-to-ligand charge-transfer triplet, which can act as a reducing agent to give $Ru(bpy)_3^{3+}$ or as an oxidizing agent to give $Ru(bpy)_3^+$. Therefore, the energy-transfer process may be coupled to either an oxidizing or a reducing agent. An oxidative coupling is shown in Scheme 7.5.

Scheme 7.5

Oxidative coupling

$$\{Ru(bpy)_3^{2+}\}^* + A \longrightarrow Ru(bpy)_3^{3+} + A^-$$

$$Ru(bpy)_3^{3+} + \text{Red. agent} \longrightarrow Ru(bpy)_3^{2+}$$

$$A^- + \text{Substrate} \xrightarrow{\text{Catalyst}} \text{Product} + A$$

Much of the interest in these systems has been concerned with the cleavage of water. For example, an oxidative system has been used[60] and studied in detail[61] in which A = $Rh(bpy)_3^{3+}$, triethanolamine is the reducing agent, H_2O is the substrate, Pt^0 is the catalyst, and H_2 is the product. A reductive scheme[62] uses A = ascorbate with $Co(bpy)_n^{2+}$ as the oxidizing agent to form $Co(bpy)_n^+$. It is proposed that the latter reacts with H^+ to form a hydride that decomposes to H_2. A major problem in these applications is the destruction of $Ru(bpy)_3^{2+}$ by photoaquation. Balzani and co-workers[63] have prepared several caged Ru(II) polypyridine complexes that are much more stable and have electronic properties similar to $Ru(bpy)_3^{2+}$.

More detailed studies[64] of the photoexcited triplet state of $Ru(bpy)_3^{2+}$ have shown that the emission is at ~625 nm, and it decays with a rate constant of 3.5 x 10^5 s^{-1} at 25°C in water. The decay occurs partially through a thermally populated d–d state with an activation energy of ~45 kJ mol^{-1}.

7.7 ORGANOMETALLIC PHOTOCHEMISTRY

Photochemical conditions are widely used in synthetic organometallic studies, and there is a steadily increasing number of quantitative studies concerned with quantum yields and elucidation of the photochemical mechanism. Most of the systematic work has been done on metal carbonyls and their derivatives, and the following discussion will be limited to this class of compounds. A substantial barrier to mechanistic studies is the uncertainty about the assignment of the bands in the electronic spectrum of even the binary carbonyls. The electronic spectra are dominated by intense metal-to-ligand charge-transfer, MLCT, bands that are transitions from the back-bonding ($d \rightarrow \pi^*CO$) π orbitals to the corresponding π^* orbitals. In addition, there are weaker, essentially d–d

transitions. The bands tend to overlap, with the latter being at somewhat longer wavelengths. The $d-s$ transitions also may be observed in the near-UV region for neutral and anionic $M(CO)_6$ species. Theoretical analyses have been given by Vanquickenborne and co-workers[65] and Veillard and co-workers,[66] but with different conclusions as to the geometry and spin state of the ligand ejection product.

7.7.a Metal Hexacarbonyls

Nasielski and Colas[67] studied the following reaction in benzene and cyclohexane with M = W:

$$M(CO)_6 + L \xrightarrow{\text{h}\nu} M(CO)_5L + CO \qquad (7.8)$$

They found that the quantum yield ($\Phi \approx 0.7$) is independent of wavelength between 254 and 366 nm and is independent of the concentration of L (= py or CH_3CN). The reaction is sensitized by $Ph_2C=O$ ($\Delta E^* = 289$ kJ mol^{-1} = 414 nm) but not by triphenylene ($\Delta E^* = 280$ kJ mol^{-1} = 427 nm). The reaction is not quenched by bibenzene ($\Delta E^* = 272$ kJ mol^{-1}) or naphthalene ($\Delta E^* - 255$ kJ mol^{-1}) and no phosphorescence was observed. The lack of quenching and phosphorescence indicates that the photoactive state is quite short-lived. The lack of concentration dependence is consistent with a dissociative mode of activation. The photoactive state was assigned to a triplet state involving metal d orbitals (see states for Co(III) in Figure 7.2) that shows a weak absorption at ~353 nm. This state presumably is reached by intersystem crossing when irradiation is at higher energies.

Flash photolysis of $Cr(CO)_6$ in cyclohexane[68] produces the transient solvent complex $Cr(CO)_5 \cdot (C_6H_{12})$, which was identified by IR. The $Cr-(C_6H_{12})$ bond strength has been estimated as ~53 kJ mol^{-1} by photoacoustic calorimetry.[69] This species reacts with CO and H_2O with rate constants of 3.6 x 10^6 and 4.5 x 10^7 M^{-1} s^{-1} at 25°C, respectively, both with $\Delta H^* \approx 22$ kJ mol^{-1}. The reaction of $Cr(CO)_5(OH_2)$ with C_6H_{12} has k = 670 s^{-1} and $\Delta H^* \approx 75$ kJ mol^{-1}. The identification of the H_2O complex is of general interest because water is a potential contaminant in many studies. Similar studies[70] in aliphatic alcohols indicate initial formation of a CH-coordinated species that rearranges to the OH-coordinated isomer with k $\approx$ 2 x 10^{10} s^{-1}.

Other work[71,72] using picosecond laser spectroscopy has shown that these reactions proceed via a solvent intermediate, $M(CO)_5$(solvent), which forms in a few picoseconds after the laser pulse and then decays to products. Lee and Harris[73] have observed formation of the solvated species $Cr(CO)_5$(cyclohexane) with a lifetime τ = 17 ps and the decay of the vibrationally excited $Cr(CO)_5$ with $\tau \approx 21$ ps (apparently at ambient temperature). These observations are at variance with those of Spears and co-workers,[74] who claim that the bare $Cr(CO)_5$ persists on the 100-ps time scale at 22°C. Hopkins and co-workers[75] have used resonance Raman detection to show that the 100-ps process is due to thermal relaxation of the excited vibrational state, probably of $Cr(CO)_5$(cyclohexane).

The kinetics of ligand substitution on $Cr(CO)_5$(heptane) was studied by Yang et al.[76] and the rate constants vary by ~20-fold for various entering groups. As noted above, the ΔH^* for CO and H_2O substitution on $Cr(CO)_5 \bullet (C_6H_{12})$ is smaller than the $Cr—(C_6H_{12})$ bond strength. These observations seem most consistent with associative activation. On the other hand, van Eldik and co-workers[77] have done several studies in mixed alkane/amine solvents and interpret the observed values of ΔV^* in terms of dissociative activation.

Photolysis of $Fe(CO)_5$ initially gives $Fe(CO)_4$, but this species is unusual because it has a triplet ground state.[78] As a result, CO addition is slower than with unsaturated singlet state analogues of other metals because the reaction is spin-forbidden.

7.7.b Substituted Metal Carbonyls: $M(CO)_5L$

The $M(CO)_5L$ systems were first studied by Wrighton et al.[79] and by Dahlgren and Zinck.[80] The quantum yields are in the range of 0.2 to 0.6. The photochemical reaction proceeds by two paths: L elimination (path **A**) and CO elimination (path **B**), as shown in Scheme 7.6.

Scheme 7.6

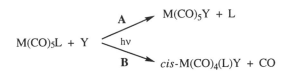

Wrighton et al. studied the system with M = Mo and L = Y = n-PrNH$_2$, so that only path **B** is observed, and found that the quantum yield decreases with increasing irradiation wavelength ($\Phi_{366} = 0.24$; $\Phi_{405} = 0.20$; $\Phi_{436} = 0.057$). When L = n-PrNH$_2$ and Y = 1-pentene, path **A** is dominant and the quantum yield is less sensitive to wavelength, even increasing slightly at longer wavelength ($\Phi_{366} = 0.60$; $\Phi_{405} = 0.65$; $\Phi_{436} = 0.73$). The results were interpreted by a ligand field model with photoactive triplet states, as shown in Figure 7.4. Longer-wavelength radiation populates the $d_{x^2-y^2}$ orbital and labilizes the cis-CO ligands.

Dahlgren and Zinck found, with M = W, that the importance of the two pathways depends on the nature of L. When L is a nitrogen-donor ligand (pyridine, piperidine, ammonia, acetonitrile), pathway **A** is dominant. When L is a phosphine, the two pathways have about equal quantum yields.

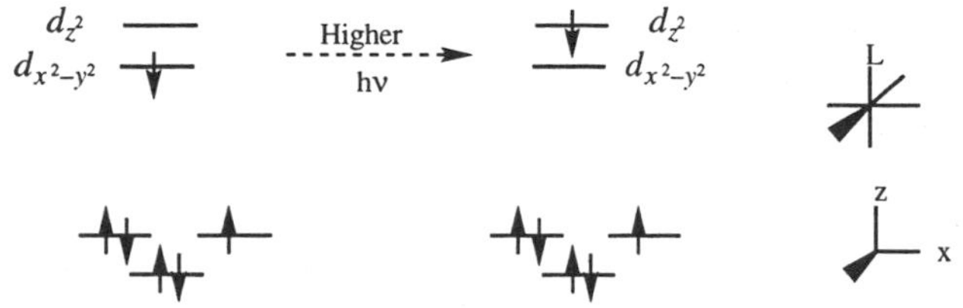

Figure 7.4. Ligand field excited states for $Mo(CO)_4(L)Y$ complexes.

The observations, as summarized in reactions (7.9) and (7.10), were rationalized in terms of the M—C bond strengths as reflected by the CO stretching frequencies. Since the N-donors do not give back π bonding, the M—C bond is stronger and CO loss is not as favored as with the phosphines, which can compete with CO for π electrons on the metal, thereby weakening the M—C bond.

$$W(CO)_5(N\text{—}R) \xrightarrow[Y]{h\nu} \begin{array}{c} W(CO)_5Y \\ + N\text{—}R \\ \Phi \approx 0.5 \end{array} + \begin{array}{c} W(CO)_4(N\text{—}R)Y \\ + CO \\ \Phi < 0.01 \end{array} \quad (7.9)$$

$$W(CO)_5(P\text{—}R) \xrightarrow[Y]{h\nu} \begin{array}{c} W(CO)_5Y \\ + P\text{—}R \\ \Phi \approx 0.3 \end{array} + \begin{array}{c} W(CO)_4(P\text{—}R)Y \\ + CO \\ \Phi \approx 0.3 \end{array} \quad (7.10)$$

Darensbourg and Murphy[81] studied the photosubstitution reactions of $Mo(CO)_5(PPh_3)$ with $Y = PPh_3$ or ^{13}CO (in THF at 313 and 366 nm). The results can be understood in terms of two five-coordinate intermediates, as shown in Scheme 7.7, where $L = PPh_3$. The interequilibration of the intermediates is indicated by the fact that *trans*-$Mo(CO)_4(PPh_3)_2$ photoisomerizes with $\Phi_{366} = 0.3$.

Scheme 7.7

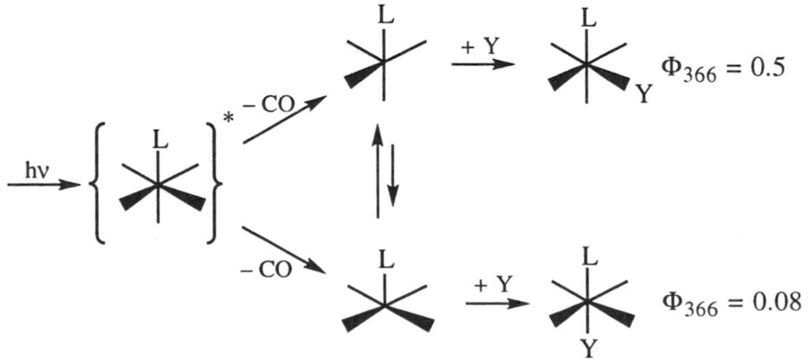

Further work has been done by Lees and co-workers[82] on the Cr and Mo systems to try to elucidate the photoactive state. It was concluded for $Mo(CO)_5L$, where L are substituted pyridines, that the lowest-energy excited state is either metal-to-ligand charge transfer or ligand field in nature, depending on the nature of L. The charge-transfer state is readily deactivated and therefore less photoactive than the ligand field state. When the latter is lower in energy, it can be populated by intersystem crossing and gives larger and less wavelength-dependent quantum yields for replacement of L by PPh_3.

Pulsed-laser studies[83] with $W(CO)_5L$ (L = pyridine or piperidine) have essentially confirmed the earlier observations. This work also shows that a solvent complex is formed within 10 ps of the laser flash, and this "intermediate" is the ultimate source of the substitution products. These authors also report that if 1-hexene is the entering ligand, then the first product has $W(CO)_5$ complexed to the "alkyl portion" of the hexene, and this species rearranges in about 10 ns to the η^2-hexene product. These observations may be relevant to those of Stoutland and Bergman on the addition of ethylene to $Ir(Cp^*)(PMe_3)$ discussed in Section 5.4.

The volumes of activation for the photosubstitution of $W(CO)_5(py)$ and 4-substituted pyridines have been determined[84] using $P(OEt)_3$ as the entering group. This work also involved a study of the effect of pressure on the emission lifetimes, and the effect was found to be small compared to the effect of pressure on the quantum yield. The ΔV^* values are positive and somewhat dependent on the nature of the pyridine (py, 5.7 cm^3 mol^{-1}; 4-cyanopy, 6.3 cm^3 mol^{-1}; 4-acetylpy, 9.9 cm^3 mol^{-1}). The results are consistent with a dissociative mode for photosubstitution.

There have been many studies of $M(CO)_4(\alpha$-diimine) systems, where the α-diimine is bpy or phen or their derivatives.[85] Recently, Vlcek and co-workers[86] have shown that photolysis of $Cr(CO)_4(bpy)$ yields only one product on the 50-ns time scale, fac-$Cr(CO)_3(bpy)(solvent)$ in the solvents dichloromethane and toluene. This conclusion has clarified a previous suggestion[87] that this might be a long-lived triplet MLCT state.

7.7.c Manganese Pentacarbonyls

Faltynek and Wrighton[88] studied the photosubstitution of $Mn(CO)_5^-$ and $Mn(CO)_4(PPh_3)^-$ in the presence of PPh_3 or $P(OMe)_3$ and found quantum yields of ~ 0.3. In $Mn(CO)_4(PPh_3)^-$, only substitution of the PPh_3 is observed.

Oxidative addition to $Mn(CO)_5^-$ is also photoactivated. A competition study of substitution versus oxidative addition gave the results shown in Scheme 7.8. The constancy of the oxidative addition yield suggests that this path may involve an ion pair. Unfortunately, the concentration of PPh_4^+ was not varied to test this hypothesis.

Scheme 7.8

$$PPh_4^+ + Mn(CO)_5^- + PPh_3 \xrightarrow[\text{THF}]{h\nu} Mn(CO)_4(PPh_3)Ph + Mn(CO)_4(PPh_3)^-$$

M	M	M	Percent	Percent
0.01	0.01	0.01	35	38
0.01	0.01	0.06	31	32
0.01	0.01	0.22	28	82

Ford and co-workers[89] studied the flash photolysis of $Mn(CO)_5(CH_3)$ at 308 nm in hydrocarbons and THF. They found that solvent reacts with the transient $\{Mn(CO)_4(CH_3)\}$ to give cis-$Mn(CO)_4(CH_3)(Solvent)$ about 5×10^3 faster than it reacts with CO to give reactant. The solvento complexes with C_6H_{12} and THF have rate constants for their reactions with CO of 2×10^6 and 1.4×10^2 M^{-1} s^{-1}, respectively.

7.7.d $Mn_2(CO)_{10}$ and Related Systems

The $Mn_2(CO)_{10}$ system has been of interest because it is a prototype for metal–metal bonded systems and because of the suggestion that $\bullet Mn(CO)_5$ radicals may be involved in the thermal substitution reactions. The photochemistry of metal–metal bonded systems has been reviewed by Meyer and Caspar.[90] Bonding theory for $Mn_2(CO)_{10}$ suggests that irradiation can populate the σ^* orbital of the Mn—Mn bond and therefore might be expected to cleave that bond and produce the $\bullet Mn(CO)_5$ radical. The $\sigma \rightarrow \sigma^*$ transition is observed at ~340 nm, and several studies[91,92] using 366-nm irradiation have found evidence for radical pathways for decomposition and substitution that are consistent with initial formation of $\bullet Mn(CO)_5$. The final products in the presence of N-donor ligands are $Mn(CO)_5^-$ and $M(N)_6^{2+}$. A radical chain process has been proposed for the decomposition. The radical $\bullet Mn(CO)_5$ can be detected by its halogen atom abstraction reaction with CCl_4 ($k \approx 10^6$ M^{-1} s^{-1}). This type of reaction with chloroalkanes is a test for the presence of radicals.

Studies by Church et al.[93] found that the recombination of $\bullet Mn(CO)_5$ occurs at a near diffusion-controlled rate ($k = 1 \times 10^9$ M^{-1} s^{-1} in heptane). The IR spectra indicate that the radical has a square pyramidal structure. A species with a bridging CO has been identified from a peak at 1760 cm^{-1} and is thought to have the following structure:

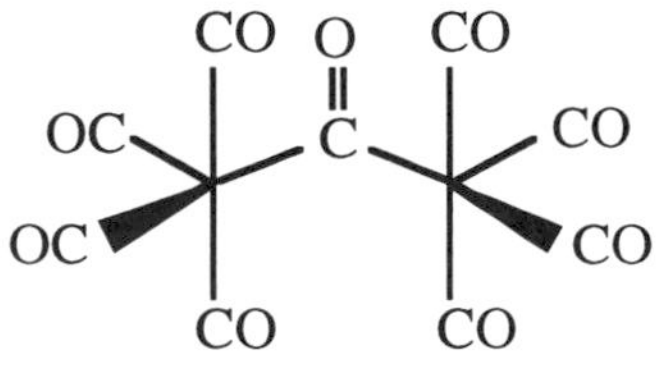

This species reacts with CO in C_7H_{16} to form $Mn_2(CO)_{10}$ with a rate constant of 2.7×10^6 M^{-1} s^{-1}. These observations have been confirmed by Seder et al.[94] in the gas phase, where the analogous rate constant is a surprisingly similar 2.4×10^6 M^{-1} s^{-1} at 323 K and the radical recombination rate constant is 4.5×10^{10} M^{-1} s^{-1} at 323 K. These observations may be relevant to the thermolysis of the Mn—Mn bond.

Wrighton and co-workers[95] first reported that photolysis at 355 nm yields substitution of PPh_3, but later modified this[96] to about 30 percent CO dissociation and subsequent substitution. Recent work[97] on the wavelength dependence of the quantum yields indicates that CO dissociation is more important as the irradiation wavelength decreases from 355 to 255 nm. This trend has been confirmed down to 193 nm in the gas phase by Seder et al. The situation is summarized in Scheme 7.9. The increase in CO dissociation at shorter wavelengths is consistent with irradiation into π^* orbitals of the M—CO bond.

Scheme 7.9

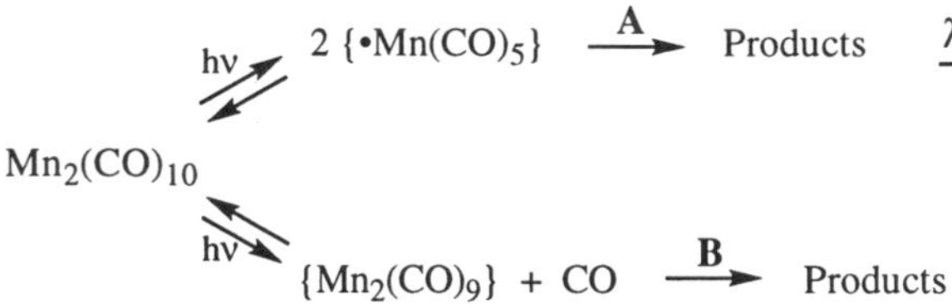

λ (nm)	Φ_A/Φ_B
355	0.74
266	0.21

Turner and co-workers[98] have compared the photochemistry of $Mn_2(CO)_{10}$, $MnRe(CO)_{10}$, and $Re_2(CO)_{10}$ in liquid argon and xenon. The former two are quite similar, but $Re_2(CO)_{10}$ is different in that it does not appear to form a bridged species. It is also known[99] that $Re_2(CO)_{10}$ gives a much greater percentage of CO dissociation at a given wavelength. The low-temperature studies in inert gases used IR spectroscopy to reveal the intermediates shown in Scheme 7.10.

Scheme 7.10

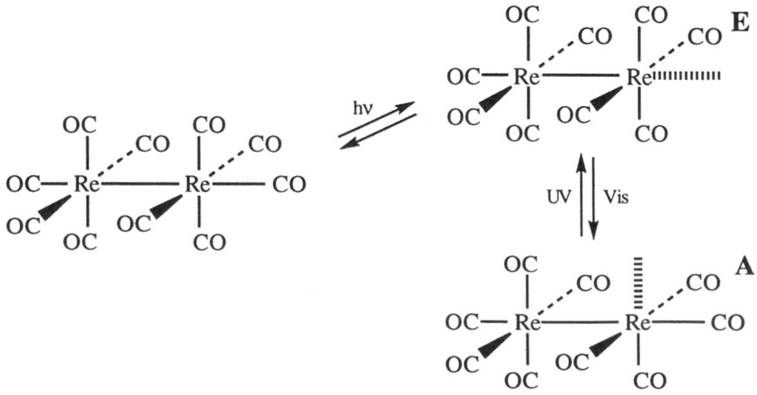

The first intermediate, **E**, with an equatorial vacancy, is quite persistent in liquid xenon and returns to reactant over a period of several hours. The second intermediate, **A**, with an axial vacancy, is formed by irradiation of **E** by visible light and returns to **E** when UV light is used. Both intermediates react with N_2 to give the corresponding dinitrogen complex. More recently, Brown and Zhang[100] found that flash photolysis of $MnRe(CO)_{10}$ in 3-methylpentane glass gives the solvento complex $MnRe(CO)_9$(solvent), which, on warming, recombines with CO to give the reactant. Continuous irradiation of the solvento complex with visible light (>450 nm) produces a new species that was assigned to the semibridging form $MnRe(CO)_8(\mu-\eta^1,\eta^2\text{-CO})$. Under the similar conditions, $Re_2(CO)_{10}$ gave no semibridging species.

7.7.e $M_3(CO)_{12}$ Systems: M = Fe, Ru, Os

These cluster compounds have been the subject of recent studies by Bentsen and Wrighton[101] and by Ford and co-workers,[102] from which some general patterns have emerged. For wavelengths below ~350 nm, the dominant photoactivated process is CO loss and substitution by solvent or added nucleophiles. For wavelengths above 400 nm, the dominant process is photofragmentation, in which the trinuclear species breaks into $M(CO)_5$, $M(CO)_4L$, $M_2(CO)_3L$, and so on. The reactions are unaffected by chlorinated organic solvents, and this is evidence against radical reaction pathways.

Bentsen and Wrighton proposed that short-wavelength irradiation of the Fe and Ru systems liberates an equatorial CO to give an intermediate that either captures solvent or added nucleophiles or rapidly rearranges to a more stable form with an axial vacancy. The latter rearranges to a bridged species identified by IR bands in the region of 1830 cm^{-1}. The Os system is similar, except that the bridged form has not been detected. The proposed structures of the intermediates are shown in Scheme 7.11.

Scheme 7.11

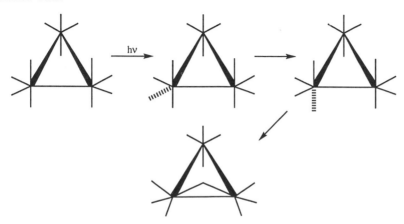

More recently, Harris and co-workers[103] have studied the flash photolysis of $Fe_3(CO)_{12}$ in cyclohexane and concluded that the major product is the bridged species that forms in ≤ 1.5 ps. It returns to the ground state with a time constant of 150 ps. However, about 20 percent of the $Fe_3(CO)_{12}$ does not return, due to fragmentation to mono- and dinuclear species. The fragmentation process has shorter time constants, in the 6- to 14-ps range, as the observation wavelength is changed from 720 to 800 nm. Harris et al. have attributed these observations to fragmentation of vibrationally excited forms of $Fe_3(CO)_{12}$, whose absorbance in the electronic spectrum shifts to longer wavelength with the increasing extent of vibrational excitation.

The photolysis of $Os_3(CO)_{12}$ at ~436 nm in benzene was studied by Poë and Sekhar.[104] The product is $Os_3(CO)_{11}(P(OEt)_3)$ when $P(OEt)_3$ is added, but $Os(CO)_4(\eta^2\text{-octene})$ and $Os_2(CO)_4(\mu\text{-}\eta^1,\eta^1\text{-octene})$ are formed in the presence of 1-octene. The quantum yield increases with increasing concentration of the nucleophile to a limiting value of ~0.04 for both nucleophiles. The observations were interpreted in terms of formation of a reactive intermediate, $Os_3(CO)_{12}$, with a bridging CO. This may undergo substitution or revert to the nonbridged ground-state structure. The mechanism is analogous to that discussed in more detail below for the longer-wavelength photolysis of $Ru_3(CO)_{12}$.

The study of Ford and co-workers on $Ru_3(CO)_{12}$ indicates that the intermediate $Ru_3(CO)_{11}$, produced by 308-nm irradiation in isooctane, reacts with CO with a rate constant of 2×10^9 M^{-1} s^{-1} at ambient temperature. Wth THF added, a THF adduct is formed, and the kinetics of the reformation of $Ru_3(CO)_{12}$ are consistent with Scheme 7.12.

Scheme 7.12

$$Ru_3(CO)_{12} \xrightarrow{h\nu} \{Ru_3(CO)_{11}\} + CO$$

$$\{Ru_3(CO)_{11}\} + THF \underset{k_{-s}}{\overset{k_s}{\rightleftarrows}} Ru_3(CO)_{11}(THF)$$

$$Ru_3(CO)_{11} + CO \xrightarrow{k_{CO}} Ru_3(CO)_{12}$$

From the dependence of the rate on [CO] and [THF] and the known value of k_{CO}, the authors obtain $k_{-s} = 2.1 \times 10^6$ s^{-1} and $k_s = 6 \times 10^9$ M^{-1} s^{-1}. The values of k_{CO} and k_s are near the diffusion-controlled limit and imply that isooctane is weakly solvating the intermediate $Ru_3(CO)_{11}$. The value of k_{CO} is much larger than that with $Mn_2(CO)_9$, in which a bridging CO is proposed to inhibit the reaction.

The long-wavelength process has been investigated by Bentsen and Wrighton[79] and by Ford and co-workers.[105] The quantum yields are low (<0.02) and dependent on the nature of the added nucleophiles and the solvent. Product studies reveal that PPh_3 and $P(OEt)_3$ favor associative substitution to give $M_3(CO)_{11}(PR_3)$. The results have been interpreted as indicating a common intermediate that is an unstable isomer of $M_3(CO)_{12}$, suggested to be the bridged isomer shown in Scheme 7.13.

Scheme 7.13

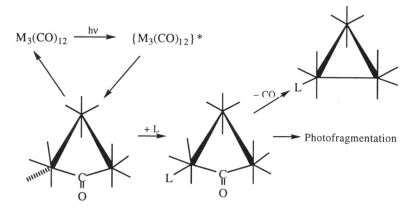

Ford has given a detailed analysis of the kinetics of this type of scheme in terms of the concentration dependence of the quantum yields for the Ru system. In cyclohexane, these results indicate that the ratio of the rate constants for substitution by L versus return to the ground state are 112, 79, and 71 for L = CO, $P(OCH_3)_3$ and PPh_3, respectively. The insensitivity of these ratios to the nature of L could mean that the unsaturated bridged species is unselectively scavenging any available ligand. Lower limits for substitution by L and return to the ground state were estimated as 5×10^6 M^{-1} s^{-1} and 5×10^4 s^{-1}, respectively.

7.8 PHOTOCHEMICAL GENERATION OF REACTION INTERMEDIATES

An increasingly important application of photochemistry is in the production of proposed intermediates in thermal reactions. Flash photolysis and pulsed-laser techniques at low temperatures can produce these species and allow for their spectroscopic characterization and studies of their reactivity. Examples of this have been seen in the areas of C—H activation and the addition of alkenes to metal centers. The only caveat is that the species must be in the electronic and thermal ground state to be relevant to the thermal mechanism.

An example of this type of application is the photochemistry of the derivative of Wilkinson's catalyst, $Rh(PPh_3)_2(CO)Cl$, studied by Wink and Ford.[106] The flash photolysis was studied in benzene for wavelengths larger than 315 nm, and a transient species was observed

L	k_L (M^{-1} s^{-1})
CO	6.9×10^7
C_2H_4	$>4 \times 10^7$
PPh_3	3.0×10^6
H_2	1.0×10^5
D_2	6.8×10^4

$$Rh(PPh_3)_2(CO)Cl$$

$$\downarrow \; -CO \Big| \; h\nu$$

$$\{Rh(PPh_3)_2Cl\}$$

$$\downarrow \; +L \Big| \; k_L$$

$$Rh(PPh_3)_2(L)Cl$$

Figure 7.5. Substitution rate constants for the intermediate photochemically produced from a derivative of Wilkinson's catalyst.

with a much higher absorbance than the reactant in the 390- to 550-nm region. The transient species, which reacts with CO to produce the reactant, has been assigned as $Rh(PPh_3)_2Cl$. The reactivity of this transient with several species was studied, and the results are summarized in Figure 7.5. It was also possible to determine the rate constant for the following dimerization reaction:

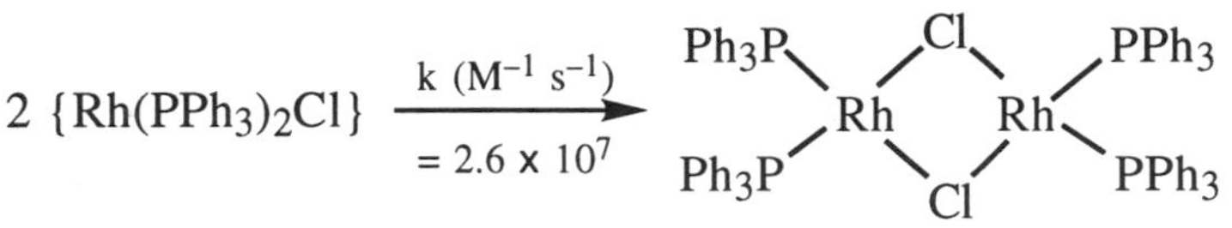

$$2\,\{Rh(PPh_3)_2Cl\} \xrightarrow[\;=\;2.6\;\times\;10^7\;]{k\;(M^{-1}\,s^{-1})} \begin{array}{c} Ph_3P \diagdown \quad {}^{Cl} \diagup \quad PPh_3 \\ \quad Rh \quad\quad Rh \\ Ph_3P \diagup \quad {}_{Cl} \diagdown \quad PPh_3 \end{array} \quad (7.11)$$

These results are in general agreement with the values or limits determined from conventional kinetic studies by Halpern and Wong[107] and Tolman et al.[108] It was found that the dimer and the dihydrogen adduct react with CO rather slowly ($k \approx 2$ s^{-1}) and by a unimolecular process. A remarkable feature of these results is the large rate constant for displacement of ethylene by CO, as shown in the following reaction:

$$\begin{array}{c} Rh(PPh_3)_2(H_2C{=}CH_2)Cl \\ +\;CO \end{array} \xrightarrow[\;=\;1\;\times\;10^8\;]{k\;(M^{-1}\,s^{-1})} \begin{array}{c} Rh(PPh_3)_2(CO)Cl \\ +\;H_2C{=}CH_2 \end{array} \quad (7.12)$$

Now, one has a quite complete picture of the reactivity of the species thought to be important in the operation of this catalytic system.

It should be noted that the closely analogous $Rh(PMe_3)_2(CO)Cl$ undergoes photoactivated oxidative addition of benzene. In the presence of CO, the benzene is carbonylated to yield benzaldehyde. The system is unusual in having two photoactivated steps, and the general reactions are shown in Scheme 7.14, which is based on the observations of Frost[109] and Goldman[110] and co-workers.

Scheme 7.14

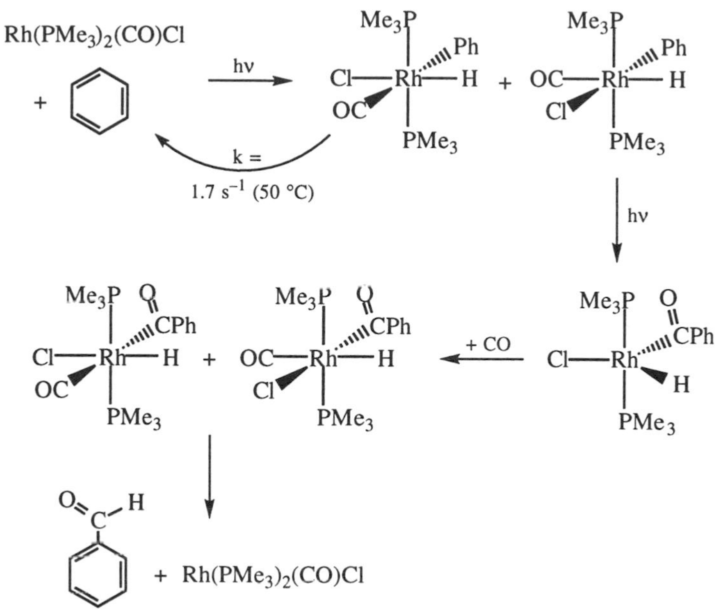

The oxidative addition yields two isomers. The one with the CO and Ph cis to each other is photoactivated to give CO insertion in competition with return to reactants. The acyl intermediate reacts by CO addition and reductive elimination to give benzaldehyde and the catalyst.

The evidence for this scheme comes from several observations. The reaction is promoted rather than inhibited by CO, so that the initial photochemical event is not CO elimination. The photochemical behavior is unusual because the quantum yield changes if the light source is pulsed on and off. Tanaka et al.[111] reported that the rate decreases by a factor of 10 when the pulse rate is decreased from 10 to 1 s^{-1}, and the phenomenon was confirmed by Goldman et al. In Scheme 7.14, this is accounted for by competition between return to reactants and the photoactivated insertion step. Goldman et al. have identified the various octahedral species by low-temperature NMR.

References
1. Balzani, V.; Carassiti, V. *Photochemistry of Coordination Compounds*; Academic Press: New York, 1970.
2. *Concepts of Inorganic Photochemistry*; Adamson, A. W.; Fleischauer, P., Eds.; Wiley: New York, 1975.

3. Geoffrey, G. L.; Wrighton, M. S. *Organometallic Photochemistry*; Academic Press: New York, 1979.
4. *Homogeneous and Heterogeneous Photocatalysis*; Pelizetti, E.; Serpone, N., Eds.; Reidel: Dordrecht, The Netherlands, 1986.
5. *Photochemistry and Photophysics of Coordination Compounds*; Yersin, H.; Vogler, A., Eds.; Springer-Verlag: Berlin, 1987.
6. Adamson, A. W.; Waltz, W. L.; Zinato, E.; Watts, D. W.; Fleischauer, P. D.; Lindholm, R. D. *Chem. Rev.* **1968**, *68*, 541.
7. Wrighton, M. S. *Top. Curr. Chem.* **1976**, *65*, 37.
8. Koerner von Gustof, E. A.; Linders, L. H. G.; Fischler, I.; Perutz, R. N. *Adv. Inorg. Chem. Radiochem.* **1976**, *19*, 65.
9. Crosby, G. A. *Acc. Chem. Res.* **1975**, *8*, 231.
10. Hollebone, B. R.; Langford, C. H.; Serpone, N. *Coord. Chem. Rev.* **1981**, *39*, 181.
11. Kirk, A. D. *Coord. Chem. Rev.* **1981**, *39*, 225.
12. Ford, P. C.; Wink, D.; DiBenedetto, J. *Prog. Inorg. Chem.* **1983**, *30*, 213.
13. Verhoeven, J. W. *Pure Appl. Chem.* **1996**, *68*, 2223.
14. Wegner, E. E.; Adamson, A. W. *J. Am. Chem. Soc.* **1966**, *88*, 394.
15. Wilson, R. B.; Solomon, E. I. *J. Am. Chem. Soc.* **1980**, *102*, 4085.
16. Scandola, F.; Scandola, M. A.; Bartocci, C. *J. Am. Chem. Soc.* **1975**, *97*, 4757.
17. Nishazawa, M.; Ford, P. C. *Inorg. Chem.* **1981**, *20*, 294.
18. Wong, C. F. C.; Kirk, A. D. *Can. J. Chem.* **1976**, *54*, 3794.
19. Kirk, A. D.; Kneeland, D. M. *Inorg. Chem.* **1989**, *28*, 4274.
20. Kirk, A. D.; Kneeland, D. M. *Inorg. Chem.* **1995**, *34*, 1536.
21. Endicott, J. F.; Ferraudi, G. J.; Barber, J. R. *J. Phys. Chem.* **1975**, *79*, 630; *J. Am. Chem. Soc.* **1975**, *97*, 219; *J. Am. Chem. Soc.* **1975**, *97*, 6406.
22. Weit, S. K.; Kutal, C. *Inorg. Chem.* **1990**, *29*, 1455.
23. Hennig, H.; Walther, D.; Thomas, P. *Z. Chem.* **1983**, *23*, 446.
24. Wang, Z.; Kutal, C. *Inorg. Chim. Acta* **1994**, *226*, 285.
25. Cope, V. W.; Hoffman, M. Z. *J. Phys. Chem.* **1978**, *82*, 2665.
26. Ferraudi, G.; Perkovic, M. *Inorg. Chem.* **1993**, *32*, 2587.
27. Poznyak, A. L.; Pavlovski, V. I. *Angew. Chem., Int. Ed. Engl.* **1988**, *27*, 789; Poznyak, A. L. *Koord. Chim.* **1991**, *17*, 1261.
28. Kawaguchi, H.; Yoshida, M.; Yonemura, T.; Ama, T.; Okamoto, K.; Yasui, T. *Bull. Chem. Soc. Jpn.* **1995**, *68*, 874.
29. Ford, P. C. *Inorg. Chem.* **1975**, *14*, 1440.
30. Bergkamp, M. A.; Watts, R. J.; Ford, P. C. *J. Am. Chem. Soc.* **1980**, *102*, 2627.
31. Weber, W.; van Eldik, R.; Kelm, H.; DiBenedetto, J.; Ducommun, Y.; Offen, H.; Ford, P. C. *Inorg. Chem.* **1983**, *22*, 623.
32. Wieland, S.; DiBenedetto, J.; van Eldik, R.; Ford, P. C. *Inorg. Chem.* **1986**, *25*, 4893.
33. Billadeau, M. A.; Wood, K. V.; Morrison, H. *Inorg. Chem.* **1994**, *33*, 5780.
34. Kelly, T. L.; Endicott, J. F. *J. Am. Chem. Soc.* **1972**, *94*, 1797; *J. Phys. Chem.* **1972**, *76*, 1937.

35. Zinato, E.; Lindholm, R. D.; Adamson, A. W. *J. Am. Chem. Soc.* **1969**, *91*, 1076.
36. Waltz, W. L.; Lillie, J.; Lee, S. H. *Inorg. Chem.* **1983**, *23*, 1768.
37. Friesen, D. A.; Lee, S. H.; Nashiem, R. E.; Mezyk, S.; Waltz, W. L. *Inorg. Chem.* **1995**, *34*, 4026.
38. Wasgestian, H. F. *Z. Phys. Chem. N. F.* **1969**, *67*, 39.
39. Angermann, K.; van Eldik, R.; Kelm, H.; Wasgestian, F. *Inorg. Chim. Acta* **1981**, *49*, 247.
40. Riccieri, P.; Zinato, E. *Inorg. Chem.* **1990**, *29*, 5035.
41. Adamson, A. W. *J. Phys. Chem.* **1967**, *71*, 798.
42. Zinck, J. I. *J. Am. Chem. Soc.* **1974**, *96*, 4464; Ibid. **1972**, *94*, 8039.
43. Vanquickenborne, L. G.; Cuelemans, A. *Coord. Chem. Rev.* **1983**, *48*, 157.
44. Kirk, A. D.; Namasivayam, C.; Ward, T. *Inorg. Chem.* **1986**, *25*, 2225.
45. Mønsted, L.; Mønsted, O. *Coord. Chem. Rev.* **1989**, *94*, 109.
46. Mønsted, L.; Mønsted, O. *Acta Chem. Scand. A* **1986**, *40*, 637.
47. Kutal, C.; Adamson, A. W. *Inorg. Chem.* **1973**, *12*, 1990.
48. Kane-Maguire, N. A. P.; Wallace, K. C.; Miller, D. B. *Inorg. Chem.* **1985**, *24*, 597.
49. Angermann, K.; van Eldik, R.; Kelm, H.; Wasgestian, F. *Inorg. Chem.* **1981**, *20*, 955.
50. Durham, B.; Caspar, J. V.; Nagle, J. K.; Meyer, T. J. *J. Am. Chem. Soc.* **1982**, *104*, 4803.
51. Durham, B.; Walsh, J. L.; Carter, C. L.; Meyer, T. J. *Inorg. Chem.* **1980**, *19*, 860.
52. Durham, B.; Wilson, S. R.; Hodgson, D. J.; Meyer, T. J. *J. Am. Chem. Soc.* **1980**, *102*, 600.
53. Tachiyashiki, S.; Ikezawa, H.; Mizumachi, K. *Inorg. Chem.* **1994**, *33*, 623.
54. Kalyanasundaram, K. *Coord. Chem. Rev.* **1982**, *46*, 159.
55. Meyer, T. J. *Pure Appl. Chem.* **1986**, *58*, 1193.
56. Juris, A.; Balzani, V.; Barigelleti, F.; Campagna, S.; Belser, P.; von Zelewsky, A. *Coord. Chem. Rev.* **1988**, *84*, 85.
57. Krausz, E.; Ferguson, J. *Prog. Inorg. Chem.* **1989**, *37*, 293.
58. DeArmond, M. K.; Myrick, M. L. *Acc. Chem. Res.* **1989**, *22*, 364.
59. Parris, T. P.; Brandt, W. W. *J. Am. Chem. Soc.* **1959**, *81*, 5001.
60. Kirch, M.; Lehn, J.-M.; Sauvage, J.-P. *Helv. Chim. Acta* **1979**, *62*, 1345; Lehn, J.-M.; Sauvage, J.-P. *Nouv. J. Chim.* **1981**, *5*, 291.
61. Chan, S.-F.; Chou, M.; Creutz, C.; Matsubara, T.; Sutin, N. *J. Am. Chem. Soc.* **1981**, *103*, 369.
62. Krishnan, C. K.; Sutin, N. *J. Am. Chem. Soc.* **1981**, *103*, 2140.
63. Barigelletti, F.; De Cola, L.; Balzani, V.; Belser, P.; von Zelewsky, A.; Vögtle, F.; Ebmeyer, F.; Grammenudi, S. *J. Am. Chem. Soc.* **1989**, *111*, 4662.
64. Sun, H.; Hoffman, M. Z. *J. Phys. Chem.* **1993**, *97*, 11956.
65. Pierloot, K.; Hoet, P.; Vanquickenborne, L. G. *J. Chem. Soc., Dalton Trans.* **1991**, 2363; Verhulst, J.; Verbeke, P.; Vanquickenborne, L. G. *Inorg. Chem.* **1989**, *28*, 3059.

66. Matsubara, T.; Daniel, C.; Veillard, A. *Organometallics* **1994**, *13*, 4905; Veillard, A. *Chem. Rev.* **1991**, *91*, 743; Daniel, C.; Veillard, A. *Nouv. J. Chimie* **1986**, *10*, 83.
67. Nasielski, J.; Colas, A. *Inorg. Chem.* **1978**, *17*, 237.
68. Church, S. P.; Grevels, F.-H.; Hermann, H.; Schaffner, K. *Inorg. Chem.* **1985**, *24*, 418; Ibid. **1984**, *23*, 3830.
69. Morse, J. M.; Parker, G. H.; Burkey, T. J. *Organometallics* **1989**, *8*, 2471.
70. Xie, X.; Simon, J. D. *J. Am. Chem. Soc.* **1990**, *112*, 1130.
71. Langford, C. H.; Moralejo, C.; Sharma, D. K. *Inorg. Chim. Acta* **1987**, *126*, L11.
72. Simon, J. D.; Xie, X. *J. Phys. Chem.* **1987**, *91*, 5538; Ibid. **1989**, *93*, 291.
73. Lee, M.; Harris, C. B. *J. Am. Chem. Soc.* **1989**, *111*, 8963.
74. Wang, L.; Zhu, X.; Spears, K. G. *J. Phys. Chem.* **1989**, *93*, 2; *J. Am. Chem. Soc.* **1988**, *110*, 8695.
75. Yu, S.-C.; Xu, X.; Lingle, R., Jr.; Hopkins, J. B. *J. Am. Chem. Soc.* **1990**, *112*, 3668.
76. Yang, G. K.; Peters, K. S.; Vaida, V. *Chem. Phys. Lett.* **1986**, *125*, 566.
77. Zhang, S.; Bajal, H. C.; Zang, V.; Dobson, G. R.; van Eldik, R. *Organometallics* **1992**, *11*, 3901, and references therein.
78. Poliakoff, M.; Weitz, E. *Acc. Chem. Res.* **1987**, *20*, 408.
79. Wrighton, M. S.; Morse, D. L.; Gray, H. B.; Ottesen, D. K. *J. Am. Chem. Soc.* **1976**, *98*, 1111.
80. Dahlgren, R. M.; Zinck, J. I. *Inorg. Chem.* **1977**, *16*, 3154.
81. Darensbourg, D. J.; Murphy, M. A. *J. Am. Chem. Soc.* **1978**, *100*, 463.
82. Lees, A. J. *J. Am. Chem. Soc.* **1982**, *104*, 2038; Kolodziej, R. M.; Lees, A. J. *Organometallics* **1986**, *5*, 450.
83. Moralejo, C.; Langford, C. H.; Sharma, D. K. *Inorg. Chem.* **1989**, *28*, 2205.
84. Wieland, S.; van Eldik, R.; Crane, D. R.; Ford, P. C. *Inorg. Chem.* **1989**, *28*, 3663.
85. Vlcek, A., Jr.; Vichova, J.; Hartl, F. *Coord. Chem. Rev.* **1994**, *132*, 167; Stufkens, D. J. *Coord. Chem. Rev.* **1990**, *104*, 39.
86. Virrels, I. G.; George, M. W.; Turner, J. J.; Peters, J.; Vlcek, A., Jr. *Organometallics* **1996**, *15*, 4089.
87. Vichova, J.; Hartl, F.; Vlcek, A., Jr. *J. Am. Chem. Soc.* **1992**, *114*, 10903.
88. Faltynek, R. A.; Wrighton, M. S. *J. Am. Chem. Soc.* **1978**, *100*, 2701.
89. Belt, S. T.; Ryba, D. W.; Ford, P. C. *Inorg. Chem.* **1990**, *29*, 3633.
90. Meyer, T. J.; Caspar, J. V. *Chem. Rev.* **1985**, *85*, 187.
91. Stiegman, A. E.; Tyler, D. R. *Inorg. Chem.* **1984**, *23*, 527.
92. McCullen, S. B.; Brown, T. L. *Inorg. Chem.* **1981**, *20*, 3528.
93. Church, S. P.; Hermann, H.; Grevels, F.-W.; Schaffner, K. J. *J. Chem. Soc., Chem. Commun.* **1984**, 785; Church, S. P.; Poliakoff, M.; Timmey, J. A.; Turner, J. J. *J. Am. Chem. Soc.* **1981**, *103*, 7515.
94. Seder, T. A.; Church, S. P.; Weitz, E. *J. Am. Chem. Soc.* **1986**, *108*, 7518.
95. Wrighton, M. S.; Ginley, D. S. *J. Am. Chem. Soc.* **1975**, *97*, 2065.

96. Hepp, A. F.; Wrighton, M. S. *J. Am. Chem. Soc.* **1983**, *105*, 5934.
97. Kobayashi, T.; Yasufuku, K.; Iwai, J.; Yesaka, H.; Noda, H.; Ohtani, H. *Coord. Chem. Rev.* **1985**, *64*, 1.
98. Firth, S.; Klotzbücher, W. E.; Poliakoff, M.; Turner, J. J. *Inorg. Chem.* **1987**, *26*, 3370.
99. Kobayashi, T.; Ohtani, H.; Noda, H.; Teratani, S.; Yamazaki, H.; Yasufuku, K. *Organometallics* **1986**, *5*, 110.
100. Brown, T. L.; Zhang, S. *Inorg. Chem.* **1995**, *34*, 1164.
101. Bentsen, J. G.; Wrighton, M. S. *J. Am. Chem. Soc.* **1987**, *109*, 4518, 4530.
102. DiBenedetto, J. A.; Ryba, D. W.; Ford, P. C. *Inorg. Chem.* **1989**, *28*, 3503.
103. Tro, N. J.; King, J. C.; Harris, C. B. *Inorg. Chim. Acta* **1995**, *229*, 469.
104. Poë, A. J.; Sekhar, C. V. *J. Am. Chem. Soc.* **1986**, *108*, 3673.
105. Desrosiers, M. F.; Wink, D. A.; Trautman, R.; Friedman, A. E.; Ford, P. C. *J. Am. Chem. Soc.* **1986**, *108*, 1917.
106. Wink, D. A.; Ford, P. C. *J. Am. Chem. Soc.* **1987**, *109*, 436.
107. Halpern, J.; Wong, S. W. *J. Chem. Soc., Chem. Commun.* **1973**, 629.
108. Tolman, C. A.; Meakin, P. Z.; Lindner, D. L.; Jesson, J. P. *J. Am. Chem. Soc.* **1974**, *96*, 2762.
109. Boyd, S. E.; Field, L. D.; Partridge, M. G. *J. Am. Chem. Soc.* **1994**, *116*, 9492.
110. Rosini, G. P.; Boese, W. T.; Goldman, A. S. *J. Am. Chem. Soc.* **1994**, *116*, 9498.
111. Moriyama, H.; Sakakura, T.; Yabe, A.; Tanaka, M. *Mol. Catal.* **1990**, *60*, L9.

8

Bioinorganic Systems

The field of bioinorganic chemistry has grown tremendously in the past 20 years. Much of the work is concerned with establishing the coordination site, ligand geometry, and metal oxidation state in biologically active systems. The field also extends to the preparation and characterization of simpler model complexes that mimic the spectroscopic properties and perhaps some of the reactivity of the biological system. Much of this characterization work must precede meaningful mechanistic studies. Williams[1] has provided an interesting overview of metal ions in biology from an inorganic perspective. There are several review series[2] and specialized journals[3] devoted to the subject.

The field is so large and the systems are so individualistic that it is necessary, for the purposes of a text such as this, to choose a few sample systems as illustrative of the mechanistic achievements and problems.

8.1 VITAMIN B_{12}

In biological systems, vitamin B_{12} consists of an enzyme–coenzyme complex that is called the *holoenzyme*. The enzyme is a peptide, whose composition depends on the biological source and typically has a molar mass in the range of 150,000 to 500,000 daltons. The coenzyme is a cobalt complex that can be isolated by denaturation of the peptide. The chemistry and biochemistry of coenzyme B_{12} are the subject of a compilation[4] and reviews.[5] Recent structural results[6,7] have implications for some of the earlier discussions.

Coenzyme B_{12}, often called adenosylcobalamin, $AdoB_{12}$ or AdoCbl, is a cobalt(III) complex whose structure is shown in Figure 8.1. In the derivative that was first characterized, the 5'-deoxyadenosyl was replaced by cyanide as a result of the isolation procedure. This Co(III) compound is commonly called vitamin B_{12} or cyanocobalamin. If the 5'-deoxyadenosyl ligand is replaced by water, the compound is called aquocobalamin or B_{12a}. Methylcobalamin has a methyl group in place of the 5'-deoxyadenosyl. The cobalt(III) in $AdoB_{12}$ can be reduced to the cobalt(II) derivative, called cob(II)alamin or B_{12r}, and further

252

reduction gives the cobalt(I) species, B_{12s}. These are all in a low-spin state and can be distinguished by their electronic spectra, and B_{12r} is EPR active. In coenzyme B_{12}, the benzimidazole may be displaced by solvent and protonated (pK_a 5–6) to give the base-off form. The benzimidazole function can be removed by hydrolysis of the nucleotide loop at the phosphate–$CH(CH_3)$ linkage to give a series of complexes known as cobinamides. These are often designated as Cbi^+, so that one has $AdoCbi^+$ and $MeCbi^+$ if the organic group is 5'-deoxyadenosyl or methyl, respectively.

The structure[8] and coordination of coenzyme B_{12} are unusual in several aspects. It is an organometallic cobalt(III) complex with a cobalt–carbon bond to the 5'-carbon of 5'-deoxyadenosine. The four

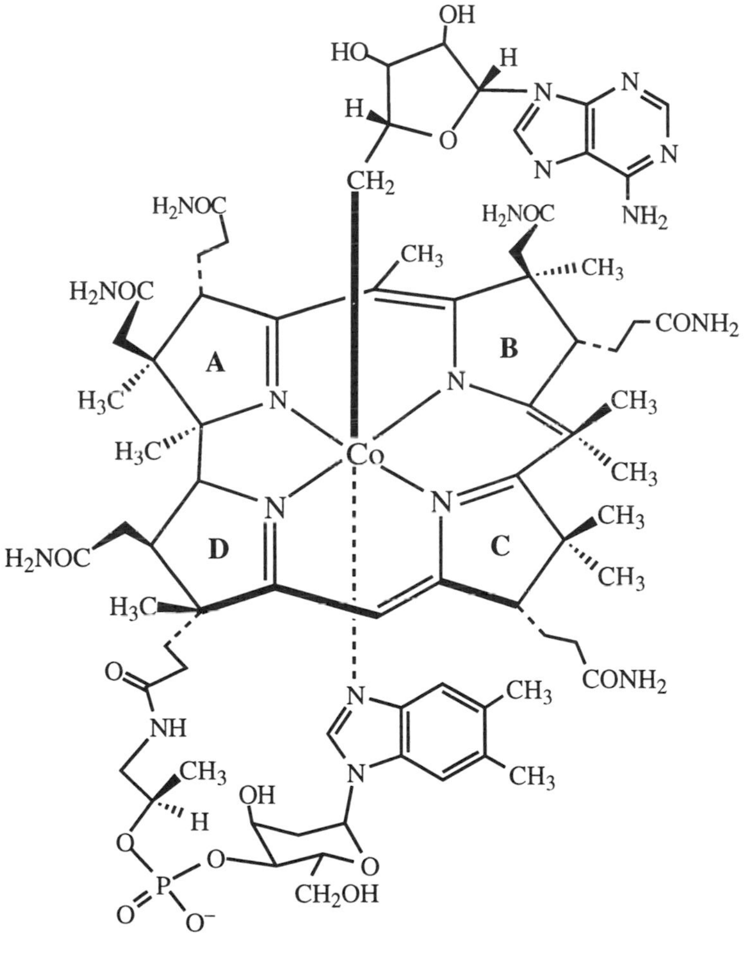

Figure 8.1. Coenzyme B_{12}.

coordination positions in the plane are occupied by nitrogens from a corrin ring. This ring differs from the more common porphyrin in that two of the pyrrole rings are directly bonded, rather than having an intervening -CH group. The pyrrole rings are designated A through D starting from the one in the upper left of Figure 8.1 and proceeding clockwise. The side chains in positions 3 and 4 of each pyrrole are designate a, b, c, etc., starting on ring A and again going clockwise. The stereochemistry of these sidechains makes the two faces of the corrin ring inequivalent. These faces are referred to as α and β, and the 5'-deoxyadenosine is on the β face in Figure 8.1. This face is generally assumed to be the less sterically crowded one because there are only three substituents at positions a, c, and g pointing in this direction, compared to four at positions b, d, e, and f pointing into the α face.

The corrin ring is somewhat flexible and puckered to varying degrees, as discussed in detail by Pett et al.[9] If the 5'-deoxyadenosyl is replaced by a methyl group,[10] then the Co—C bond length shortens from 2.05 to 1.99 Å and the Co—N(benzimidazole) also shortens from 2.24 to 2.19 Å. The crystal structure[11] indicates that the cobalt(II) derivative is five-coordinate, with the cobalt 0.12 Å below the plane of the corrin nitrogens toward the benzimidazole. The bond to the benzimidazole nitrogen is shorter in the cobalt(II) derivative (2.13 Å) than in coenzyme B_{12} (2.24 Å). An EXAFS study[12] has suggested that the benzimidazole-N—Co(II) bond is even shorter (1.99 Å).

There are a number of simple model complexes of cobalt that mimic various aspects of the chemistry of coenzyme B_{12}, such as $Co(DMGH)_2$,[13] $Co(C_2(DO)(DOH))_{pn}$,[14] and $Co(salen)$,[15] which are shown in Figure 8.2.

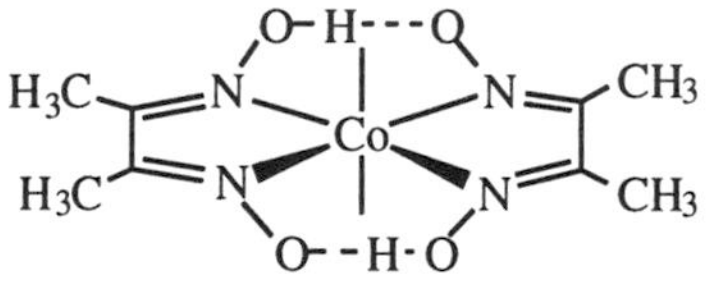

Co(DMGH)₂ 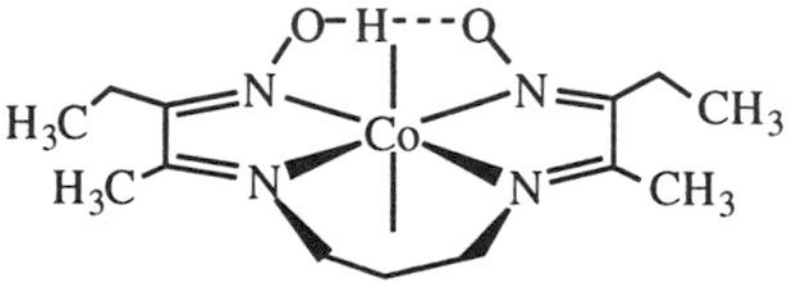   Co(C₂(DO)(DOH))ₚₙ

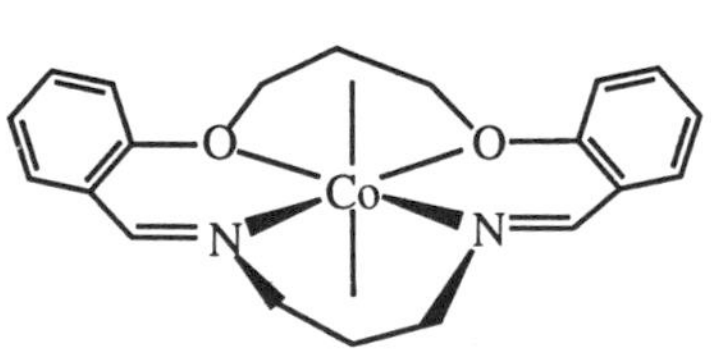

Co(salen)

Figure 8.2. Some coenzyme B_{12} model systems.

These complexes have derivatives with cobalt in the (III), (II), and (I) oxidation states analogous to coenzyme B_{12}. They form a wide range of organometallic cobalt(III) species in which both the organic group and the sixth ligand can be varied. The models have been particularly useful for the characterization of the electronic and EPR spectral features of coenzyme B_{12} and its derivatives.

Some enzymic reactions that occur with coenzyme B_{12} are given in Figure 8.3. The first five examples have as a common feature the interchange of H and another substituent X on an adjacent carbon atom (X = OH, NH_2, $CH(NH_2)CO_2H$, and (CoA)—S represents coenzyme A). Tritium and deuterium labeling experiments have established that a hydrogen of the substrate is transferred to the 5'-carbon of the deoxyadenosine during the reactions of diol dehydrase.[16] This system

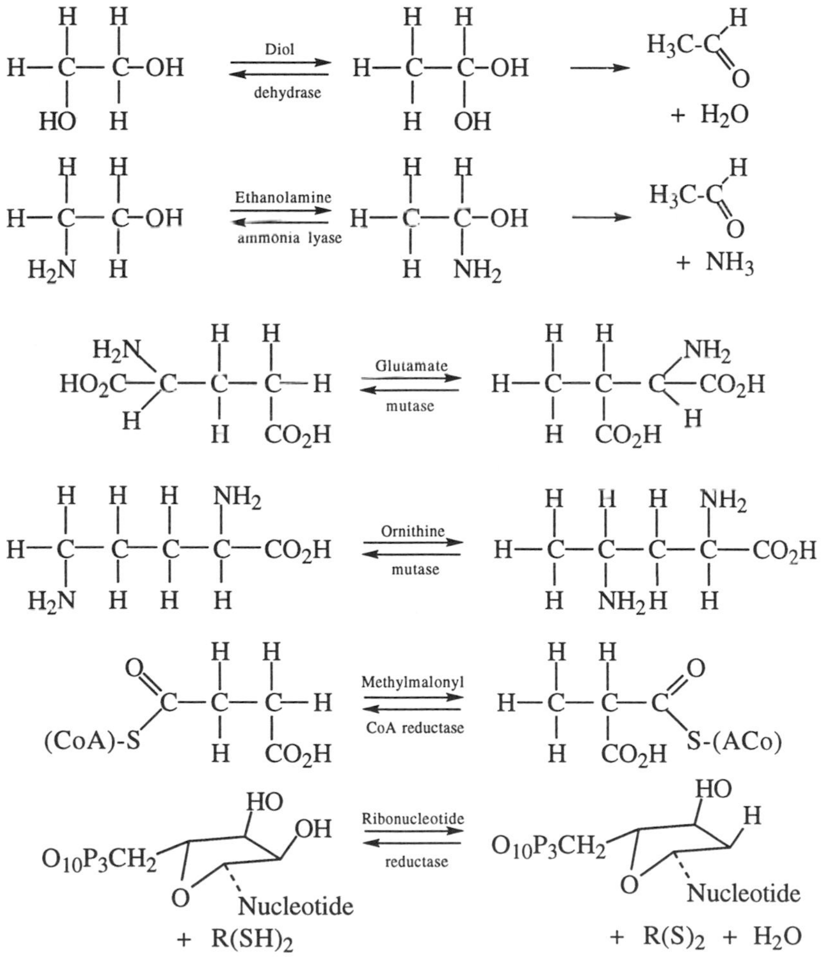

Figure 8.3. Some coenzyme B_{12} reactions.

gives inversion at the C-2 carbon when 1,2-propanediol is the substrate, and glutamate mutase also produces inversion. The reactions of methylmalonyl–CoA mutase and ribonucleotide reductase proceed with retention. The labeling experiments indicate that the initial reaction with a substrate RH may occur by a radical mechanism, as in (8.1), or by an ionic mechanism, as in (8.2).

$$Co^{III}\text{---}CH_2Ado \; \rightleftharpoons \; \begin{array}{c} Co^{II} \\ + \; \cdot CH_2Ado \end{array} \; \overset{RH}{\rightleftharpoons} \; \begin{array}{c} Co^{II} \; + \; R\cdot \\ + \; CH_3Ado \end{array} \qquad (8.1)$$

$$Co^{III}\text{---}CH_2Ado \; \rightleftharpoons \; \begin{array}{c} Co^{I} \\ + \; {}^-CH_2Ado \end{array} \; \overset{RH}{\rightleftharpoons} \; \begin{array}{c} Co^{I} \; + \; R^- \\ + \; CH_3Ado \end{array} \qquad (8.2)$$

The radical mechanism involves homolysis of the Co—C bond and the ionic mechanism involves heterolysis of this bond. Since coenzyme B_{12} has both Co(II) and Co(I) derivatives, either of these mechanisms is chemically reasonable. In addition, the substrate radical or anion might combine with the cobalt to give an organocobalt complex as the product of the second step in both (8.1) and (8.2). Several lines of evidence favor the radical mechanism. The presence of organic radicals and cobalt(II) as B_{12r} has been established by EPR when substrate is added to the ethanol ammonia lyase,[17] diol dehydrase,[18] ribonucleotide reductase,[19] glutamate mutase,[20] and methylmalonyl-CoA mutase.[21] Stubbe and co-workers[22] have given a detailed analysis of the EPR spectra for ribonucleotide reductase and described the effect that can result from long-range coupling of Co(II) and the organic radical.

Methionine synthase uses methylcobalamin as the cofactor that undergoes heterolysis to transfer a methyl group to homocysteine. The structure of this enzyme has been reported by Ludwig et al.[6] with the surprising result that the 5,6-dimethylbenzimidazole loop serves to anchor the cofactor to the protein and the axial position on cobalt is occupied by a histidine nitrogen from the protein. Banerjee and co-workers[21] suggested an analogous structure for methylmalonyl-CoA by using ^{15}N labeled protein and observing the superhyperfine coupling in the EPR of the Co(II) derivative. The crystal structure[23] of the Co(III) form has shown a histidine nitrogen coordination. This structural aspect may turn out to be a general feature of the B_{12} enzymes.[24]

Much of the recent work has concentrated on the radical pathway because it seems to have solid support for some systems, as noted earlier. There are two chemical problems with the radical mechanism: the first is the thermal stability of the Co—C bond compared to the high reactivity of the enzymic reactions; the second is the very limited precedent for the type of radical rearrangements that are required. Various aspects and approaches to these problems are summarized in recent reviews.[25,26]

There have been a number of studies of the Co—C homolysis rate and the bond energy in model compounds and more recently in coenzyme B_{12} itself. Halpern et al.[27] studied the temperature dependence of the equilibrium constant for reaction (8.3) and determined that ΔH° is 22.1 kcal mol^{-1}.

$$\underset{\underset{\displaystyle CH(CH_3)(C_6H_5)}{|}}{(py)(DMGH)_2Co} \quad \xrightarrow{\text{Toluene}} \quad \underset{+ \ CH_2{=}CHC_6H_5 + 1/2\,H_2}{Co^{II}(DMGH)_2(py)} \tag{8.3}$$

They combined this value with the ΔH° of -2.2 kcal mol^{-1} for the reaction

$$C_6H_5CH{=}CH_2 \ + \ 1/2\,H_2 \ \rightleftharpoons \ C_6H_5\overset{\bullet}{C}H(CH_3) \tag{8.4}$$

to obtain a ΔH° of 19.9 kcal mol^{-1} for reaction (8.5), which defines the Co—C bond energy of the reactant in reaction (8.3).

$$\underset{\underset{\displaystyle CH(CH_3)(C_6H_5)}{|}}{(py)(DMGH)_2Co} \quad \longrightarrow \quad \underset{+ \ C_6H_5\overset{\bullet}{C}H(CH_3)}{Co^{II}(DMGH)_2(py)} \tag{8.5}$$

A kinetic method also was devised to determine the bond energy. The rate of reaction (8.3) in the absence of H_2 is first order in the reactant, and it was assumed that the rate-controlling step is reaction (8.5), with a rate constant $k_f = 7.8 \times 10^{-4}$ s^{-1} (25°C), $\Delta H_f^* = 21.2$ kcal mol^{-1}, and $\Delta S_f^* = -1.4$ cal mol^{-1} K^{-1}. Other work[28] indicates that the reverse of reaction (8.5) will be near the diffusion-controlled limit and therefore should have a low value of $\Delta H_r^* \sim 2$ kcal mol^{-1}. These results predict the ΔH° for reaction (8.5) as $21.2 - 2 = 19.2$ kcal mol^{-1}, in agreement with the value from the equilibrium measurements. This provides the basis for the determinations of a number of Co—C bond energies by measurement of the homolysis kinetics, especially if a radical scavenger can be added to drag the reaction to completion.

Variations in the values of ΔH° for reactions such as (8.5) with different axial bases and alkyl groups may be due to other bond energy differences in reactants and products and not just to the Co—C bond energy changes. Measurements[29] of the Co—C stretching frequency (~ 500 cm^{-1}) in L(DMGH)$_2$Co—CH$_3$ systems have been interpreted to indicate that apparent bond energy variations with L are more reflective of changes in the stability of the Co(II) product than of the Co(III) reactant. Spiro and co-workers[30] have come to a similar conclusion for AdoCbl, based on resonance Raman studies of the base-on and base-off forms. The applications and complications of the kinetic method for these bond energy determinations are the subject of several recent discussions.[31,32] In the absence of trapping agents, the radical from thermolysis or photolysis may react with the planar N_4 ligand system to give a stable product, as observed[33] with the benzyl derivative of the $Co(C_2(DO)(DOH))_{pn}$ model (see Figure 8.2), for example. Various equilibrium, competition, and kinetic methods for determining Co—C bond energies have been described by Wayland and co-workers.[34]

The temperature dependence of the thermal decomposition of coenzyme B_{12} was studied first in ethylene glycol by Finke and Hay,[35] then in water by Halpern et al.,[36] and most recently again in water by Hay and Finke.[37] The latter authors have criticized the earlier aqueous study for its failure to separate all the reaction pathways and in particular the heterolysis reaction, which is significant for the base-off form of the coenzyme and is found to contribute between 77 percent (85°C) and 45 percent (110°C) of the product at the pH of 4.3 of the earlier study. From results between 85°C and 110°C at pH 7, after correction for heterolysis and the base-off form, Finke and Hay find the kinetic parameters for the homolysis to be $\Delta H_h^* = 33 \pm 2$ kcal mol^{-1} and $\Delta S_h^* = 11 \pm 3$ cal mol^{-1} K^{-1}, which give $k_h = 1 \times 10^{-9}$ s^{-1} (25°C). The Co—C bond dissociation energy is then estimated to be ~30 kcal mol^{-1} after correction for the activation energy for the reverse reaction. Martin and Finke[38] used a similar method to estimate a Co—C bond dissociation energy of 37 kcal mol^{-1} for methylcobalamin. These studies have been summarized by Waddington and Finke.[39]

The small value of $k_h \approx 10^{-9}$ s^{-1} for the coenzyme is noteworthy in comparison to values of ~2 x 10^2 s^{-1} estimated for the rate-determining step in the diol dehydrase and ethanolamine ammonia lyase systems. This shows one of the problems in using the radical mechanism. This difficulty usually is explained by assuming that the enzyme somehow distorts the coenzyme so that the homolysis is about 15 kcal mol^{-1} more favorable in the holoenzyme. This distortion may destabilize the reactant and/or stabilize the transition state in order to hasten homolysis in the holoenzyme. For some time, it was thought that the axial base might promote these distortions, but Hay and Finke[40] found that replacement of the 5,6-dimethylbenzimidazole in AdoCbl by water only decreased the homolysis rate 10^2 times. Garr et al.[41] recently have given further examples of the effect of the axial base on the thermal homolysis rate and products of AdoCbl$^+$.

Finke and co-workers have summarized the current information on these systems and suggest that the dimethylbenzimidazole loop may serve two functions: to anchor the cofactor to the protein and to inhibit homolysis before the cofactor–protein complex is formed. The activation for homolysis may be provided by distortions of the flexible corrin ring in an upward "butterfly" motion, induced by the protein.

The problem of precedents for the organic rearrangements required by the radical mechanism has been addressed and discussed by Wollowitz and Halpern.[42] They reacted $(n\text{-Bu})_3\text{SnH}$ with various bromide derivatives to generate the radicals and obtained some rearranged product, as shown for typical conditions in Scheme 8.1.

Scheme 8.1

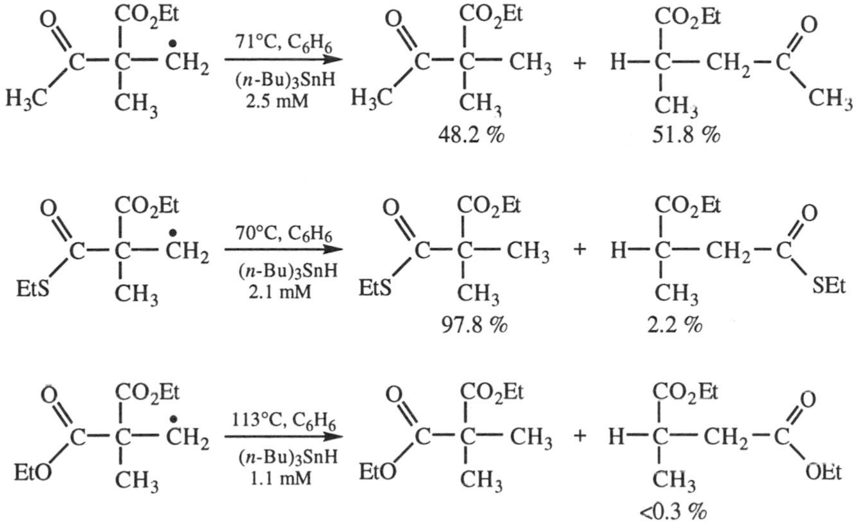

The product ratios were determined as a function of $(n\text{-Bu})_3\text{SnH}$ concentration in order to determine the relative rates of rearrangement and H• abstraction from $(n\text{-Bu})_3\text{SnH}$ by the parent radical. It is clear from the product amounts in Scheme 8.1 that the migratory aptitudes of the groups are $\text{O=CCH}_3 > \text{O=CSEt} > \text{O=COEt}$. The rearrangement is envisaged as proceeding through the following cyclopropyloxy intermediate or transition state:

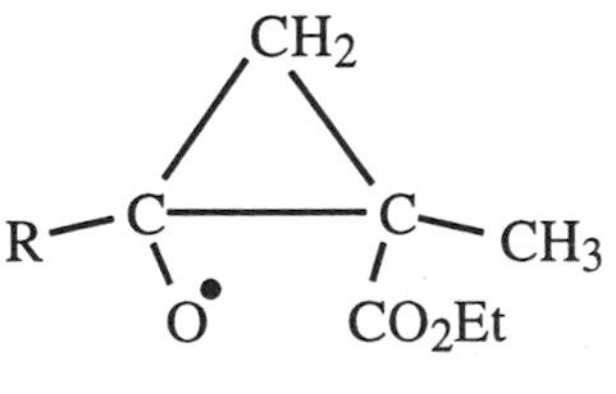

Rearrangement with the O=CSEt group seems relevant to the methylmalonyl–CoA reductase system, but there are still no precedents for the OH and NH_2 migration. In addition, Wollowitz and Halpern observed rearrangements in the corresponding anions that would result from heterolysis [see reaction (8.2)].

Choi and Dowd[43] have reported substantial amounts of rearranged products for reactions (8.6) to (8.8):

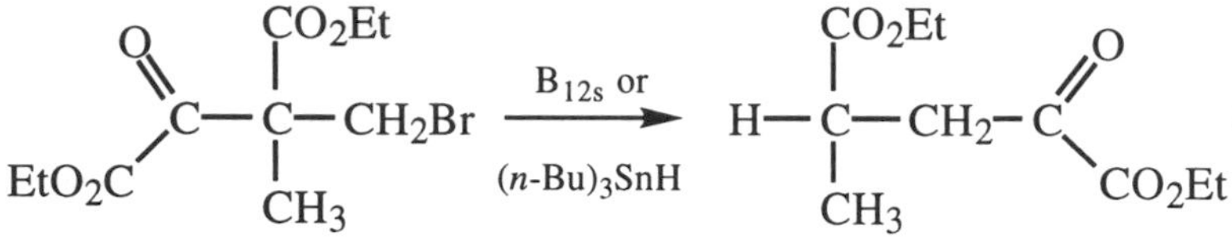

$$(8.6)$$

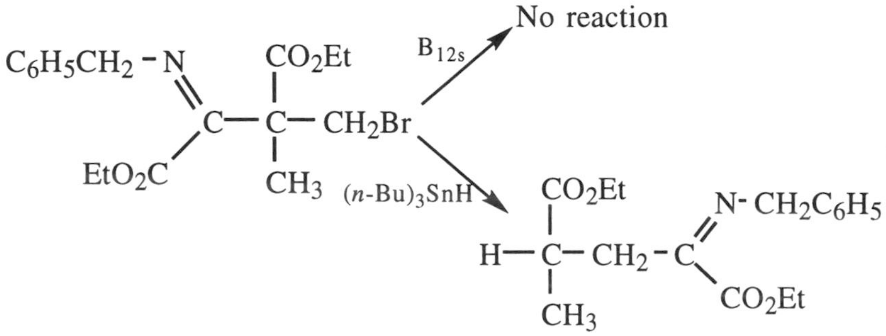

$$(8.7)$$

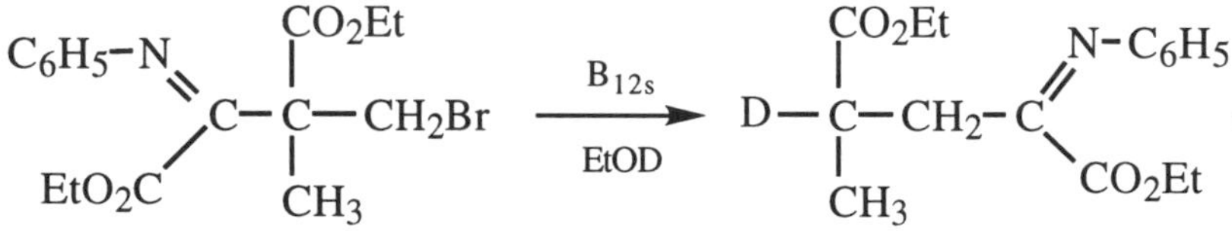

$$(8.8)$$

These reactions are expected to proceed by nucleophilic attack of B_{12s} on the halide to give the organo-B_{12} derivative that would undergo homolysis to products. However, the reactivity patterns with B_{12s} and the formation of the deuterated product in EtOD are more easily rationalized if the reactions are proceeding through an anionic organic intermediate, which might be produced by reduction of the organic radical by excess B_{12s} or other reducing agents used to prepare the B_{12s}. Murakami et al.[44] have noted that anionic rearrangements are more facile than those involving radicals. They observed that addition of cyanide to organo-B_{12} derivatives yields the organic anion under photolysis conditions because the NC—Co^{II} homolysis product reduces the radical to the anion, which then is rearranged.

Dowd et al.[45] found that no rearranged product was obtained for the B_{12} derivative Co—$CH_2CH(CO_2H)CH(CO_2^-)(NH_3^+)$ under thermal or photochemical decomposition. However, Murakami et al.[46] observed substantial rearrangement in the organic products with the same and several related species when photolysis is done while the complex is bound to an anionic surfactant (octopus azaparacyclophane). They proposed that the surfactant is acting in a way analogous to the enzyme in the biological system. These results indicate that studies on the coenzyme alone may not be appropriate models for the enzyme.

To test for radical intermediates, Dowd and co-workers have used a pendant 4-pentenyl side chain trap[47] and a cyclopropyl substituent on the substrate.[48] If radicals are formed, the trap should give cyclization through the 5-hexenyl radical and the cyclopropyl ring should open. Substantial amounts of unrearranged product were found in both cases, and it was concluded that, if any radicals do form, they must have very short lifetimes, in the microsecond range.

Finke et al.[49] found that, with the model $Co(C_2(DO)(DOH))_{pn}$ in Figure 8.2, the $Co—CH_2CH(=O)$ complex was too stable to be an intermediate in the diol to acetaldehyde reaction in their system. The stability of the aldehyde complex is consistent with earlier work of Silverman and Dolphin,[50] who found that the analogous B_{12} complex decomposed by an acid-catalyzed path with $k_{obsd} = 2.1 \times 10^3[H^+]$ s^{-1} at 25°C. Finke and co-workers appear to have found a synthetic scheme that proceeds at least in part through the organometallic diol complex, as shown in Scheme 8.2.

Scheme 8.2

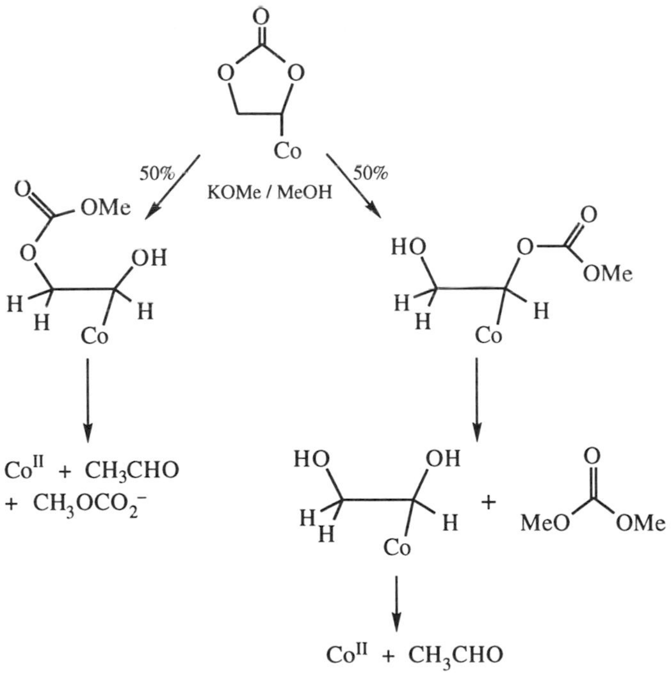

The intermediates in Scheme 8.2 apparently are too unstable to be isolated and have been identified through the nature and stoichiometry of their decomposition products $(Co(II):CH_3CHO:CH_3OCO_2^- = 2:2:1)$

and radical trapping experiments. The decomposition is too fast to be consistent with the formation of the relatively stable aldehyde complex as an intermediate. The observations have been interpreted by assuming that the radical formed after homolysis escapes from the solvent cage and undergoes a normal radical decomposition, as shown in reaction (8.9), without the intervention of a cobalt complex.

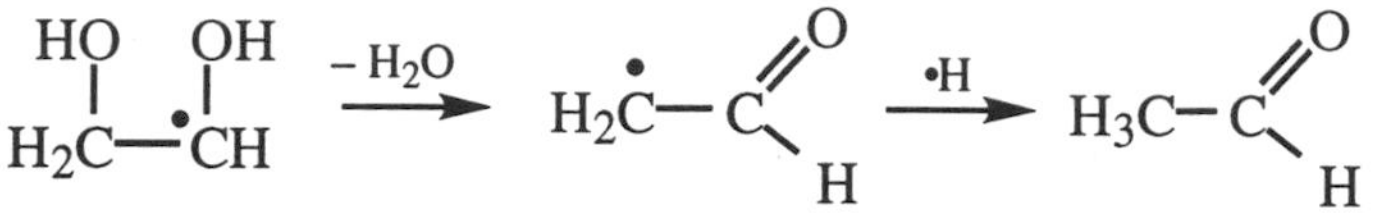

$$
\begin{array}{c}
\text{HO} \quad \text{OH} \\
| \quad \bullet | \\
\text{H}_2\text{C}\!-\!\text{CH}
\end{array}
\xrightarrow{-\text{H}_2\text{O}}
\;
\text{H}_2\overset{\bullet}{\text{C}}\!-\!\text{C}\!\!\underset{\text{H}}{\overset{\text{O}}{<}}
\;\xrightarrow{\;\bullet\text{H}\;}\;
\text{H}_3\text{C}\!-\!\text{C}\!\!\underset{\text{H}}{\overset{\text{O}}{<}}
\qquad (8.9)
$$

However, this mode of decomposition does not involve the 1,1-diol intermediate shown to be required by the elegant stereochemical and labeling studies of Arigoni and co-workers[51] on diol dehydrase. Finke et al.[26] proposed a bound radical mechanism for the enzyme in which formation of the 1,1-diol assists in binding the radical to the enzyme by hydrogen bonding. The bound radical mechanism involves rehydration of the radical intermediate in reaction (8.9) to give the 1,1-diol.

For systems such as glutamate mutase, 2-methyleneglutarate mutase, and methylmalonyl-CoA, Golding and co-workers[52] have proposed the rather different radical rearrangment mechanism illustrated in Scheme 8.3 for 2-methyleneglutarate mutase.

Scheme 8.3

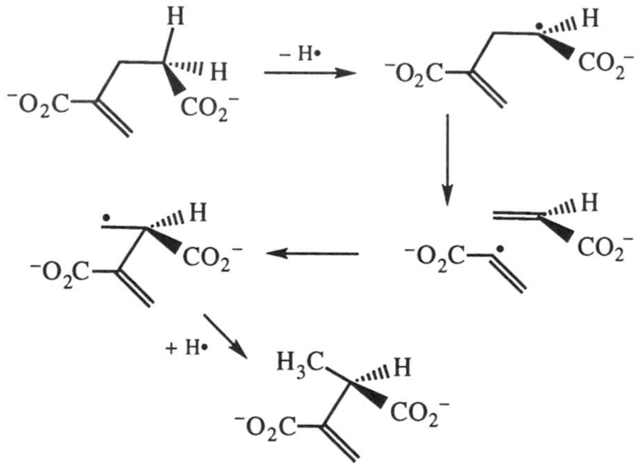

The kinetics of radical interactions with coenzyme B_{12} have been the subject of several flash photolysis studies.[53] The general picture is that photolysis yields a geminate radical pair, which may undergo recombination or escape the solvent cage and then recombine or undergo other reactions. In the recent study of Grissom and co-workers,[54] it was concluded that the geminate pair of cobalt(II) and the adenosyl radical recombine with a rate constant of 1×10^9 s^{-1} in water, but the value for the methyl radical is much smaller, with an upper limit of ~2 x 10^8 s^{-1}. The difference was attributed to the planar geometry of the methyl radical. The natural system may counteract the rapid recombination of the adenosyl radical if the protein serves to twist the adenosyl radical away from the cobalt(II).

8.2 A ZINC(II) ENZYME: CARBONIC ANHYDRASE

Zinc is the second most abundant metal, after iron, in humans. There are a number of enzyme systems in which zinc(II) is bound to a peptide (apoenzyme) and is at the active site in the enzyme. Examples of these enzymes are carbonic anhydrase, carboxypeptidase A and B, alkaline phosphatase, alcohol dehydrogenase, and RNA polymerase. In most cases, these systems bring about the making or breaking of covalent bonds in an organic substrate, but they may also use an oxidizing or reducing coenzyme to produce a redox change of the substrate. The area has been the subject of a recent review[55] and a compilation.[56] Studies on carbonic anhydrase are representative of the methods used to elucidate a considerable range of enzyme mechanisms.

Most of the kinetic work in this area has been done by biochemists, and they have evolved their own terminology for describing the results. The following section provides an outline of some of these terms and the general methodology.

8.2.a Introduction to Terms and Methods of Enzyme Kinetics

The simplest picture of an enzymic reaction is given by

$$E + S \underset{k_2}{\overset{k_1}{\rightleftharpoons}} ES \underset{k_4}{\overset{k_3}{\rightleftharpoons}} E + P \qquad (8.10)$$

where E is the enzyme, S is the substrate, ES is the enzyme–substrate complex, and P is the product. These systems are often studied under steady-state conditions in which the enzyme (usually $<10^{-6}$ M) and substrate (usually $>10^{-4}$ M) are mixed in a buffered solution and the initial rate of loss of S or formation of P is determined. The formation of ES is assumed to be a rapid pre-equilibrium and only the initial rate is studied, so that P does not accumulate and the k_4 step is ignored. This system was analyzed in Section 2.2 and the rate is given by the *Michaelis–Menten* equation:

$$\frac{d[P]}{dt} = v_i = \frac{k_3 [E]_T [S]}{[S] + \dfrac{k_2}{k_1}} = \frac{k_3 [E]_T [S]}{[S] + K_m} \qquad (8.11)$$

where v_i is the initial rate, $[E]_T = [E] + [ES]$ is the total concentration of enzyme, and K_m is the *Michaelis constant*. Since $K_m = [E][S]/[ES]$, it would be called a dissociation constant by inorganic chemists, and smaller values of K_m indicate stronger enzyme to substrate binding. This rate law gives what inorganic kineticists call saturation behavior and the rate reaches a maximum when $[S] \gg K_m$, so that $V_{max} = k_3[E]_T$. The ratio $V_{max}/[E]_T = k_3$ is called the *turnover number*. Since the total enzyme concentration may not be precisely known, it is useful to substitute V_{max} into Eq. (8.11) to obtain the more conventional form of the Michaelis–Menten equation. This equation can be rearranged by taking the reciprocal of both sides to obtain

$$\frac{1}{v_i} = \frac{K_m}{V_{max}\,[S]} + \frac{1}{V_{max}} \qquad (8.12)$$

which predicts that a plot of v_i^{-1} versus $[S]^{-1}$, called a *Lineweaver–Burke plot*, will be linear with an intercept of $-K_m^{-1}$ when $v_i^{-1} = 0$.

If a steady-state approximation for ES is used (see Section 2.1) instead of the rapid-equilibrium assumption, then one obtains Eq. (8.13), known as the *Briggs–Haldane equation*:

$$\frac{d\,[P]}{d\,t} = v_i = \frac{k_3\,[E]_T\,[S]}{[S] + \dfrac{k_2 + k_3}{k_1}} \qquad (8.13)$$

This equation has the same mathematical form as Eq. (8.11), except that $K_m = (k_2 + k_3)/k_1$. The K_m will be the same as defined previously only if $k_2 \gg k_3$; otherwise, K_m is always larger than k_2/k_1.

The variation of the rate of the enzyme-catalyzed reaction with pH can provide information about the ionization of functional groups important for the reaction. For example, the system might be described by Scheme 8.4, which has been simplified from more general possibilities in order to give a more tractable result.

Scheme 8.4

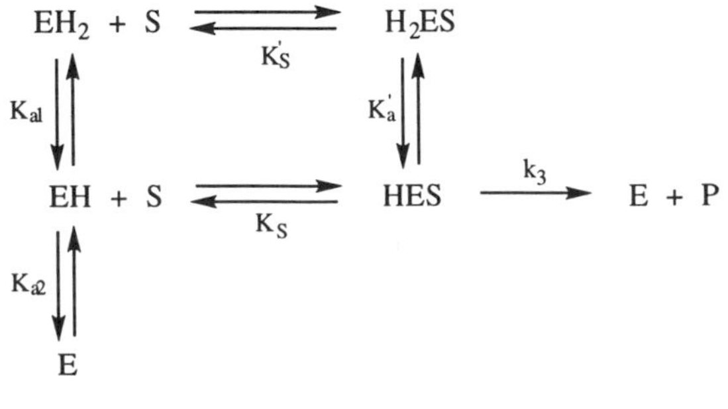

If it is assumed that the substrate binding and proton equilibria are rapidly maintained, then the initial rate can be put in the standard Michaelis–Menten form:

$$\frac{v_i}{[E]_T} = \frac{k_{cat}[S]}{[S] + K_m} \tag{8.14}$$

where

$$k_{cat} = \frac{k_3}{\left(1 + \frac{[H^+]}{K_a'}\right)} \quad \text{and} \quad K_m = K_s \frac{\left(1 + \frac{[H^+]}{K_{a1}} + \frac{K_{a2}}{[H^+]}\right)}{\left(1 + \frac{[H^+]}{K_a'}\right)}$$

If the conditions are such that $[S] \gg K_m$ (called swamping conditions), then the variation of k_{cat} with pH can be used to determine K_a'. If $K_m \gg [S]$, then k_{cat}/K_m is independent of K_a' and the values of K_{a1} and K_{a2} can be determined.

Many enzyme systems are affected by inhibitors. The structure of the inhibitors may be helpful in defining the size and binding requirements of the active site, but they may bind elsewhere and cause distortions of the peptide that affect the active site. The kinetic effect of inhibitors can be described by three possibilities, and mixtures of these possibilities may be observed.

A system with a *competitive inhibitor*, I, is described by the sequence

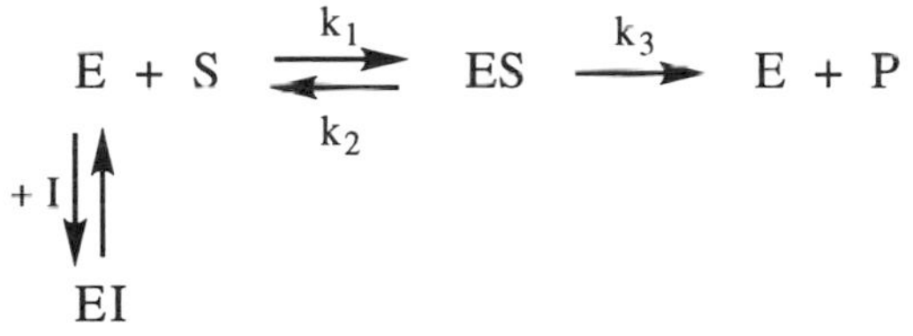

$$E + S \underset{k_2}{\overset{k_1}{\rightleftharpoons}} ES \xrightarrow{k_3} E + P$$
$$+ I \updownarrow$$
$$EI \tag{8.15}$$

The inhibitor competes with substrate for the enzyme. If a steady state is assumed for ES and the inhibitor reaction with E is a rapidly maintained equilibrium with $K_I = [E][I]/[EI]$, then

$$v_i = \frac{k_3 [E]_T [S]}{[S] + \frac{(k_2 + k_3)}{k_1}\left(\frac{K_I + [I]}{K_I}\right)} = \frac{k_3 [E]_T [S]}{[S] + K_m \left(\frac{K_I + [I]}{K_I}\right)} \tag{8.16}$$

An *uncompetitve inhibitor* is one that prevents the enzyme–substrate complex from following the main reaction pathway, as described by the sequence

$$E + S \underset{k_2}{\overset{k_1}{\rightleftharpoons}} ES \overset{k_3}{\longrightarrow} E + P$$

$$+ I \updownarrow$$

$$ESI \tag{8.17}$$

If the same kinetic assumptions as before are used, then

$$v_i = \frac{k_3 [E]_T [S] \left(\dfrac{K_I}{K_I + [I]}\right)}{[S] + \dfrac{(k_2 + k_3)}{k_1}\left(\dfrac{K_I}{K_I + [I]}\right)} = \frac{k_3 [E]_T [S] \left(\dfrac{K_I}{K_I + [I]}\right)}{[S] + K_m \left(\dfrac{K_I}{K_I + [I]}\right)} \tag{8.18}$$

A *noncompetitive inhibitor* binds to both E and ES. If it is assumed that K_I is the same for both species, then

$$v_i = \frac{k_3 [E]_T [S] \left(\dfrac{K_I}{K_I + [I]}\right)}{[S] + \dfrac{(k_2 + k_3)}{k_1}} \tag{8.19}$$

The various types of inhibition can be distinguished by studies of v_i for different S and I concentrations. Then, Lineweaver–Burke plots of v_i^{-1} versus $[S]^{-1}$ at constant [I] or *Dixon plots* of v_i^{-1} versus [I] at constant [S] can be used to determine K_I and the type of inhibition.

8.2.b Mechanism of Carbonic Anhydrase Action

The carbonic anhydrase enzymes catalyze the hydration of CO_2, shown by

$$CO_2 + H_2O \rightleftharpoons HCO_3^- + H^+ \tag{8.20}$$

They also catalyze the hydrolysis of organic esters and the hydration of aldehydes, but with such low efficiency that this does not seem to be a significant biological function. Various aspects have been the subject of several reviews.[57]

In mammals, there are three distinct isozymes of carbonic anhydrase, designated as CA I, CA II, and CA III. Their maximum turnover numbers at 25°C are 2×10^5, 1×10^6, and 3×10^3 s^{-1}, respectively, and CA II is one of the most efficient of all known enzymes. Although they have different reactivities and amino acid sequences, the structures of

CA I and II are known to be homologous. The active site is typically a conical cleft, about 15 Å wide at the base and 12 Å deep, with a zinc(II) atom at the apex coordinated to three imidazole nitrogens from histidines and probably a water molecule, giving a distorted tetrahedral coordination about the zinc. The zinc can be removed by chelating agents, and the resulting apoenzyme reacts with various metals, such as cobalt(II), copper(II), cadmium(II), and nickel(II). The cobalt(II) derivative has about 50 percent of the catalytic activity of the zinc enzyme, and various spectroscopic measurements on the cobalt analogue have been helpful in elucidating the coordination chemistry.

The pH profile for the CO_2 hydration is consistent with the reactions in Scheme 8.4, with $pK_{a1} \approx 7$ and $pK_{a2} \approx 9$ (values vary somewhat with the isozyme and ionic medium). The assignment of these ionizations to particular subunits has been an area of controversy. Simpler systems would lead to the expectation that pK_{a1} is due to an imidazole nitrogen of histidine and that pK_{a2} is due to ionization of $(Im)_3Zn-OH_2^{2+}$. However, the opposite assignment is now generally favored. The problem with this is that many model zinc(II) complexes have values of $pK_a > 8$, but these models are five- or six-coordinate species,[58,59] and it can be argued that four-coordinate zinc(II) in the enzyme will be more acidic than the models. Brown et al.[60] found that a trisimidazole chelate complex of cobalt(II) has a pK_a of 7.6, but the ionization is accompanied by a coordination change from six to either five or four, based on changes in the electronic spectrum. The closest model so far appears to be the trinitrogen chelate, [12]aneN$_3$, of zinc(II), which Kimura et al.[61] have reported to have a pK_a of 7.3 and which mimics the aldehyde hydration and ester hydrolysis activity of carbonic anhydrase. The biomimetic chemistry of this system has been reviewed recently.[62]

Spectrophotometric titration[63] of the cobalt(II) derivative of human CA I gives two pK_a values of 6.9 and 8.7, and both ionizations have a similar effect on the electronic spectrum. The cadmium(II) derivatives have pK_{a1} in the range of 9.5 to 10.[64] The variation of pK_a with the metal shows that the metal ion is involved in the first ionization to give the active species, which is widely accepted to be $(Im)_3Zn-OH$.

The catalytic cycle is shown in Scheme 8.5, with the first step being the formation of $(Im)_3Zn-OH$. The next step is the hydration of the CO_2, which is taken to be attack of the zinc-bound OH^- on CO_2, for which there are ample precedents[65] in the chemistry of carbonate complexes of inert metal ions. This may proceed through the intermediate at the bottom of Scheme 8.5 since zinc(II) can be five-coordinate. This rearranges to give zinc-bound bicarbonate, for which two possibilities have been suggested,[66,67] as shown in Scheme 8.5. Both proposals suggest stabilization through hydrogen bonding to threonine-199, but there is little consensus on which one is correct.

Scheme 8.5

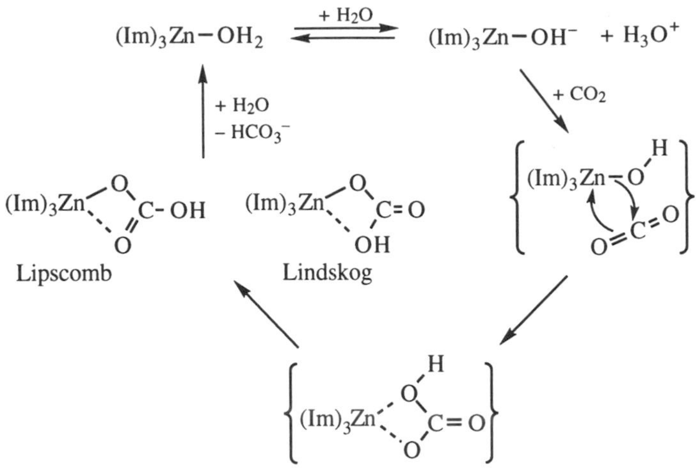

The final step for CO_2 hydration is replacement of bicarbonate by water in the coordination sphere of the zinc. This will be a facile process because of the substitution lability of zinc(II). The $(Im)_3Zn—OH_2$ then is converted to $(Im)_3Zn—OH$ by proton transfer to some base. The base most often suggested is an imidazole from histidine(64), which is connected to the $Zn—OH_2$ by a water-bridged hydrogen bonding network. Other proton relay networks have been suggested involving different peptide groups[68] or just water molecules in the active site cleft.[69] It has been noted that the replacement of histidine(64) by other amino acids in mutant enzymes reduces the activity only by 1.5 to 3 times. Pocker and Janjić[70] argued that these mutants may permit other sidechain amino acids to be active and may open the cavity to allow other basic buffer components to enter. These suspicions seem to be confirmed by the report of Tu et al.[71] that the alanine mutant has about 20 times lower activity in the absence of buffer, and the activity can be restored if imidazole or 1-methylimidazole buffers are used. Buffer effects on the rate of H_2O^{18}–CO_2 exchange by the less reactive human CA III also have been reported.[72]

The most recent theoretical work by Merz and Banci[73] essentially follows the pathway in Scheme 8.5 and finds that the bicarbonate is bound in the Lipscomb model, with the Lindskog structure being an intermediate or transition state on the pathway to the more stable form. In either case, the calculated $Zn—O=C$ or $Zn—OH$ bonds are long, 3.2 and 2.6 Å, respectively, and it is debatable whether these should be considered as five-coordinate zinc complexes. It should be noted that the theoretical conclusions have changed and evolved with time[74] and must be viewed accordingly.

Model systems may provide some indication of the coordination state of Zn(II) in the bicarbonate complex. Kinetic studies by van Eldik and co-workers of the Zn(II) complexes of three-coordinate [12]aneN$_3$ and four-coordinate [12]aneN$_4$ indicate that the k_{cat} of the latter is ~5 times larger than that of the former.[75] The four-coordinate macrocycle would be expected to disfavor chelation by bicarbonate, so that these results imply that chelation is not important. Other model studies indicate that bicarbonate chelation is detrimental to reactivity.[76]

Carbonic anhydrase is inhibited by a wide range of anions. Pocker and Diets[77] found that the inhibition is of the competitive type at pH 6.6 but is uncompetitive at pH 9.9. The uncompetitive inhibition implies that Zn—OH is not complexed by inhibitor, but zinc(II) in the enzyme–substrate complex can expand its coordination number to five with the ligands being three imidazoles, substrate, and either water or inhibitor. Alkyl or aryl sulfonamides are strong inhibitors for carbonic anhydrase and bind as the anion, RSO_2NH^- or $ArSO_2NH^-$, respectively. Dugad et al.[78] found that p-fluorobenzenesulfonamide forms a bis complex, indicating that the zinc is quite capable of becoming five-coordinate. Liang and Lipscomb[79] have modeled the formation and bonding of acetamide and sulfonamide inhibitors. Phenol is a competitive inhibitor of CAII for the hydration reaction. The structure of the phenol–CAII complex[80] reveals that two phenols are bound: one in the active-site pocket but not coordinated to the zinc, and the other in a hydrophobic patch, about 15 Å from the zinc. The complex was prepared at pH 10 and the phenol may be present in the neutral or anionic form.

Bertini et al.[81] showed that nitrate binds to the EH$_2$ and EH forms of the cobalt(II)-substituted enzyme. They attribute this to the EH form having tautomers E(histidine-H)(Zn—OH) and E(histidine)(Zn—OH$_2$), with water replacement in the latter being the source of nitrate binding to EH. This maintains consistency with the standard interpretation that anions cannot compete with OH$^-$ in the Zn(OH) species.

8.3 ENZYMIC REACTIONS OF DIOXYGEN

There are a number of metalloenzymes, EM, that bring about reactions of dioxygen and its derivatives, hydrogen peroxide and the superoxide ion. They may act as reversible oxygen carriers, such as hemoglobin and myoglobin in the following reaction:

$$EM + O_2 \rightleftharpoons EM(O_2) \qquad (8.21)$$

The enzyme may introduce either one or two oxygens into the substrate, as shown in the following reactions:

$$O_2 + SH + AH_2 \xrightarrow[\text{oxygenases}]{\text{Mono-}} SOH + H_2O + A \qquad (8.22)$$

$$O_2 + SH \xrightarrow{\text{Dioxygenases}} SHO_2 \qquad (8.23)$$

Other enzymes protect the system[82] from reactions with superoxide or hydroxy radicals by catalyzing the disproportionation of O_2^- and the decomposition of hydrogen peroxide, as in reactions (8.24) and (8.25).

$$2\,O_2^- + 2\,H^+ \xrightarrow[\text{dismutase}]{\text{Superoxide}} O_2 + H_2O_2 \qquad (8.24)$$

$$H_2O_2 + AH_2 \xrightarrow{\text{Peroxidases}} H_2O + A \qquad (8.25)$$

The oxidases and peroxidases are the subject of a recent, succinct review.[83] The following sections will discuss two specific examples of these metalloenzymes.

8.3.a Oxygen Carriers: Myoglobin

Myoglobin consists of one iron–protoporphyrin complex (heme prosthetic group) with an imidazole nitrogen of a histidine also coordinated to the iron and a single peptide chain. Since this system is simpler than hemoglobin, which has four heme units, much of the mechanistic work has been done on myoglobin. There are several sources of myoglobin that differ in the peptide composition, but the work discussed in this section is on the most common source, sperm whale myoglobin, unless otherwise indicated. The thermodynamic and structural effects of dioxygen binding to myoglobin and hemoglobin have been reviewed recently.[84]

The structural features of the iron coordination in the iron(II) deoxy form and the dioxygen complex have been determined by Takano[85] and Phillips,[86] respectively, and are shown in Figure 8.4. The iron(II) in deoxymyoglobin is displaced by 0.42 Å out of the plane of the four porphyrin nitrogens toward the imidazole nitrogen. In the dioxygen complex, this displacement is reduced to 0.18 Å. The dioxygen is bonded in an end-on fashion with an Fe—O—O angle of 115°. The complexation of myoglobin by dioxygen appears to be a simple process accompanied by some structural changes. However, the iron(II) is initially in the high-spin state and reacts with paramagnetic dioxygen to yield a diamagnetic product and a lot is happening electronically. The dioxygen complex can be pictured as $Fe^{II}(O_2)$, $Fe^{III}(O_2^-)$ or $Fe^{IV}(O_2^{2-})$. The $Fe^{III}(O_2^-)$ formalism is now favored as best describing the spectroscopic properties of the system. In oxymyoglobin, MbO_2, the terminal oxygen is hydrogen bonded to an imidazole of a distal histidine(64). Mutants in which this histidine is replaced by other amino acids show a lower discrimination between O_2 and CO.[87]

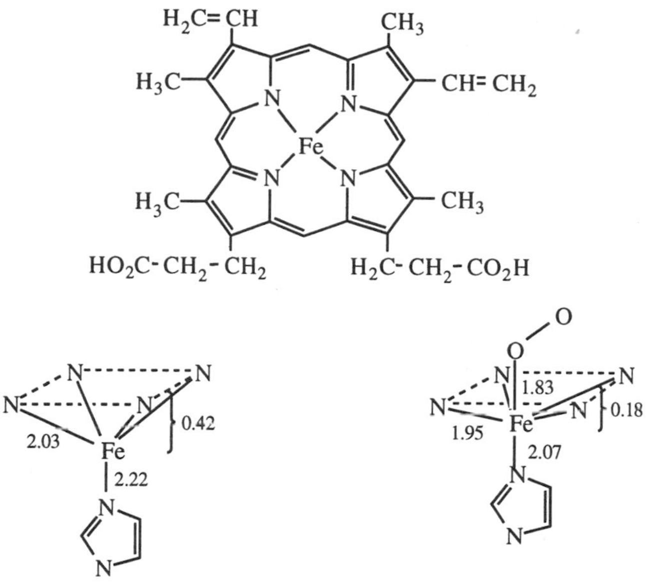

Figure 8.4. Iron protoporphyrin IX, the prosthetic group in myoglobin, and the coordination geometry of iron in myoglobin and oxymyoglobin.

Takano[88] reported the structure of metmyoglobin, the iron(III) complex with axial water, and imidazole ligands. The same structural features also are found in the model system described by Collman and co-workers,[89] where the displacement of iron depends on the steric requirements of the axial ligand, being 0.086 Å with 2-methylimidazole and 0.03 Å with 1-methylimidazole.

In the development of model systems, a major problem has been that the simple iron(II) porphyrins form diiron μ-peroxo complexes:

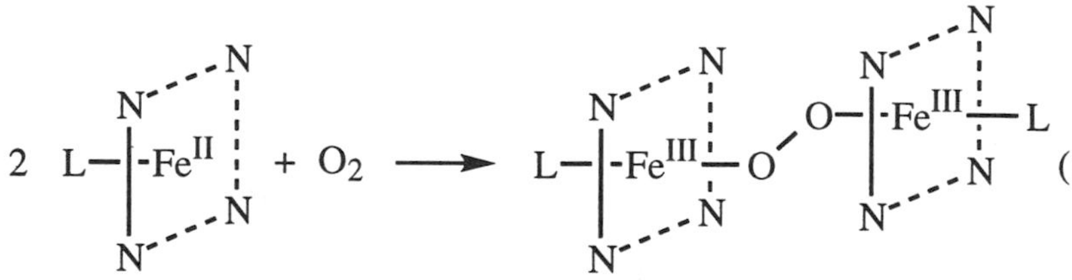

$$2 \ L-|-Fe^{II} \ + \ O_2 \ \longrightarrow \ L-|-Fe^{III}-O-O-|-Fe^{III}-L \quad (8.26)$$

This can be overcome by introducing appropriate steric hindrance to dimer formation. This was achieved by Collman et al.[90] by using the bulky *o*-pivalamidophenyl substituent on *meso*-tetraphenylporphyrin to prepare a model oxygen carrier. In the biological system, the peptide may serve a similar purpose.

Table 8.1. Kinetic Parameters for Complexing of Myoglobin, Mb

Reaction	k (M^{-1} s^{-1})	ΔH^* (kJ mol^{-1})	ΔS^* (J mol^{-1} K^{-1})	ΔV^* (cm^3 mol^{-1})
Mb + O_2 [a]	2.5×10^7	26 [b]	−15	5.2
Mb + O_2 [c]	1.3×10^7	23.0	−30	7.8
Mb + O_2 [d]	2.4×10^7			4.6
Mb + CO [c]	3.8×10^5	17.1	−81.1	−8.9
Mb + CO [d]	6.7×10^5			−9.2
Mb + CO [e]	5.2×10^5			−10.0
Mb + MeNC	1.2×10^5 [f]			8.8 [e]
Mb + n-BuNC [f]	3.0×10^4			

[a] Projahn, H.-D.; Dreher, C.; van Eldik, R. *J. Am. Chem. Soc.* **1990**, *112*, 17; in 5 x 10^{-3} M Tris buffer, 0.1 M NaCl, pH 8.5, 25°C.

[b] Calculated from the dissociation rate and the equilibrium constant parameters.

[c] Hasinoff, B. B. *Biochemistry* **1974**, *13*, 3111; in 0.1 M phosphate buffer, pH 7.0, 25°C.

[d] Adachi, S.; Morishima, I. *J. Biol. Chem.* **1989**, *264*, 18896; in 0.1 M Tris buffer, pH 7.8, 20°C.

[e] Taube, D. J.; Projahn, H.-D.; van Elaik, R.; Magde, D.; Traylor, T. G. *J. Am. Chem. Soc.* **1990**, *112*, 6880; in 0.05 M bis-Tris buffer, 0.1 M NaCl, pH 7.0, 25°C.

[f] Rohlfs, R. J.; Olson, J. S.; Gibson. Q. H. *J. Biol. Chem.* **1988**, *263*, 1803; in 0.1 M phosphate buffer, pH 7, 20°C.

The kinetic parameters for the binding and dissociation of some small molecules to myoglobin are given in Table 8.1. The relatively small rate constants for the isocyanides have been attributed to steric effects imposed by the peptide as these larger molecules move through the peptide cleft toward the iron. However, the kinetic difference between O_2 and CO is more remarkable. The ΔV^* for CO binding is quite negative, whereas that for O_2 is positive, and the CO rate constant is smaller, largely because of the less favorable ΔS^*. The binding of CO gives a low-spin iron(II) complex with the iron 0.1 Å below the porphyrin plane,[91] compared to 0.18 Å in the O_2 complex., The O_2 is bent and hydrogen bonded to a distal histidine, whereas the CO is linear but still rather immobile according to ^{17}O NMR studies.[92] The binding of O_2 is independent of pH, and CO binding is faster at lower pH.[93]

For mutants in which histidine(64) is replaced, the dissociation rates of O_2 are increased by 50 to 1500 times, while the association rates for O_2 and CO increase 5 to 15 times. Resonance Raman studies[94] of MbCO indicate that histidine(64) also affects the Fe—C stretching frequency, and the pH effect indicates that protonation of the histidine occurs with a pK_a of ~4.5. These studies also revealed that removal of histidine(64) increases the autooxidation rate of MbO_2, so that the histidine is serving some protective role as well.

It has been possible to obtain a detailed picture of the complexation process by using laser flash photolysis to dissociate the bound ligand and then watch its recombination or movement out of the active site pocket. Martin et al.[95] studied CO dissociation from several hemes using a 250-fs laser pulse at 307 nm and concluded that the high-spin iron(II) state is formed within 0.35 ps. Finsden at al.[96] followed the Raman spectrum of the Fe—N(imidazole) stretch, after dissociation of CO and O_2 from hemoglobin, and found that the deoxy form was produced within the 10-ns pulse length. This is rather surprising since low-spin to high-spin changes for octahedral iron(II) complexcs[97,98] have rate constants of ~10^7 s^{-1}. Traylor and co-workers[99] observed the time evolution of difference spectra in the 400- to 500-nm range after flash photolysis and found one process on the nanosecond and another on the picosecond time scale. Olson et al.[100] have done similar experiments but using essentially nanosecond time scale detection. Both groups interpret the results in terms of a four-state model, described in Scheme 8.6, where B is called a geminate pair and the leaving group has moved further away and probably rotatcd somewhat in C.

Scheme 8.6

$$Mb-XY \underset{k_{BA}}{\overset{h\nu}{\rightleftarrows}} \{Mb\text{---}XY\} \underset{k_{CB}}{\overset{k_{BC}}{\rightleftarrows}} \left\{ Mb \begin{matrix} X \\ | \\ Y \end{matrix} \right\} \underset{k_{DC}}{\overset{k_{CD}}{\rightleftarrows}} Mb + XY$$

$$\quad A \qquad\qquad\qquad B \qquad\qquad\qquad C \qquad\qquad\qquad D$$

The kinetic observations combincd with quantum yields have been used to determine rate constants for the various steps. Some data from both studies are given in Table 8.2. The results show some variation, especially with regard to k_{BA} and k_{BC}. A transient time-dependent signal was not observed for CO dissociation, although CO dissociation has a high quantum yield.[101] This could be explained by a larger k_{BC} for CO, but the value for O_2 is at the diffusion limit according to the data of Traylor et al., so that k_{BA} must be smaller to explain the observations. Since k_{BA} and k_{CB} do not show any correlation with the steric bulk of XY, Traylor and co-workers suggested that the slower overall complexation of isocyanides is due to changes in k_{DC} caused by steric effects as the ligand enters the protein pocket. This is consistent with a crystal structure of the ethyl isocyanide complex.[102]

In a study on myohemerythrin,[103] a diferrous species, it was noted that the ratio of the overall on and off rate constants for O_2 does not give the correct value of the equilibrium constant. This might indicate that photolysis does not completely liberate O_2 or does not leave the protein in its ground state before O_2 undergoes recombination.

Table 8.2. Rate Constants Following Flash Photolysis of Myoglobin–XY Complexes Analyzed According to Scheme 8.4

XY	k_{BA} (s^{-1})	k_{BC} (s^{-1})	k_{CB} (s^{-1})	k_{CD} (s^{-1})
O_2 [a]	1.4×10^{10}	1.5×10^{10}	4.6×10^6	8.8×10^6
O_2 [b]	4.9×10^8	1.2×10^8	8.5×10^6	1.4×10^7
MeNC [a]	1.3×10^{10}	4.0×10^{10}	7.8×10^7	5.9×10^6
MeNC [b]	2.4×10^7	4.9×10^6		
n-BuNC [b]	1.6×10^7	1.4×10^7	8.0×10^5	1.2×10^6
t-BuNC [a]	2.6×10^{10}	8.0×10^9	5.2×10^6	3.6×10^6

[a] Jongeward, K. A.; Magde, D.; Taube, D. J.; Marsters, J. C.; Traylor, T. G.; Sharma, V. S. *J. Am. Chem. Soc.* **1988**, *110*, 380; in 0.1 M Tris buffer, 0.1 M NaCl, pH 7, unspecified temperature.

[b] Rohlfs, R. J.; Olson, J. S.; Gibson. Q. H. *J. Biol. Chem.* **1988**, *263*, 1803; in 0.1 M phosphate buffer, pH 7, 20°C.

Magde and co-workers[104] have measured k_{CD} and its Arrhenius activation energy for several complexes of horse heart myoglobin and for the O_2 sperm whale myoglobin system. For the latter, they report $k_{CD} = 7.7 \times 10^6$ s^{-1} (25°C, 0.1 M Tris buffer, 0.1 M NaCl, pH 7.0) and $E_a = 7.4$ kcal mol^{-1}. The range of the E_a values for all the systems is 6 to 9 kcal mol^{-1}. They also suggest that a five-state model may be required, involving either two escape routes from the peptide pocket or a second bound site within the pocket.

The observed steady-state rate constant for Scheme 8.6 is given by

$$k_{obsd} = \frac{k_{BA}\,k_{CB}\,k_{DC}}{k_{BA}\,k_{CB} + k_{BA}\,k_{CD} + k_{BC}\,k_{CD}} \qquad (8.27)$$

If k_{BA} is unusually small for CO, then the first two terms in the denominator of Eq. (8.27) are small relative to the third term, and the equation simplifies to $k_{obsd} = k_{BA}(k_{CB}/k_{BC})(k_{DC}/k_{CD}) = k_{BA}K_{CB}K_{DC}$. This corresponds to step B to A being rate-controlling and preceded by fast equilibria of D to C and C to B. For O_2, the first two terms in the denominator are somewhat larger than the third, so that the k_{BA} term approximately cancels in the numerator and denominator. Then, $k_{obsd} = k_{CB}(k_{DC}/k_{CD}) = k_{CB}K_{DC}$ because k_{CD} is larger than k_{CB}, and step C to B is rate-controlling. This would explain the kinetic differences between O_2 and CO binding.

Adachi and Morishima[105] have determined the pressure dependence for some of the steps in Scheme 8.4 for several myoglobins. Their results, given in Table 8.3, show that changes in the peptide cause small changes in the rate constants but substantial differences in ΔV^*.

Table 8.3. Rate Constants (20°C) and Activation Volumes for the Reactions of O_2 and CO with Different Myoglobins[a]

	Sperm Whale	Horse	Dog
k_{CA} (s^{-1})	3.9×10^6	3.5×10^6	3.6×10^6
ΔV_{CA}^* (cm^3 mol^{-1})	3.5	-8.4	-17.8
k_{CD} (s^{-1})	5.6×10^6	6.2×10^6	7.4×10^6
ΔV_{CD}^* (cm^3 mol^{-1})	16.7	11.2	-2.1
k_{DC} (M^{-1} s^{-1})	5.7×10^7	7.9×10^7	9.1×10^7
ΔV_{DC}^* (cm^3 mol^{-1})	10.4	12.3	8.6
k_{obsd} [c] (M^{-1} s^{-1})	2.4×10^7	2.9×10^7	3.0×10^7
ΔV_{obsd}^* [c] (cm^3 mol^{-1})	4.6	3.8	0
k_{obsd} (CO) [c] (M^{-1} s^{-1})	6.7×10^6	6.8×10^6	8.5×10^6
ΔV_{obsd}^* (CO) [c] (cm^3 mol^{-1})	-9.2	-12.7	-18.8

[a] In 0.1 M Tris buffer, pH 7.8; values are for O_2 unless otherwise indicated.
[b] $k_{CA} = k_{BA}k_{CB}/(k_{BA} + k_{BC})$ from Scheme 8.4.
[c] Values from steady-state conditions.

The systems in Table 8.3 all have the histidine(64) residue that is involved in hydrogen bonding to the bound O_2, but they vary in the residues that may be affecting the size and structure of the pocket leading to the active site. Adachi and Morishima have given a more detailed rationalization of the effects of varying amino acid residues. Traylor and co-workers[106] subsequently reported activation volumes for k_{CD} with sperm whale myoglobin of 11.7, 12.6, and 9.1 cm^{-3} M^{-1} for CO, O_2, and MeNC, respectively. For the overall formation reaction, the more negative values of ΔV_{obsd}^*(CO) can be attributed to contractions occurring at the B to A stage, which follows the rate-limiting step for O_2 binding and therefore does not influence ΔV_{obsd}^*(O_2). Van Eldik and Projahn have summarized the overall kinetic and equilibrium parameters for the CO and O_2 systems.[107]

An interesting recent development is the use of X-ray methods at liquid-helium temperatures to identify reaction intermediates. Schlichting et al. found that the intermediate for CO dissociation from myoglobin has the CO almost parallel to the heme plane.[108]

8.3.b Cytochrome P-450

There is a large and wide-ranging family of monooxygenase enzymes, called cytochrome P-450s, whose chemistry and biochemistry are the subjects of a recent book.[109] Their name derives from their characteristic absorbance at 450 nm. They all contain a heme prosthetic group and catalyze the reactions of O_2 and a reducing agent to bring about transformations, such as those in (8.28) to (8.31). These are only a few examples of the many reactions catalyzed by P-450 enzymes.

$$R\text{-}CH_3 + O_2 + 2\,H^+ + 2\,e^- \xrightarrow{\text{P-450}} R\text{-}CH_2OH + H_2O \qquad (8.28)$$

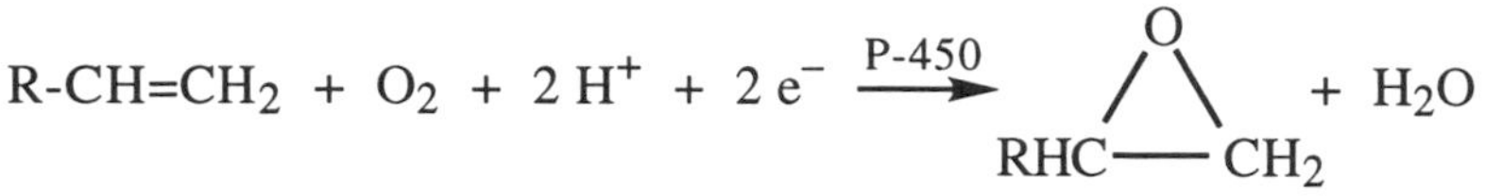

$$R\text{-}CH{=}CH_2 + O_2 + 2\,H^+ + 2\,e^- \xrightarrow{\text{P-450}} \underset{RHC \longrightarrow CH_2}{\overset{O}{\triangle}} + H_2O \quad (8.29)$$

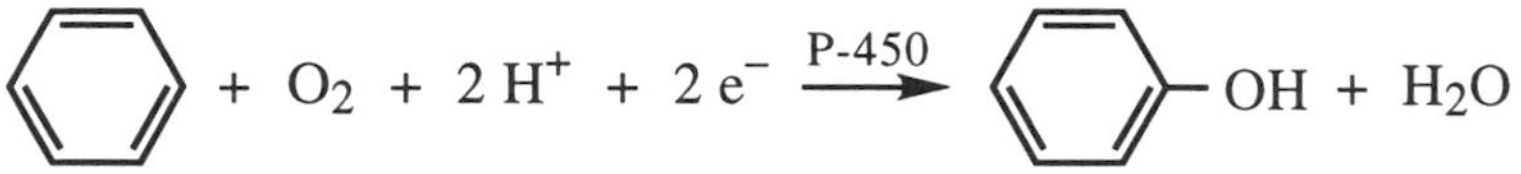

$+ O_2 + 2\,H^+ + 2\,e^- \xrightarrow{\text{P-450}}$ $-OH + H_2O \qquad (8.30)$

$+ O_2 + 2\,H^+ + 2\,e^- \xrightarrow{\text{P-450}}$ $+ H_2O \; (8.31)$

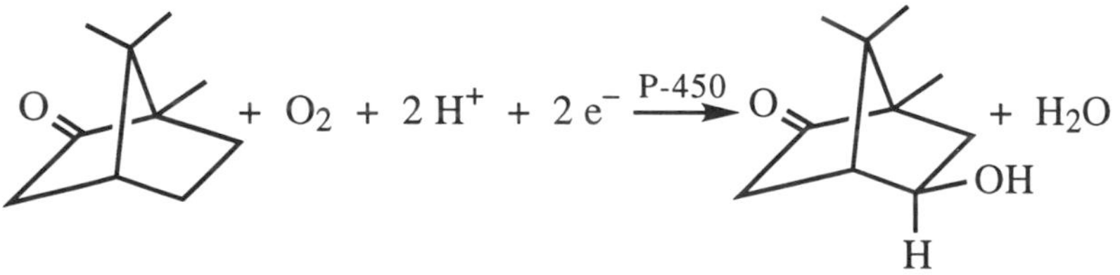

Reaction (8.31) represents the transformation of camphor to the exo-alcohol by P-450-CAM, one of the best characterized of these enzymes. The crystal structure of P-450-CAM and its camphor adduct have been determined by Poulos and co-workers.[110,111] The solution-state coordination of iron has been characterized by Dawson and co-workers[112,113] using extended X-ray absorption fine structure, EXAFS. The resting-state prosthetic group is a six-coordinate low-spin iron(III) heme with axial S and O ligands. The S is derived from cysteine, and Poulos has argued from bond lengths (Fe—S = 2.2 Å) that it is a thiolate cys-S⁻—Fe linkage. The O ligand is OH_2, based on electron spin-echo envelope modulation, ESEEM, results.[114] The iron(III) is displaced 0.29 Å from the plane of the porphyrin nitrogens toward the S⁻. The active-site pocket contains five hydrogen-bonded waters and is lined with hydrophobic amino acids with no acid or base residues. If camphor is added to P-450-CAM under anaerobic conditions, then camphor displaces water and moves into the active site pocket. This complex contains five-coordinate high-spin iron(III), with the iron atom 0.43 Å out of the porphyrin plane and an essentially unchanged Fe—S bond length. The camphor is poised over the heme, with the C to be hydroxylated closest to the iron and the C=O of camphor hydrogen bonded to an -OH of a tyrosine. An EXAFS study by Dawson and co-workers of the low-spin iron(II)–camphor–O_2 adduct, prepared at −40°C in water–ethylene glycol, indicates bond lengths very similar to those of a model thiolate–Fe^{II}(porphyrin)–O_2 complex.[115]

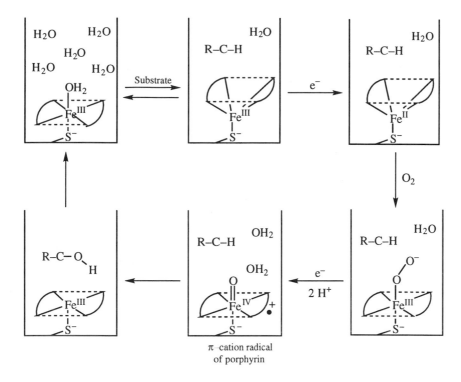

Figure 8.5. A catalytic cycle for alkane hydroxylation by cytochrome P-450 in which the first four intermediates have been identified and the π-cation radical is speculative.

Since the operation of P-450 requires the enzyme plus three reagents (substrate, O_2, and reducing agent), it is possible to control the reaction conditions in order to identify intermediates and to establish the reaction pathway.[116] Current evidence is consistent with the sequence in Figure 8.5. When the substrate (R–C–H) enters the active site pocket, a high-spin iron(III) complex forms due to small changes in the peptide conformation or loss of the water ligand. This form is 0.12 V[117] more easily reducible than the resting enzyme and therefore favors the reduction to high-spin iron(II), which then forms the O_2 complex similar to myoglobin.

Precedent for the π-cation radical Fe=O species comes from the more stable and well-characterized intermediate, HRP I,[118,119] of horseradish peroxidase. Furthermore, it has been shown that the iron(III) resting state of P-450 reacts with iodosylbenzene to produce a species that hydroxylates alkanes. This can be understood if oxygen transfer from iodosylbenzene proceeds to give the π-cation radical Fe=O species, as shown in (8.32).

$$PhIO + (porphyrin)Fe^{III} \longrightarrow PhI + (porphyrin\bullet^+)Fe^{IV}=O \quad (8.32)$$

This synthetic version causes the same hydroxylations as the natural system. However, the iodosylbenzene product gives ^{18}O exchange with $^{18}OH_2$, while the natural system shows no exchange.[120] Groves et al.[121] found that the reaction of P-450$_{LM2}$ with its reductase gives exchange of the trans 1-H of propylene with solvent D_2O, but the enzyme with iodosylbenzene gave no exchange. The iodosylbenzene reactions may be different because the substrate enters the active site only after the phenyl iodide leaves.

Model porphyrin systems show the same type of reaction and catalyze the hydroxylation of alkanes[122] and the epoxidation of alkenes[123] by iodosylbenzene. There are many examples of high-oxidation-state M=O (oxene) species[124] that act as oxygen transfer agents. Groves et al.[125] have shown that a Ru^{III}(porphyrin) with 2,6-dichloropyridine-N-oxide can hydroxylate alkanes by a mechanism that appears to involve a Ru^V=O species. The details of the oxygen insertion step, often called the rebound step, remain open to speculation. Champion[126] has suggested that the (porphyrin$\bullet^+$)Fe^{IV}=O species is more reactive in P-450 than in horseradish peroxidase because of better back π bonding from the -S$^-$ ligand in P-450 compared to the nitrogen of imidazole in horseradish peroxidase. This makes the reactive (porphyrin)Fe^{IV}(O$\bullet$) resonance form more favorable with the P-450 systems. Then, the reaction with the alkane could proceed as shown in Figure 8.6.

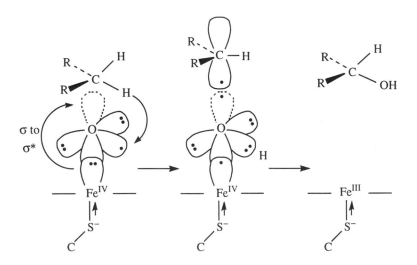

Figure 8.6. Possible product formation steps for P-450 hydroxylation starting from the (porphyrin)Fe^{IV}(O$\bullet$) resonance form favored by back π bonding from S$^-$.

Recently, the intermediacy of an organic free radical in these reactions has been cast into doubt through observations on radical clock reactions. Evidence for a radical came from the observation of Ortiz de Montellano and Stearns[127] that hydroxylation of bicyclo[2.1.0]pentane gave bicyclo[2.1.0]pentan-2-ol and rearranged 3-cyclopentenol, the latter being indicative of a radical intermediate. Atkinson and Ingold[128] used a series of five calibrated radical clocks to estimate the radical lifetime, and thereby the value of the rate constant of the oxygen transfer or rebound step. They found rather variable rate constants in the range of 1.4×10^{10} to 7×10^{12} s^{-1}. More recently, the observations of Newcomb and co-workers[129] indicate a seemingly impossibly high value of 1.4×10^{13} s^{-1}. These are all with different substrates and could indicate substrate specificity, but it has been argued that this is unlikely because the P-450 systems are so unselective with regard to substrate. Newcomb also developed the clock substrate in Scheme 8.7 that gives different rearrangement products shown from radical and cation routes.

Scheme 8.7

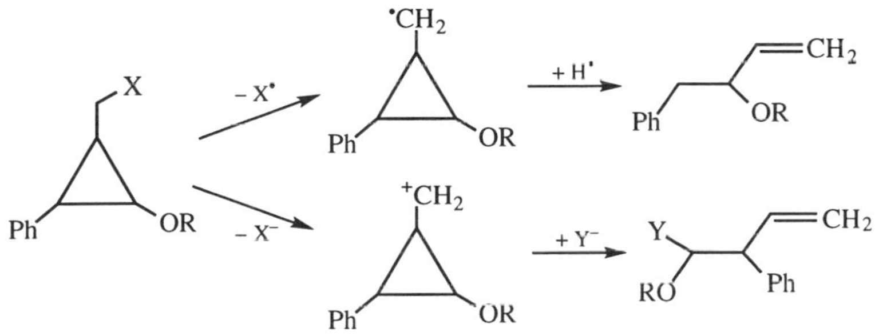

With P-450 from rat liver,[130] this substrate (X = H) gave predominantly unrearranged product with hydroxylation at the methyl group and similar amounts of Ph hydroxylation products. The remaining product (5 to 15 percent) is derived from the cationic pathway. It was suggested that the latter may be the source of rearranged product in previous radical clock studies. These and further observations suggest that the oxygen transfer reaction is essentially concerted through a triangular transition state in which H transfer to O precedes O transfer to C.[131]

The epoxidation of olefins, given in (8.29), presents some new mechanistic problems. Ortiz de Montellano and co-workers[132] have shown that the enzyme gives both epoxidation and alkylation of a pyrrole nitrogen, the latter occurring with alkenes that have a terminal $=CH_2$. Alkenes and alkynes alkylate a different pyrrole nitrogen. The epoxidation seems to be stereospecific, since retention is found for the reaction of *trans*-1-[1-^{2}H]octene. The simplest explanation for these results would seem to be that the peptide controls the orientation of the substrate over the iron and the reactions proceed as shown in Figure 8.7.

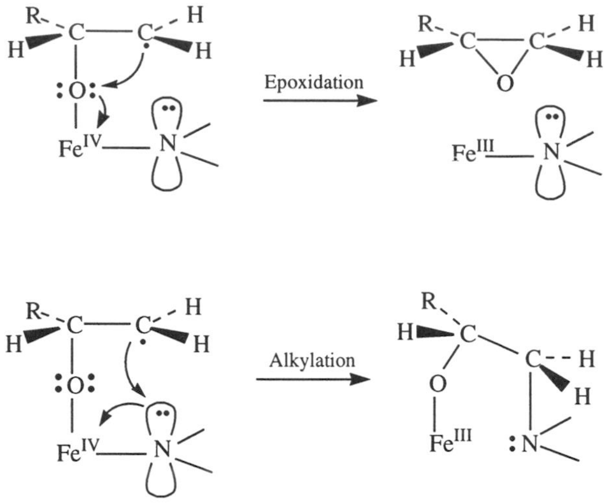

Figure 8.7. Possible routes to epoxide and N-alkylation products in the reaction of alkenes with P-450 systems.

A concerted mechanism, such as that in Figure 8.7, would not predict the deuterium exchange results of Groves and co-workers, and models involving organo-iron species have been invoked[133,134] based on known model chemistry. The deuterium exchange was originally explained as proceeding through the metallocycle in Scheme 8.8.

Scheme 8.8

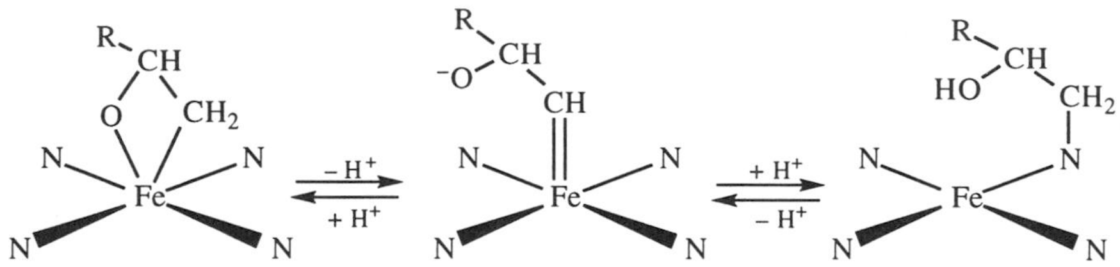

Dolphin et al.[135] have suggested, based on model studies with tetra-(2,6-dichlorophenyl)porphyrin, that a ß-hydroxyalkyl species can be formed from the N-alkyl derivative under reducing conditions, as shown in Scheme 8.9. This provides an alkyl derivative that gives proton exchange through deprotonation to the carbene. Mansuy and co-workers[136] have reported efficient preparations of iron *meso*-(tetraphenyl)porphyrin carbenes such as $(N)_4Fe^{IV}(=C(CH_3)COPh)$. Sterically crowded alkenes are reactive with both P-450 and model systems, yet they do not show N-alkylation and are unlikely to form carbenes. Therefore, the latter processes appear to be side reactions of the main epoxidation process.

Scheme 8.9

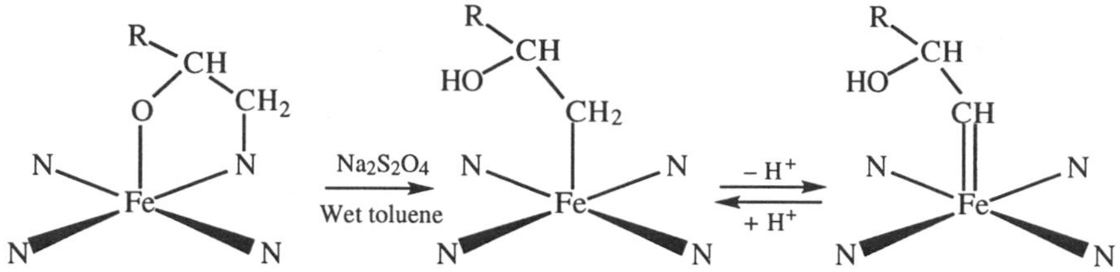

Collman et al.[137] have summarized and discussed the competition between epoxide formation and N-alkylation and the numerous mechanistic possibilities. They conclude in part that "Mechanisms ranging from fully concerted reaction to a stepwise reaction involving an initial electron transfer may all be possible depending on the nature of the system." Such a conclusion is encouraging for those seeking challenging mechanistic problems, but discouraging to those hoping for a unified mechanism to explain all observations.

8.4 ENZYMIC REACTIONS OF NITRIC OXIDE

In 1987 it was announced by two groups[138,139] that the endothelium-derived relaxing factor, EDRF, is nitric oxide. Previous work[140] had shown that this factor is produced in the endothelium and diffuses into the underlying muscle, where it causes smooth muscle relaxation by triggering the conversion of guanosine triphosphate to cyclic guanosine monophosphate. Smooth muscle surrounds blood vessels and controls the resistance to blood flow in the arterial system; the relaxation of this muscle tissue can give relief from hypertension and angina pain and can assist the recovery from heart attack. Since this discovery, there has been a greatly renewed interest in the chemistry and biological functions of NO; several reviews[141] and books[142] give further details.

With the benefit of hindsight, it is not surprising that NO is involved in smooth muscle relaxation. Angina pain has been treated for many years by vasodilators, such as glyceryl trinitate, amyl nitrite, and isosorbide dinitrate, all of which are potential sources of NO. There is still some question as to whether NO itself or some derivative is the actual EDRF, and nitrosothiols (RS—NO) are most often invoked as possibilities.[143]

In the biological system, NO is produced by nitric oxide synthase, NOS, which occurs in at least two forms. The constitutive form is always present in endothelial cells, while the inductive form in macrophages appears when triggered by the calcium-binding protein, calmodulin. The NOS enzymes contain iron and appear to be closely analogous to cytochrome P-450 reductase. The substrate is L-arginine, and the products are NO and citrulline, as shown in (8.33).

$$
\underset{\substack{\displaystyle \text{HN--C} \\ \displaystyle (CH_2)_3 \\ \displaystyle H_3\overset{+}{N}\text{--HC--CO}_2^-}}{\overset{\displaystyle NH}{\underset{\displaystyle NH_3^+}{}}} + {}^{18}O_2 \xrightarrow[\text{H}_2\text{O}]{\text{NOS}} \underset{\substack{\displaystyle H_3\overset{+}{N}\text{--HC--CO}_2^-}}{} + N^{18}O \quad (8.33)
$$

As indicated, it has been shown[144] that both NO and citrulline derive their oxygen from O_2. The mechanism of action may be analogous to that shown in Figure 8.5. There is a problem with the details, however, because NO is known to bind to iron in hemes, yet the enzyme is not inhibited by NO.[145] Inhibitors of NOS may be useful in injury cases were blood vessel contraction is desirable.

The binding of NO to various iron(III)-heme proteins was studied recently by Ford and co-workers.[146] For the equilibrium reaction

$$(Heme)Fe^{III} + NO \rightleftharpoons (Heme)Fe^{III}\text{---NO} \quad (8.34)$$

they determined equilibrium constants for myoglobin, horse heart cytochrome c, and methemyoglobin in the narrow range of 1.6×10^4 to $1.3 \times 10^4 \ M^{-1}$. However, the system is complicated by the following reductive hydroxylation and complexation reactions:

$$(Heme)Fe^{III}\text{---NO} + OH^- \longrightarrow (Heme)Fe^{II} + HNO_2 \quad (8.35)$$

$$(Heme)Fe^{II} + NO \longrightarrow (Heme)Fe^{II}\text{---NO} \quad (8.36)$$

Reaction (8.36) has rate constants in the 10^2 to $10^3 \ M^{-1} \ s^{-1}$ range and becomes a serious complication for pH >7.5. At pH 6.4, the rate constants for $(Heme)Fe^{III}$ + NO recombination after flash photolysis were determined as 1.9×10^5 and $7.2 \times 10^2 \ M^{-1} \ s^{-1}$ for myoglobin and cytochrome c, respectively.[147] The differences in rate constants were attributed to the different ligands coordinated to iron(III). The $(Heme)Fe^{II}$ systems in reaction (8.36) are very strongly complexed by NO. Flash photolysis of such complexes[148] has indicated recombination rate constants close to the diffusion-controlled limit and analogous to those for recombination with O_2.

The aqueous chemistry of NO has been extensively studied. Nitric oxide has a modest solubility of 1.9 mM in water under 1 atmosphere of NO. Descriptions of its reactivity have tended to be highly variable. Recent work indicates that NO is not highly reactive, despite having an unpaired electron, but reacts with O_2 to form reactive species. Traces of

O_2 are the probable source of different conclusions about the reactivity of NO.

A series of recent papers by Goldstein and Czapski[149] have served to clarify the pathways for the reaction of NO with O_2 and the reactivity of intermediates. The overall reaction can be described by

$$4\,NO + O_2 + 2\,H_2O \longrightarrow 4\,H^+ + 4\,NO_2^- \qquad (8.37)$$

There is general agreement that the rate is first order in O_2 and second order in NO, and the specific rate constant, k, can be defined from $-d[O_2]/dt = k[O_2][NO]^2$. The reaction pathways are described in Scheme 8.10.

Scheme 8.10

$$^\bullet NO + O_2 \; \underset{k_{-1}}{\overset{k_1}{\rightleftharpoons}} \; (ON:O_2)$$

$$(ON:O_2) + {}^\bullet NO \; \overset{k_2}{\longrightarrow} \; ONOONO$$

$$ONOONO \; \overset{k_3}{\longrightarrow} \; 2\,{}^\bullet NO_2$$

$$^\bullet NO_2 + {}^\bullet NO \; \underset{k_{-4}}{\overset{k_4}{\rightleftharpoons}} \; N_2O_3$$

$$N_2O_3 + H_2O \; \overset{k_5}{\longrightarrow} \; 2\,NO_2^- + 2\,H^+$$

The product of the k_1 step is uncertain and the suggested weak $ON:O_2$ complex is based on theoretical work by McKee.[150] The values of k_4, k_{-4}, and k_5 are known[148] to be 1.1×10^9 M^{-1} s^{-1}, 8.1×10^4 s^{-1}, and 5.3×10^2 s^{-1}, respectively. A steady-state treatment of this mechanism gives $k = k_1 k_2 / k_{-1} = 2.5 \times 10^6$ M^{-2} s^{-1}.

For reactions with other substrates, S, it is proposed that these involve the following reactions with the intermediates in Scheme 8.10, and it is found that the rate is independent of the substrate.[151]

$$S + ONOONO \longrightarrow S^+ + NO_2^- + {}^\bullet NO_2$$

$$S + {}^\bullet NO_2 \longrightarrow S^+ + NO_2^- \qquad (8.38)$$

$$S + N_2O_3 \longrightarrow S^+ + NO_2^- + {}^\bullet NO$$

In general, the amount of reaction proceeding by these pathways will depend on the substrate and lead to variable stoichiometry for the overall substrate-oxidation reaction. For the important case of thiols, the observations suggest that $\bullet NO_2$ and/or N_2O_3 are the reactive oxidants.

Goldstein and Czapski have estimated that, under physiological conditions, the nitrosation of a thiol would have a half-life of > 7 min and would be too slow to produce the nitrosothiols that have been suggested as alternatives to NO as the EDRF.

Since the discovery of the involvement of NO in smooth muscle relaxation, intensive investigations have revealed additional functions of NO. The brain contains more NOS than any other organ, and NO has been suggested as the chemical messenger at the synapses. The antibacterial action of macrophages appears to involve NO production. Since superoxide ion also is produced in these systems, it has been suggested[152] that the active agent is peroxynitrite, $ONOO^-$. Lymar and Hurst[153] found that $ONOO^-$ reacts rapidly with CO_2 to give $ONO_2CO_2^-$ and this will be the dominant pathway for removal of $ONOO^-$ under physiological conditions. However, the product $ONO_2CO_2^-$ or its decomposition products might be active oxidants.

Lest one think that, if a little is good then a lot of NO is better, it must be noted that NO can be a serious air pollutant and a carcinogen, although it is toxic to early tumor cells. It may be involved in mutagenesis through the conversion of 5-methycytosine to thymine.

References

1. Williams, R. J. P. *Coord. Chem. Rev.* **1987**, *79*, 175.
2. *Metal Ions in Biological Systems*; Sigel, H., Ed.; Marcel Dekker: New York; *Advances in Inorganic Biochemistry*; Eichorn, G. L.; Marzilli, L. G., Eds.; Elsevier: New York; *Advances in Inorganic and Bioinorganic Mechanisms*; Sykes, A. G., Ed.; Academic Press: New York; *Metal Ions in Biology*; Spiro, T. G., Ed.; Wiley-Interscience: New York.
3. *J. Inorg. Biochem.*; *Bioinorg. Chem.*; *Inorg. Chim. Acta*; *Bioinorg. Chem. Art. Lett.*
4. *B₁₂*; Dolphin, D., Ed.; Wiley: New York, 1982; Vols. I and II.
5. Halpern, J. *Science* **1985**, *227*, 869; Pratt, J. M. *Chem. Soc. Rev.* **1985**, 161; Finke, R. G. In *Molecular Mechanisms in Bioorganic Processes*; Royal Society of Chemistry: Thomas Graham House, Science Park, Cambridge, U.K., 1990; pp 244-279.
6. Ludwig, M. L.; Drennan, C. L.; Mathews, R. G. *Structure*, **1996**, *4*, 505; Drennan, C. L.; Huang, S.; Drummond, J. T.; Mathews, R. G.; Ludwig, M. L. *Science*, **1994**, *266*, 1699.
7. Mancia, F.; Rasmussen, B.; Boscke, P.; Diat, O.; Evans, P. R. *Structure* **1996**, *4*, 339.
8. Lenhert, P. G.; Hodgkin, D. C. *Nature (London)* **1961**, *192*, 937; Lenhert, P. G. *Proc. R. Soc. London* **1968**, *A303*, 45; Savage, H. F.; Lindley, P. F.; Finney, J. L.; Timmins, P. A. *Acta Crystallogr.* **1987**, *B43*, 296; Bouquiere, J. P.; Finney, J. L.; Lehmann, M. S.; Lindley, P. F.; Savage, H. F. *Acta Crystallogr.* **1993**, *B49*, 79; Kraütler, B.; Konrat, R.; Stupperich, E.; Gerald, F.; Gruber, K.; Kratky, C. *Inorg. Chem.* **1994**, *33*, 4128.

9.   Pett, V. B.; Liebman, M. N.; Murray-Rust, P.; Prasad, K.; Glusker, J. P. *J. Am. Chem. Soc.* **1987**, *109*, 3207.

10. Rossi, M.; Glusker, J. P.; Randaccio, L.; Summers, M. F.; Toscano, P. J.; Marzilli, L. G. *J. Am. Chem. Soc.* **1985**, *107*, 1729.

11. Kräutler, B.; Keller, W.; Kratky, C. *J. Am. Chem. Soc.* **1989**, *111*, 8936.

12. Sagi, I.; Wirt, M. D.; Chen, E.; Frisbie, S.; Chance, M. R. *J. Am. Chem. Soc.* **1990**, *112*, 8639.

13. Shrauzer, G. N. *Acc. Chem. Res.* **1968**, *1*, 97.

14. Finke, R. G.; Smith, B. L.; McKenna, W.; Christian, P. A. *Inorg. Chem.* **1981**, *20*, 687.

15. Costa, G. *Coord. Chem. Rev.* **1972**, *8*, 63.

16. Rétey, J.; Umani-Ronchi, A.; Seibl, J.; Arigoni, D. *Experientia* **1966**, *22*, 502.

17. Babior, B. M.; Moss, T. A.; Orme-Johnson, W. H.; Beinert, H. *J. Biol. Chem.* **1974**, *249*, 4537.

18. Valinsky, J. E.; Abeles, J. E.; Fee, J. A. *J. Am. Chem. Soc.* **1974**, *96*, 4709.

19. Padmakumar, R.; Banerjee, R. *J. Biol. Chem.* **1995**, *270*, 1; Babior, B. M. *Acc. Chem. Res.* **1975**, *8*, 376, and references therein.

20. Zelder, O.; Beatrix, B.; Leytbecher, U.; Buckel, W. *Eur. J. Biochem.* **1994**, *226*, 577.

21. Padmakumar, R.; Taoka, S.; Padmakumar, R.; Banerjee, R. *J. Am. Chem. Soc.* **1995**, *117*, 7033; Padmakumar, R.; Banerjee, R. *J. Biol. Chem.* **1995**, *270*, 9295; Zhao, Y.; Abend, A.; Kunz, M.; Such, P.; Rétey, J. *Eur. J. Biochem.* **1994**, *225*, 891; Keep, N. H.; Smith, G. A.; Evans, M. C. W.; Diakun, G. P.; Leadlay, P. F. *Biochem. J.* **1993**, *295*, 387.

22. Gerfen, G. J.; Licht, S.; Willems, J.-P.; Hoffman, B. M.; Stubbe, J. *J. Am. Chem. Soc.* **1996**, *118*, 8192.

23. Mancia, F.; Keep, N. H.; Nakagawa, A.; Leadlay, P. F.; McSweeny, S.; Rasmussen, B.; Boescke, P.; Diat, O.; Evans, P. R. *Stucture* **1996**, *4*, 339.

24. Stubbe, J. *Science* **1994**, *266*, 1663; Drennan, C. L.; Mathews, R. G.; Ludwig, M. L. *Curr. Opin. Struct. Biol.* **1994**, *4*, 919.

25. Halpern, J. *Science* **1985**, *227*, 869.

26. Finke, R. G.; Schiraldi, D. A.; Mayer, B. J. *Coord. Chem. Rev.* **1984**, *54*, 1.

27. Halpern, J.; Ng, F. T. T.; Rempel, G. L. *J. Am. Chem. Soc.* **1979**, *101*, 7124.

28. Endicott, J. F.; Ferraudi, G. J. *J. Am. Chem. Soc.* **1977**, *99*, 243.

29. Nie, S.; Marzilli, P. A.; Marzilli, L. G.; Yu, N.-T. *J. Am. Chem. Soc.* **1990**, *112*, 6084.

30. Dong, S.; Padmakumar, R.; Banerjee, R.; Spiro, T. *J. Am. Chem. Soc.* **1996**, *118*, 9182.

31. Halpern, J. *Polyhedron* **1988**, *7*, 1483.

32. Koenig, T. W.; Hay, B. P.; Finke, R. G. *Polyhedron* **1988**, *7*, 1499; Koenig, T. W.; Finke, R. G. *J. Am. Chem. Soc.* **1988**, *110*, 2657.

33. Daikh, B. E.; Huthchison, J. E.; Gray, N. E.; Smith, B. L.; Weakley, J. R.; Finke, R. G. *J. Am. Chem. Soc.* **1990**, *112*, 7830.

34. Woska, D. C.; Xie, Z. D.; Gridnev, A. A.; Ittel, S. D.; Fryd, M.; Wayland, B. B. *J. Am. Chem. Soc.* **1996**, *118*, 9102.
35. Finke, R. G.; Hay, B. P. *Inorg. Chem.* **1984**, *23*, 3043; Ibid. **1985**, *24*, 1278.
36. Halpern, J.; Kim, S. H.; Leung, T. W. *J. Am. Chem. Soc.* **1984**, *106*, 8317; Ibid. **1985**, *107*, 2199.
37. Hay, B. P.; Finke, R. G. *J. Am. Chem. Soc.* **1986**, *108*, 4820.
38. Martin, B. D.; Finke, R. G. *J. Am. Chem. Soc.* **1990**, *112*, 2419.
39. Waddington, M. D.; Finke, R. G. *J. Am. Chem. Soc.* **1993**, *115*, 4629.
40. Hay, B. P.; Finke, R. G. *J. Am. Chem. Soc.* **1987**, *109*. 8012.
41. Garr, C. D.; Sirovatka, J. M.; Finke, R. G. *J. Am. Chem. Soc.* **1996**, *118*, 11142.
42. Wollowitz, S.; Halpern, J. *J. Am. Chem. Soc.* **1988**, *110*, 3112.
43. Choi, S.-C.; Dowd, P. *J. Am. Chem. Soc.* **1989**, *111*, 2313, and references therein.
44. Murakami, Y.; Hisaeda, Y.; Ozaki, T.; Ohno, T.; Fan, S.-D.; Matsuda, Y. *Chem. Lett.* **1988**, 839.
45. Dowd, P.; Choi, S.-C.; Duak, F.; Kaufman, C. *Tetrahedron* **1988**, *44*, 2137.
46. Murakami, Y.; Hisaeda, Y.; Kikuchi, J.; Ohno, T.; Suzuki, M.; Matsuda, Y.; Matsura, T. *J. Chem. Soc., Perkin Trans. 2* **1988**, 1237.
47. Choi, G.; Choi, S.-C.; Galan, A.; Wilk, B.; Dowd, P. *Proc. Natl. Acad. Sci. U.S.A.* **1990**, *87*, 3174.
48. He, M.; Dowd, P. *J. Am. Chem. Soc.* **1996**, *118*, 711.
49. Finke, R. G.; McKenna, W. P.; Schiraldi, D. A.; Smith, B. L.; Pierpont, C. *J. Am. Chem. Soc.* **1983**, *105*, 7592; Finke, R. G.; Schiraldi, D. A. *J. Am. Chem. Soc.* **1983**, *105*, 7605.
50. Silverman, R. B.; Dolphin, D. *J. Am. Chem. Soc.* **1976**, *98*, 4633.
51. Arigoni, D. In *Vitamin B₁₂, Proceedings of the Third European Symposium on Vitamin B₁₂ and Intrinsic Cofactor*; Zagalak, B.; Friedrich, W., Eds.; Walter de Gruyter: Berlin, 1979; p. 389.
52. Beartix, B.; Zelder, O.; Kroll, F. K.; Örlygsson, G.; Golding, B. T.; Buckel, W. *Angew. Chem., Int. Ed. Engl.* **1995**, *34*, 2398.
53. Chen, E.; Chance, M. R. *J. Biol. Chem.* **1990**, *265*, 12987; Endicott, J. F.; Ferraudi, G. J. *J. Am. Chem. Soc.* **1977**, *99*, 243.
54. Lott, W. B.; Chagovetz, A. M.; Grissom, C. B. *J. Am. Chem. Soc.* **1995**, *117*, 12194.
55. Valle, B. L.; Auld, D. S. *Acc. Chem. Res.* **1993**, *26*, 543.
56. *Zinc Enzymes, Prog. Inorg. Biochem. Biophys.*; Bertini, I.; Luchinat, C.; Maret, W.; Zeppezauer, M., Eds.; Birkhauser: Boston, 1986, Vol. 1.
57. Christianson, D. W.; Fierke, C. A. *Acc. Chem. Res.* **1996**, *29*, 331; Silverman, D. N. *Methods Enzymol.* **1995**, *249*, 479; Lindskog, S.; Liljas, A. *Curr. Opin. Struct. Biol.* **1993**, *3*, 915; Silverman, D. N.; Lindskog, S. *Acc. Chem. Res.* **1988**, *21*, 30.
58. Woolley, P. *Nature (London)* **1975**, *258*, 677.
59. Chaberek, S.; Courtney, R. C.; Martell, A. E. *J. Am. Chem. Soc.* **1952**, *74*, 5057.

60. Brown, R. S.; Salmon, D.; Curtis, N. J.; Kusuma, S. *J. Am. Chem. Soc.* **1982**, *104*, 3188.
61. Kimura, E.; Shiota, T.; Koike, T.; Shiro, M.; Kodama, M. *J. Am. Chem. Soc.* **1990**, *112*, 5805.
62. Kimura, E. *Prog. Inorg. Chem.* **1994**, *41*, 443.
63. Bertini, I.; Dei, A.; Luchinat, C.; Monnanni, R. *Inorg. Chem.* **1985**, *24*, 301.
64. Bauer, R.; Limkilde, P.; Johansen, J. T. *Biochemistry* **1976**, *15*, 334; Tibell, L.; Lindskog, S. *Biochim. Biophys. Acta* **1984**, *778*, 110.
65. Palmer, D. A.; van Eldik, R. *Chem. Rev.* **1983**, *83*, 651.
66. Lindskog, S. In *Zinc Enzymes*; Spiro, T. G. Ed.; Wiley: New York, 1983; pp 77 - 121.
67. Lipscomb, W. N. *Ann. Rev. Biochem.* **1983**, *52*, 17; Liang, J.-Y.; Lipscomb, W. N. *Int. J. Quantum Chem.* **1989**, *36*, 299.
68. Kannan, K. K.; Ramanadham, M.; Jones, T. A. *Ann. N.Y. Acad. Sci.* **1984**, *429*, 49.
69. Vedani, A.; Huhta, D. W.; Jacober, S. P. *J. Am. Chem. Soc.* **1989**, *111*, 4075.
70. Pocker, Y.; Janji´c, N. *J. Am. Chem. Soc.* **1989**, *111*, 731.
71. Tu, C.; Silverman, D. N.; Forsman, C.; Jonsson, B.-H.; Lindskog, S. *Biochemistry* **1989**, *28*, 7913.
72. Tu, C.; Paranawithana, S. R.; Jewell, D. A.; Tanhauser, S. M.; LoGrasso, P. V.; Wynns, G. C.; Laipis, P. J.; Silverman, D. N. *Biochemistry* **1990**, *29*, 6400.
73. Merz, K. M., Jr.; Banci, L. *J. Am. Chem. Soc.* **1997**, *119*, 863.
74. Aqvist, J.; Fothergill, M.; Warshel, A. *J. Am. Chem. Soc.* **1993**, *115*, 631; Zheng, Y.-J.; Merz, K. M., Jr. *J. Am. Chem. Soc.* **1992**, *114*, 10498; Merz, K. M., Jr. *J. Am. Chem. Soc.* **1991**, *113*, 406; Merz, K. M., Jr.; Hoffmann, R.; Dewar, M. J. S. *J. Am. Chem. Soc.* **1989**, *111*, 5636.
75. Zhang, X.; van Eldik, R.; Koike, T.; Kimura, E. *Inorg. Chem.* **1993**, *32*, 5749; Zhang, X.; van Eldik, R. *Inorg. Chem.* **1995**, *34*, 5606.
76. Looney, A.; Han, R.; McNeil, K.; Parkin, G. *J. Am. Chem.Soc.* **1993**, 115, 703.
77. Pocker, Y.; Diets, T. L. *J. Am. Chem. Soc.* **1982**, *104*, 2424.
78. Dugad, L. B.; Cooley, C. R.; Gerig, J. J. *Biochemistry* **1989**, *28*, 3955.
79. Liang, J.-Y.; Lipscomb, W. N. *Biochemistry* **1989**, *28*, 9724.
80. Nair, S. K.; Ludwig, P. A.; Christianson, D. W. *J. Am. Chem. Soc.* **1994**, *116*, 3679.
81. Bertini, I.; Dei, A.; Luchinat, C.; Monnanni, R. *Prog. Inorg. Biochem. Biophys.* **1986**, *1*, 371.
82. Imlay, J. A.; Linn, S. *Science* **1988**, *240*, 1302.
83. Dawson, J. H. *Science*, **1988**, *240*, 433.
84. Perutz, M. F.; Fermi, G.; Luisi, B.; Shaanan, B.; Liddington, R. C. *Acc. Chem. Res.* **1987**, *20*, 309.
85. Takano, T. *J. Mol. Biol.* **1977**, *110*, 569.
86. Phillips, S. E. V. *J. Mol. Biol.* **1980**, *142*, 531.

87. Springer, B. A.; Egeberg, K. D.; Sligar, S. G.; Rohlfs, R. J.; Mathews, A. J.; Olson, J. S. *J. Biol. Chem.* **1989**, *264*, 3057.
88. Takano, T. *J. Mol. Biol.* **1977**, *110*, 537.
89. Jameson, G. B.; Molinaro, F. S.; Ibers, J. A.; Collman, J. P.; Brauman, J. I.; Rose, E.; Suslick, K. S. *J. Am. Chem. Soc.* **1980**, *102*, 3224.
90. Collman, J. P.; Gagne, R. R.; Halbert, T. R.; Marchon, J.-C.; Reed, C. A. *J. Am. Chem. Soc.* **1973**, *95*, 7868.
91. Norvell, J. C.; Nunes, A. C.; Schoenborn, B. P. *Science* **1975**, *190*, 569.
92. Lee, C. L.; Oldfield, E. *J. Am. Chem. Soc.* **1989**, *111*, 1584.
93. Coletta, M.; Ascenzi, P.; Traylor, T. G.; Brunori, M. *J. Biol. Chem.* **1985**, *260*, 4151.
94. Morikis, D.; Champion, P. M.; Springer, B. A.; Sligar, S. G. *Biochemistry* **1989**, *28*, 4791; Ramsden, J.; Spiro, T. G. *Biochemistry* **1989**, *28*, 3125.
95. Martin, J. L.; Migus, A.; Poyart, C.; Lecarpentier, Y.; Astier, R.; Antonetti, A. *Proc. Natl. Acad. Sci. U.S.A.* **1983**, *80*, 173.
96. Finsden, E. W.; Friedman, J. M.; Ondrias, M. R.; Simon, S. R. *Science*, **1985**, *229*, 661.
97. Beatie, J. K.; Binstead, R. A.; West, R. J. *J. Am. Chem. Soc.* **1978**, *100*, 3046; McGarvey, J. J.; Lawthers, I.; Heremans, K.; Toftlund, H. *Inorg. Chem.* **1990**, *29*, 252.
98. Beattie, J. K. *Adv. Inorg. Chem.* **1988**, *32*, 2.
99. Jongeward, K. A.; Magde, D.; Taube, D. J.; Marsters, J. C.; Traylor, T. G.; Sharma, V. S. *J. Am. Chem. Soc.* **1988**, *110*, 380.
100. Olson, J. S.; Rohlfs, R. J.; Gibson. Q. H. *J. Biol. Chem.* **1987**, *262*, 12930.
101. Gibson, Q. H.; Olson, J. S.; McKinnie, R. E.; Rohlfs, R. J. *J. Biol. Chem.* **1986**, *261*, 10228.
102. Johnson, K. A.; Olson, J. S.; Phillips, G. N., Jr. *J. Mol. Biol.* **1989**, *207*, 459.
103. Lloyd, C. R.; Eyring, E. M.; Ellis, W. R., Jr. *J. Am. Chem. Soc.* **1995**, *117*, 11993.
104. Chatfield, M. D.; Walda, K. N.; Magde, D. *J. Am. Chem. Soc.* **1990**, *112*, 4680.
105. Adachi, S.; Morishima, I. *J. Biol. Chem.* **1989**, *264*, 18896.
106. Taube, D. J.; Projahn, H.-D.; van Eldik, R.; Magde, D.; Traylor, T. G. *J. Am. Chem. Soc.* **1990**, *112*, 6880.
107. Projahn, H.-D.; van Eldik, R. *Inorg. Chem.* **1991**, *30*, 3283.
108. Schlichting, I.; Berendzen, J.; Phillips, G. N., Jr.; Sweet, R. M. *Nature* **1994**, *371*, 808.
109. *Cytochrome P-450: Structure, Mechanism and Biochemistry*; Ortiz de Montellano, P. R., Ed.; Plenum Press: New York, 1986.
110. Poulos, T. L.; Howard, A. J. *Biochemistry* **1987**, *26*, 8165; Poulos, T. L.; Finzel, A. J.; Howard, J. *J. Mol. Biol.* **1987**, *195*, 687.
111. Poulos, T. L. *Adv. Inorg. Biochem.* **1987**, *7*, 1.
112. Dawson, J. H.; Kau, L.-S.; Penner-Hahn, J. E.; Sono, M.; Eble, K. S.; Bruce, G. S.; Hager, L. P.; Hodgson, K. O. *J. Am. Chem. Soc.* **1986**, *108*, 8114.

113. Dawson, J. H.; Sono, M. *Chem. Rev.* **1987**, *87*, 1255.
114. Thomann, H.; Bernardo, M.; Goldfarb, D.; Kroneck, P. M. H.; Ullrich, V. *J. Am. Chem. Soc.* **1995**, *117*, 8243.
115. Ricard, L.; Schappacher, M.; Weiss, R.; Montiel-Montoya, R.; Bill, E.; Gonser, U.; Trautwein, A. *Nouv. J. Chim.* **1983**, *7*, 405.
116. Gunsalus, I. C.; Meeks, J. R.; Lipscomb, J. D.; Debnner, P.; Munck, E. In *Molecular Mechanisms of Oxygen Activation*; Hayaishi, O., Ed.; Academic Press: New York, 1974; p. 559.
117. Sligar, S. G.; Gunsalus, I. C. *Proc. Nat. Acad. Sci. U.S.A.* **1976**, *73*, 1078.
118. Dunford, H. B. *Adv. Inorg. Biochem.* **1982**, *4*, 41.
119. Penner-Hahn, J. E.; Eble, K. S.; McMurry, T. J.; Renner, M.; Balch, A. L.; Groves, J. T.; Dawson, J. H.; Hodgson, K. O. *J. Am. Chem. Soc.* **1986**, *108*, 7819.
120. McDonald, T. L.; Burka, L. T.; Wright, S. T.; Guengerich, F. P. *Biochem. Biophys. Res. Commun.* **1982**, *104*, 620.
121. Groves, J. T.; Avaria-Neisser, G. E.; Fish, K. M.; Imachi, M.; Kuczkowsli, R. L. *J. Am. Chem. Soc.* **1986**, *108*, 3837.
122. Groves, J. T.; Nemo, T. E.; Myers, R. S. *J. Am. Chem. Soc.* **1979**, *101*, 1032; Chang, C. K.; Kuo, M.-S. *J. Am. Chem. Soc.* **1979**, *101*, 3413.
123. Ostovi`c, D.; Bruice, T. C. *J. Am. Chem. Soc.* **1989**, *111*, 6511.
124. Holm, R. H. *Chem. Rev.* **1987**, *87*, 1401.
125. Groves, J. T.; Bonchio, M.; Carofiglio, T.; Shalyaev, K. *J. Am. Chem. Soc.* **1996**, *118*, 8961.
126. Champion, P. M. *J. Am. Chem. Soc.* **1989**, *111*, 3433.
127. Ortiz de Montellano, P. R.; Stearns, R. A. *J. Am. Chem. Soc.* **1987**, *109*, 3415.
128. Atkinson, J. K.; Ingold, K. U. *Biochemistry* **1993**, *32*, 9209.
129. Newcomb, M.; Le Tadic, M.-H.; Putt, D. A.; Hollenberg, P. F. *J. Am. Chem. Soc.* **1995**, *117*, 3312.
130. Newcomb, M.; Le Tadic-Biadatti, M.-H.; Chestney, D. L.; Roberts, E. S.; Hollenberg, P. F. *J. Am. Chem. Soc.* **1995**, *117*, 12085.
131. Choi, S.-Y.; Eaton, P. E.; Hollenberg, P. F.; Liu, K. E.; Lippard, S. J.; Newcomb, M.; Putt, D. A.; Upadhyaya, S. P.; Xiong, Y. *J. Am. Chem. Soc.* **1996**, *118*, 6547.
132. Ortiz de Montellano, P. R.; Mangold, B. L. K.; Wheeler, C.; Kunze, K. L.; Reich, N. O. *J. Biol. Chem.* **1983**, *258*, 4208; Kunze, K. L.; Mangold, B. L. K.; Wheeler, C.; Beilan, H. S.; Ortiz de Montellano, P. R. *J. Biol. Chem.* **1983**, *258*, 4202.
133. Mansuy, D. *Pure Appl. Chem.* **1987**, *59*, 759.
134. Brothers, P. J.; Collman, J. P. *Acc. Chem. Res.* **1986**, *19*, 209.
135. Dolphin, D.; Matsumoto, A.; Shortman, C. *J. Am. Chem. Soc.* **1989**, *111*, 411.
136. Artaud, I.; Gregoire, N.; Battioni, J.-P.; Dupre, D.; Mansuy, D. *J. Am. Chem. Soc.* **1988**, *110*, 8714.
137. Collman, J. P.; Hampton, P. D.; Brauman, J. I. *J. Am. Chem. Soc.* **1990**, *112*, 2986.

138. Palmer, R. M. J.; Ferrige, A. G.; Moncada, S. *Nature* **1987**, *327*, 524.
139. Ignarro, L. J.; Buga, G. M.; Wood, K. S.; Byrns, R. E.; Chaudhuri, G. *Proc. Natl. Acad. Sci. U.S.A.* **1987**, *84*, 9265.
140. Furchgott, R. F.; Zawadzki, J. V. *Nature* **1980**, *288*, 373.
141. Williams, R. J. P. *Chem. Soc. Rev.* **1996**, 78; Butler, A. R.; Williams, D. L. H. *Chem Soc. Rev.* **1993**, 233; Clarke, M. J.; Gaul, J. B. *Structure and Bonding*, Vol. 81; Springer-Verlag: Berlin, 1993; pp. 47-181; Feldman, P. L.; Griffith, O. W.; Stuehr, D. J. *Chem. Eng. News* **1993** (December 20), 26; Snyder, S. H.; Bredt, D. S. *Scientific American* **1992**, *226* (May), 28; Moncada, S.; Palmer, R. M. J.; Higgs, E. A. *Pharmacol. Rev.* **1991**, 43, 109.
142. *The Biology of Nitric Oxide*; Moncada, S.; Marletta, M. A.; Hibbs, J. B.; Higgs, E. A., Eds.; Portland Press: London, 1992; Vols. 1 and 2; *Nitric Oxide in the Nervous System*; Vincent, S. R.; Academic Press: New York, 1995.
143. Myers, P. R.; Minor, R. L.; Guerra, R.; Bates, J. N.; Harrison, G. D. *Nature* **1990**, *345*, 161; Stamler, J. S.; Simon, D. I.; Osborne, J. A.; Mullins, M. E.; Jaraki, O.; Michel, T.; Singel, D. J.; Loscalzo, J. *Proc. Natl. Acad. Sci. U.S.A.* **1992**, *89*, 444.
144. Leone, A. M.; Palmer, R. M. J.; Knowles, R. G.; Francis, P. L.; Ashton, D. S.; Moncada, S. *J. Biol. Chem.* **1991**, *226*, 23790.
145. Wang, J.; Rousseau, D. L.; Abu-Soud, H. M.; Stuehr, D. J. *Proc. Natl. Acad. Sci. U.S.A.* **1994**, *91*, 10512.
146. Hoshino, M.; Maeda, M.; Konishi, R.; Seki, H.; Ford, P. C. *J. Am. Chem. Soc.* **1996**, *118*, 5702.
147. Hoshino, M.; Ozawa, K.; Seki, H.; Ford, P. C. *J. Am. Chem. Soc.* **1993**, *115*, 9568.
148. Traylor, T. G.; Magde, D.; Marsters, J.; Jongeward, K.; Wu, G.-Z.; Walda, K. *J. Am. Chem. Soc.* **1993**, *115*, 4808.
149. Goldstein, S.; Czapski, G. *J. Am. Chem. Soc.* **1995**, *117*, 12078.
150. McKee, M. L. *J. Am. Chem. Soc.* **1995**, *117*, 1629.
151. Goldstein, S.; Czapski, G. *J. Am. Chem. Soc.* **1996**, *118*, 3419, 6806.
152. Beckman, J. S.; Beckman, T. W.; Chen, J.; Marshall, P. A.; Freeman, B. A. *Proc. Natl. Acad. Sci. U.S.A.* **1990**, *87*, 1620.
153. Lymar, S. V.; Hurst, J. K. *J. Am. Chem. Soc.* **1995**, *117*, 8867.

9

Experimental Methods

A kinetic study generally proceeds after the reactants, products, and stoichiometry of the reaction have been satisfactorily characterized. The more one knows about the chemistry of the reaction, the better the conclusions that one can draw from a kinetic study. The discussion here describes techniques often used in inorganic studies, emphasizes their time range and general area of applicability, and gives some examples of their use. Further details can be found in other sources.[1,2]

Any experimental kinetic method must somehow monitor change of concentration with time. Many studies are done under pseudo-first-order conditions, and then one must monitor the deficient reactant or product(s) because the other species undergo small changes in concentration. The kinetic method(s) of choice often will be dictated by the time scale of the reaction. The detection method(s) will be determined by the spectroscopic properties of the species to be monitored. The efficient use of materials can be a significant factor in the choice of method because a kinetic study generally involves a number of runs at different concentrations and temperatures, and conservation of difficult to prepare or expensive reagents may be a critical factor.

The detection method should be as species specific as possible, and ideally one would like to measure both reactant disappearance and product formation. The method must not be subject to interference from other reactants and should be applicable under a wide range of concentration conditions so that the rate law can be fully explored. Often there is a practical trade-off between specificity, sensitivity, and reaction time. For example, NMR is quite specific but rather slow and has relatively low sensitivity, unless the system allows time for signal accumulation. Spectrophotometry in the UV and visible range often has good sensitivity and speed, but the specificity may be poor because absorbance bands are broad and intermediates may have chromophoric properties similar to those of the reactant and/or product. Vibrational spectrophotometry can be better if the IR bands are sharp, as in the case of metal carbonyls, but the solvent must be chosen to provide an approriate spectral window. Conductivity change can be very fast but is rather unspecific.

9.1 FLOW METHODS

In these methods, the reagent solutions are brought together by flowing them through a mixer from which the reaction solution emerges to be analyzed. The flow may simply be driven by gravity or by mechanical pressure applied to syringes containing the reagents. The minimum time scale depends on various factors such as the reagent flow rate, the efficiency of the mixer, and the response time of the analyzer. This general process has been adapted in various ways to minimize the amount of reagents used, optimize the detection sensitivity, and shorten the accessible reaction time. Some of these adaptations are described in the following sections.

9.1.a Quenched Flow

This method involves driving the reagent solutions through a mixer and then having some means of stopping (quenching) the reaction as the solution emerges from the mixer. The reaction time can be controlled by changing the length of tubing between the mixer and quencher. Calibration with a reaction of known rate is necessary. The main trick is to find an effective quenching method, and this will depend on the chemical reaction; adding acid, base, or precipitating agents, and rapid cooling are common methods. The short time limit is ~20 ms, but this depends on the effectiveness of the quenching method.

The advantages of this method are that the apparatus is simple and that analysis of the quenched solution can be done without time constraints. The disadvantages are the sometimes tedious analysis of many samples and the consumption of substantial amounts of reagents for each kinetic run. The method has been used especially for isotope exchange reactions where the subsequent analysis of isotopic content is a slow process.

9.1.b Stopped Flow

For reactions with half-times in the 10-ms to ~60-s range, stopped flow is the most popular technique, and several commercial instruments are available.[3]

In a typical instrument, the reagent solutions are contained in two drive syringes whose plungers can be advanced by activating an air pressure or electrical drive system. This moves the solutions through a mixer into an observation cell and then to a stopping syringe. A mechanical stop on the stopping syringe or drive mechanism stops the flow and triggers the observation and data recording system. The standard system mixes equal volumes (~0.2 to 0.5 mL) of each reagent solution and uses single-wavelength, single-beam UV–visible spectrophotometry as the detection method. A number of variations have been described using other detection methods (conductivity, fluorimetry, NMR,[4] ESR,[5] EXAFS[6]), and multiwavelength detection.[7] Instruments have been described for measurements at high pressure,[8] at subzero temperatures,[9] and for multiple mixing,[10] first of two reagents followed by a third after some time interval. In general, the method is quite adaptable and widely applicable.

The time range limitations are determined on the short end by the deadtime of the system (the time for mixing of reagents and transfer to the observation cell) and on the long end by the diffusion of reagents into the observation cell. The experimental first-order rate constant, k_{exp}, can be corrected for the mixing time effect, k_{mix}, to obtain the true pseudo-first-order rate constant, k_{true}, from the relationship suggested by Dickson and Margerum[11] that $(k_{true})^{-1} = (k_{exp})^{-1} - (k_{mix})^{-1}$. The rearranged version of this expression, given by

$$k_{true} = k_{exp}\left(1 - \frac{k_{exp}}{k_{mix}}\right)^{-1} \tag{9.1}$$

shows that the true rate constant is always larger than the experimental value, but the correction will be insignificant if $k_{exp}/k_{mix} \ll 1$. To determine k_{mix}, measurements can be done on a well-characterized system under pseudo-first-order conditions with $k_{true} = k_1[R]$, where k_1 is known and R is the excess reagent whose concentration can be varied to change k_{true}. Then, the variation of k_{exp} with [R] is used to determine k_{mix}. Margerum and co-workers[12] have reported values of k_{mix} for Durrum and Hi-Tech instruments of 1.7×10^3 and 2.9×10^3 s^{-1}, respectively.

The deadtime is due primarily to the physical separation of the mixer and observation cell and also depends on the flow velocity. Typical deadtimes are in the 1- to 5-ms range and can be determined by extrapolation of the observable back to the known value at true time zero, as shown by the dashed lines in Figure 9.1.

The actual time, t, is related to the experimentally recorded time, t_{exp}, and the deadtime, t_d, by

$$t = t_{exp} + t_d \tag{9.2}$$

For spectrophotometric detection, Figure 9.1 illustrates the relationship between these times and A_0^{pred}, the predicted absorbance at true zero time, and A_0^{obsd}, the initial absorbance at the start of the detection system. The reaction has a pseudo-first-order rate constant $k_{exp} = k_1[R]$, so that the time dependence of the absorbance change is given by

$$\left(A_\infty - A\right) = \left(A_\infty - A_0^{pred}\right)\exp\left(-k_{exp}\,t\right) \tag{9.3}$$

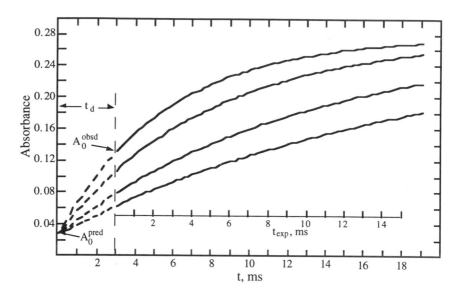

Figure 9.1. Schematic stopped-flow recordings with varying first-order rate constants, showing the variation of the observed initial absorbance, A_0^{obsd}, with the rate constant.

where A_∞ is the final absorbance and A is the absorbance at any time t. Both sides of Eq. (9.3) can be multiplied by $\exp(k_{exp}t_d)$, and noting that $t_{exp} = t - t_d$, one obtains

$$\left(A_\infty - A \right) \exp\left(k_{exp}\, t_d \right) = \left(A_\infty - A_0^{pred} \right) \exp\left(-k_{exp} \left(t - t_d \right) \right)$$

$$= \left(A_\infty - A_0^{pred} \right) \exp\left(-k_{exp}\, t_{exp} \right) \qquad (9.4)$$

Substitution of the limiting condition that $A = A_0^{obsd}$ when $t_{exp} = 0$ into Eq. (9.4) and rearrangement yields the expression

$$\left(A_\infty - A_0^{obsd} \right) = \left(A_\infty - A_0^{pred} \right) \exp\left(-k_{exp}\, t_d \right) \qquad (9.5)$$

which can be used to determine t_d from the measurable quantities k_{exp}, A_0^{obsd}, and A_0^{pred}. Since $k_{exp} = k_1[R]$, then R can be varied in a series of experiments and A_0^{obsd} should change, as shown in Figure 9.1, so that an average t_d, which should be independent of [R], can be calculated.

Substitution from Eqs. (9.2) and (9.5) into Eq. (9.3) gives

$$\left(A_\infty - A \right) = \left(A_\infty - A_0^{obsd} \right) \exp\left(-k_{exp}\, t_{exp} \right) \qquad (9.6)$$

which shows that k_{exp} is independent of t_d and is determined from the dependence of $(A_\infty - A)$ on t_{exp}.

The situation is more complex for studies under second-order conditions because the reagent concentrations at the true zero time must be known. Meagher and Rorabacher[13] have analyzed the second-order system of reactants A + B coming to equilibrium with products C + D and have given the integrated expression for the time dependence of the concentration change. The deadtime could be adjusted to give the A_0^{pred}, if this value is known, and Meagher and Rorabacher have suggested an empirical way of dealing with the mixing problem.

Another limitation of stopped-flow mixing is that the reagent solutions should be of similar composition in order to avoid spurious effects due to inhomogeneous mixing. It is always advisable to do blank observations to ensure that no apparent reaction is observed in the absence of each reactant.

There are numerous applications of this method in which the collection and analysis of the experimental rate constants are entirely straightforward. The few examples described below illustrate some special aspects, or address problems of longstanding interest.

The hydrolysis of cobalt(III) carbonato chelates in acidic solution has been the subject of numerous studies and may be a model for the carbonic anhydrase catalyzed dehydration of CO_2. The results have been reviewed and reanalyzed.[14] Buckingham and Clark[15] have provided new insight into this reaction by taking advantage of the multiwavelength observation capabilities of modern instruments. As a result, it has been possible to show that the reactions are often biphasic and to identify the optimum wavelength to observe the biphasic character. The elements of the mechanism are shown in the Scheme 9.1.

Scheme 9.1

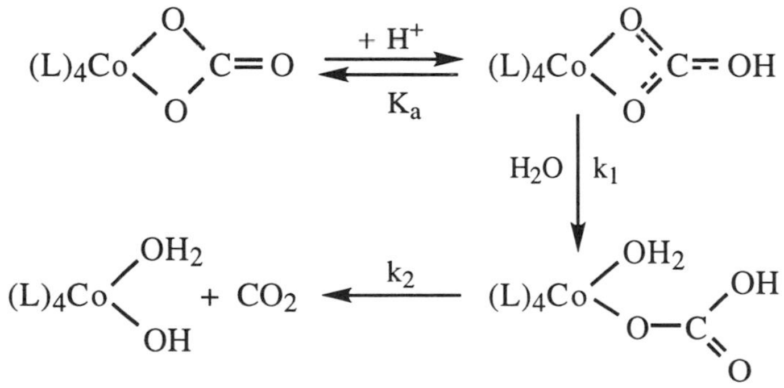

For some systems the protonated species has been detected and the K_a determined. The two steps in the reaction are assigned to chelate ring opening, k_1, and decarboxylation, k_2, of the monodentate bicarbonate complex. This work provides an example of the ambiguity in assigning each rate constant to the correct reaction in biphasic systems. The $[H^+]$ dependence was used to assign the larger rate constant to the chelate ring opening for $Co(NTA)(O_2CO)^{2-}$ and $Co(tren)(O_2CO)^+$, but to the decarboxylation for $Co(gly)_2(O_2CO)^-$ and $Co(NH_3)_4(O_2CO)^+$. It was found that $K_a \sim 1$ M and $k_2 \sim 1$ s^{-1} (25°C, 1.0 M NaClO$_4$) for all the systems studied. This insensitivity to the ancilliary ligands is consistent with protonation at a site remote from the Co(III), and for the O—C rather than the Co—O bond breaking for the decarboxylation. However, k_1 changes by $\sim 10^4$ s^{-1} for different $(L)_4$ systems because the Co—O bond is broken. Buckingham and Clark[15] suggest a detailed mechanism that involves an intermediate or minor equilibrium species with a proton transferred to a carbonate oxygen bound to Co(III).

Reactions of aqueous iron(III) with various ligands often yield highly colored products. The classic example of this is the deep red thiocyanate complex $Fe(OH_2)_5(NCS)^{2+}$. This system is ideal for study because of the large absorbance change and the ready availability of reagents. There is some mechanistic interest because reactivity arguments have suggested that the substitution mechanism is associative for $Fe(OH_2)_6^{3+}$ and dissociative for $Fe(OH_2)_5(OH)^{2+}$. A number of pressure-dependent studies have been done in the expectation that ΔV^* values would help to validate the reactivity arguments. There is general agreement that the reaction proceeds by the following two pathways:

$$Fe(OH_2)_6^{3+} + SCN^- \xrightarrow{k_1} Fe(OH_2)_5(NCS)^{2+} + H_2O$$

$$(9.7)$$

$$Fe(OH_2)_5(OH)^{2+} + SCN^- \xrightarrow{k_2} Fe(OH_2)_4(OH)(NCS)^+ + H_2O$$

Earlier studies,[16] mainly using pressure- and temperature-jump relaxation methods, obtained k_1 of (1.2 to 1.5) x 10^2 M^{-1} s^{-1} and k_2 of (1 to 4) x 10^4 M^{-1} s^{-1}, at 25°C. But stopped-flow methods[17,18] have given values of 70 to 90 M^{-1} s^{-1} for k_1 and (4 to 7) x 10^3 M^{-1} s^{-1} for k_2. Furthermore, the activation volumes from six studies range from -6.1 to $+6.7$ cm^3 M^{-1} for k_1, and from 0 to $+16.5$ cm^3 M^{-1} for k_2. The most recent study by Grace and Swaddle[19] used high-pressure stopped flow and low concentrations and second-order conditions for Fe(III) and SCN$^-$, in order to avoid higher-order thiocyanate complexes. They obtained ΔV^* values of -5.7 cm^3 M^{-1} and $+9.0$ cm^3 M^{-1} for k_1 and k_2, respectively. These results agree with the earlier stopped-flow study of Funahashi et al.[18] with a large excess of SCN$^-$, but not with results from temperature-jump and pressure-jump relaxation methods. Funahashi et al. suggested that some of the earlier work was affected by nitrate ion complexation from the NaNO$_3$/HNO$_3$ medium used, but the results of Capitan et al.[17] show rather small differences in rate constants at 25°C between nitrate and perchlorate media. Grace and Swaddle[19] noted that some relaxation studies might be affected by incorrect speciation, but reanalysis did not remove the disparity. They proposed that electric discharge effects in the temperature-jump measurements may affect the observations. It may be relevant that Betts and Dainton[20] have observed the oxidation of SCN^- by aqueous iron(III).

9.1.c Continuous Flow

Historically, continuous flow preceded stopped flow as a method for studying moderately fast reactions. The reactant solutions are made to flow continuously through the mixer and observation chamber, and the time dependence of the reaction can be obtained by changing the flow rate or moving the observation point to various distances from the mixer. The apparatus can be quite simple, but large amounts of reagents are consumed. Pulsed continuous flow,[21] a method in which continuous flow is established for a short time, can reduce reagent consumption to ~5 mL, and fast jet mixers have lowered the accessible reaction half-time to the 10-µs range. The concentration of the reagent being monitored can be lowered if integrating observation is used in which the flowing solution is viewed down the length of the observation cell. A combined continuous flow with integrating observation and stopped-flow system has been described.[22]

9.1.d Pulsed Accelerated Flow

This method may be viewed as an adaptation of pulsed continuous flow in which the flow rate through the mixer and observation chamber is varied during the course of one run. Most recent applications of this method have been from Margerum and co-workers.[23] The advantage of the method is that it can be used for half-times down to ~10 µs, compared to ~10 ms for stopped flow. Because of the complexity of the analysis, the method is limited to first-order reaction conditions.

9.2 RELAXATION METHODS

In these methods, a system at equilibrium is subjected to a perturbation and the kinetics of the system relaxing to the new equilibrium condition is followed. The perturbation normally is a change in temperature, pressure, or concentration of one of the reagents and the methods are known as temperature jump, pressure jump, and concentration jump, respectively. The advantage of these methods is that the perturbation, especially of temperature and pressure, can be applied very quickly and reactions with half-times in the microsecond range can be observed. The pioneering work on these methods by Eigen and co-workers[24] greatly extended the time scale for solution kinetic studies. The major limitation is that the equilibrium position of the reaction must involve significant concentrations of both reactants and products; therefore, relaxation methods are not applicable to essentially irreversible reactions. These methods are especially useful for Lowry–Brønsted acid–base reactions in which the equilibrium position can be adjusted simply by changing the pH of the solution and for ligand substitution reactions that involve proton production or consumption.

It is a noteworthy feature of relaxation methods that the changes in concentration caused by the perturbation should not be too large, so that the mathematical analysis can be simplified. This poses some limitations for the detection method, in that it must be fast but also sensitive enough to detect these small concentration changes. However, it is possible to repeat the perturbation and improve the signal to noise ratio through signal averaging.

The standard mathematical analysis may be illustrated for the following system:

$$A + B \underset{k_r}{\overset{k_f}{\rightleftharpoons}} C \qquad (9.8)$$

After the perturbation, the system comes to a new equilibrium with final concentrations $[A_e]$, $[B_e]$, and $[C_e]$, and these may be related to the concentration at any time through the concentration change variable Δ, so that $[A] = [A_e] + \Delta$, $[B] = [B_e] + \Delta$, and $[C] = [C_e] - \Delta$. Simple differentiation shows that $d\Delta/dt = d[A]/dt = d[B]/dt = -d[C]/dt$, and one can write the usual differential equation for the system as follows:

$$\frac{d\Delta}{dt} = \frac{d[A]}{dt} = -k_f[A][B] + k_r[C]$$

$$= -k_f\Big([A_e] + \Delta\Big)\Big([B_e] + \Delta\Big) + k_r\Big([C_e] - \Delta\Big) \qquad (9.9)$$

Expansion and collection of terms gives

$$\frac{d\Delta}{dt} = -k_f\big([A_e] + [B_e]\big)\Delta - k_r\Delta - k_f[A_e][B_e] + k_r[C_e] - k_f\Delta^2 \qquad (9.10)$$

At equilibrium, $k_f[A_e][B_e] = k_r[C_e]$, and these terms cancel. Next, the assumption is made that Δ is very small, so that the term in Δ^2 can be neglected and Eq. (9.10) simplifies to

$$\frac{d\Delta}{dt} = -\big\{k_f\big([A_e] + [B_e]\big) + k_r\big\}\Delta = \frac{\Delta}{\tau} \qquad (9.11)$$

where τ is defined as the relaxation time.

If experiments are done with varying positions of the equilibrium, then a plot of τ^{-1} versus $[A_e] + [B_e]$ should have a slope of k_f and an intercept of k_r. This is different from the normal pseudo-first-order system coming to equilibrium where the experimental rate constant is equal to $k_f + k_r$.

9.2.a Temperature Jump

For a system at equilibrium, if the temperature is changed by ΔT, then the equilibrium concentrations will change because of the thermodynamic relationship $(\delta \ln K / \delta T)_P = -\Delta H^o_{rxn} / RT^2$ between the equilibrium constant and the enthalpy change for the reaction, ΔH^o_{rxn}. In early applications and commercial instruments, the temperature was quickly changed, typically by ~5°C, by an electrical discharge. The sample was contained between two electrodes and subjected to a voltage of ~10 to 100 kV across the electrodes to produce the discharge. This requires that the solution be electrically conducting and clearly is somewhat invasive on the sample. It has been noted[25] that ion polarization at the charged electrodes can cause spurious but reproducible signals.

For nonconducting solutions, applications of pulsed microwave heating have been described.[26,27] More recently, heating by an iodine laser has been used[28]; laser photons at 1315 nm are absorbed by overtone vibrations of OH bonds in the solvent to cause the heating.

Hague and Martin[29] used T-jump to study the complexation of aqueous manganese(II) by 2,2'-bipyridine, shown by the reaction in (9.12). The work was extended by Doss and van Eldik[30] to determine the pressure dependence (21°C, 0.3 M $NaClO_4$, pH ~6.8).

$$Mn(OH_2)_6{}^{2+} + bpy \underset{k_r}{\overset{k_f}{\rightleftharpoons}} Mn(OH_2)_4(bpy)^{2+} + 2\,H_2O \quad (9.12)$$

From both studies, $k_f \approx 2 \times 10^5$ M^{-1} s^{-1} and the pressure dependence gave $\Delta V_f{}^* = -3$ cm^3 M^{-1}. This negative value was taken as evidence for an I_a mechanism for substitution on $Mn(OH_2)_6{}^{2+}$.

It has been suggested[31] that, with proper calibration, the magnitude of the temperature-jump change can be used to determine K and ΔH^o_{rxn}. Secco and co-workers[32] have done such a study on the reaction of iron(III) with thiocyanate ion. Their analysis gave $K = 1.2 \times 10^2$ M^{-1}, $\Delta H^o_{rxn} = -1.6$ kcal mol^{-1}, and $k_1 = 6 \times 10^2$ M^{-1} s^{-1} (25°C, 0.5 M $HClO_4$). Unfortunately, these values are not in agreement with current stopped-flow values of $K \approx 2 \times 10^2$ M^{-1} and $k_1 \approx 1 \times 10^2$ M^{-1} s^{-1}. Secco and co-workers overlooked the earlier studies of Funahashi et al.[18] and others,[16] and omitted the contribution of the k_2 pathway to the observations.

9.2.b Pressure Jump

This method requires a finite volume change for the reaction, ΔV^o_{rxn}, so that the equilibrium constant will change with pressure due to the relationship $(\delta \ln K/\delta P)_T = -\Delta V^o_{rxn}/RT$. The experiment is done by putting the sample under high pressure and then suddenly reducing the pressure by piercing a diaphragm. High-pressure equipment and observation cells are required, but the perturbation seems less invasive on the sample than T-jump by electrical discharge. A P-jump system with conductivity detection has been described recently.[33]

9.2.c Concentration Jump

The system is perturbed by adding a small amount of one of the species in the equilibrium reaction. Generally the apparatus is much simpler compared to the T-jump or P-jump methods, but the perturbation can not be done as quickly so that the short-time limit is in the millisecond range.

9.3 ELECTROCHEMICAL METHODS

Various electrochemical methods, such as cyclic voltammetry, polarography, chronoamperometry, and chronopotentiometry, can be used to measure homogeneous reaction rates. It is beyond the scope of this text to explore all the variations and intricacies of electrochemical methods, but they are described in several sources.[34] The purpose here is to give some basic background and to provide some examples that illustrate the technique.

In general, electrochemical observations can give information about homogeneous reaction rates when an electrode reaction is coupled to a homogeneous chemical reaction, and the rate of the latter becomes rate limiting for the process at the electrode. Sometimes the chemical rate constant can be extracted fairly directly from the observations, or it may require curve matching of experimental and simulated curves computed with various rate constants. Since the size and composition of the electrode and the diffusion coefficients of reagents affect the kinetics of the electrode reaction, these factors will influence the observations and the effective time range for these methods.

The field has a well-developed nomenclature and symbolism. The one-electron electrode reaction is designated by **E** and a chemical reaction by **C**. There are extensions of this system, such as **E+E** for a two-electron electrode reaction, $\bar{\mathbf{E}}$ and $\bar{\bar{\mathbf{E}}}$ for reduction and oxidation, **C1** and **C2** for first- and second-order reactions, and **C1'** for a pseudo-first-order reaction. The examples described below rely heavily on cyclic voltammetry. This is the most widely used technique because of the availability of appropriate instrumentation, and the number of applications is likely to increase with the recent availability of software[35] to simulate cyclic voltammograms. Such simulations generally are essential for the determination of meaningful kinetic parameters.

An idealized cyclic voltammogram, CV, and some terminology of this technique are shown in Figure 9.2. The experiment is carried out by changing the voltage, E, of the working electrode at some constant sweep rate, v, and measuring the current, i. Then, the sweep rate and reagent concentrations are changed and the changes in cathodic and anodic peak potentials, E_{pc} and E_{pa}, and peak current, i_p, are analyzed.

The quantitative analysis requires knowledge of the rate(s) of the heterogeneous electrode reaction(s), reagent diffusion coefficients, and the transfer coefficient. If the electrode reaction is reversible, most of these parameters can be determined from the CV experiments. The formal reduction potential, $E^{o'}$, differs from the standard potential, E^{o}, because the latter is obtained by extrapolation to infinite dilution, while the former refers to the actual experimental conditions of ionic strength and temperature. For a fast, reversible process, $E^{o'} \approx E_{1/2} \pm 10$ mV if the diffusion coefficients of the oxidized and reduced forms are within a factor of two. Potentials are often reported relative to some standard electrode, such as ferrocene/ferrocinium ion, saturated calomel, SCE, or Ag/AgCl, and this must be taken into account in comparing results from different sources.

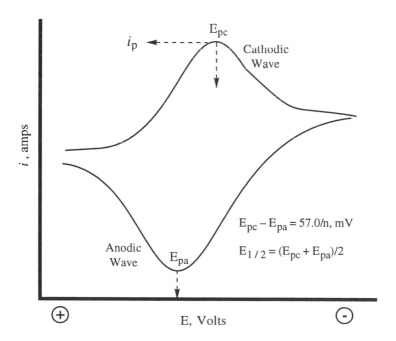

Figure 9.2. Sketch of a cyclic voltammogram for a reversible electrode reaction.

One restriction on these methods is that the medium must contain an "inert" electrolyte to maintain electrical conductivity. Typically, 0.1 M tetraalkylammonium salts of PF_6^-, $CF_3SO_3^-$, or ClO_4^- are used. Problems can arise due to adsorption of reagents on the electrodes and uncertainties in the chemical characterization of the product of the electrode reaction. The experiment can give the number of electrons, n, involved and the reduction potential, and then the nature of the electrochemically generated reagent often is inferred by chemical reasoning and analogy. It is possible to couple the system to some spectroscopic technique, such as EPR[36] or IR spectroscopy[37] (using transparent electrodes) to provide further characterization.

The electrochemical behavior of aqueous Cu(II) and Cu(I) complexed by 2,9-dimethyl-1,10-phenanthroline, DMP, has been studied by Lei and Anson.[38] The measurements involved cyclic and rotating-disk voltammetry with glassy carbon electrodes at pH 5.2 in a buffer containing 0.04 M aqueous acetic, phosphoric, and boric acids at ambient temperature. Glassy carbon electrodes were used to minimize adsorption of electroactive species on the electrode. If the initial ratio of DMP to Cu(II) is ≥ 2, then a normal CV is observed and assigned to the following reaction:

$$Cu(DMP)_2^{2+} + e^- \rightleftharpoons Cu(DMP)_2^+ \quad E^{o\prime} = 0.35 \text{ V (vs SCE)} \quad (9.13)$$

With equal concentrations of DMP and Cu(II), the electrochemical response is more complicated. Cathodic and anodic peaks appear at the same positions as in the $Cu(DMP)_2^{2+}/Cu(DMP)_2^+$ system, and a new cathodic peak is observed at ~0.1 V and assigned to the reaction

$$Cu(DMP)^{2+} + e^- \rightleftharpoons Cu(DMP)^+ \quad E^{o\prime} = 0.08 \text{ V (vs SCE)} \quad (9.14)$$

There is no anodic peak, however, and the sum of the cathodic peak currents is smaller than expected for reduction of all the Cu(II) present. This may be accounted for by the ligand redistribution reaction

$$Cu(DMP)^{2+} + Cu(DMP)^+ \underset{}{\overset{k_1}{\rightleftharpoons}} Cu(DMP)_2^+ + Cu^{2+} \quad (9.15)$$

which removes $Cu(DMP)^+$ and forms Cu^{2+}. Uncomplexed Cu^{2+} was not detected because it would appear below the experimental scan range, at –0.1 V. When the scan rate is increased from 50 to 250 mV s^{-1}, the anodic part of the $Cu(DMP)^{2+}/Cu(DMP)^+$ reaction appears, indicating that the scanning is faster than the ligand redistribution reaction (9.15) that removes $Cu(DMP)^+$ at low scan rates. This reaction must be faster than the redistribution of the Cu(II) species, shown by the following reaction:

$$2 \; Cu(DMP)^{2+} \; \underset{\longleftarrow}{\overset{k_2}{\longrightarrow}} \; Cu(DMP)_2{}^{2+} \; + \; Cu^{2+} \tag{9.16}$$

Otherwise, no $Cu(DMP)^{2+}/Cu(DMP)^+$ wave would be observed.

Quantitative analysis, based on comparison of experimental and simulated voltamograms, indicates values of $k_1 = 1.1 \times 10^4 \; M^{-1} \; s^{-1}$ and $k_2 = 5 \times 10^2 \; M^{-1} \; s^{-1}$, with equilibrium constants for the same reactions of 3×10^3 and 7.9×10^{-2}, respectively. Further analysis gave complex formation constants for $Cu(DMP)^+$ and $Cu(DMP)_2{}^+$ of 6.7×10^8 and $1.5 \times 10^{10} \; M^{-1}$, respectively, and it is the large value of the latter that leads to much of the complexity of the system.

The substitution inertness of Cr(III) and the lability of Cr(II) have allowed Hecht, Schultz, and Speiser[39] to observe some ring-opening and ring-closing reactions of amino-carboxylate ligands. The experiments used a stationary Hg drop electrode in 1.0 M Na_2SO_4 at pH 8.5, apparently at ambient temperature. Their observations can be explained by Scheme 9.2, where the aliphatic substituent R gives the *trans*-N,N geometry shown.

Scheme 9.2

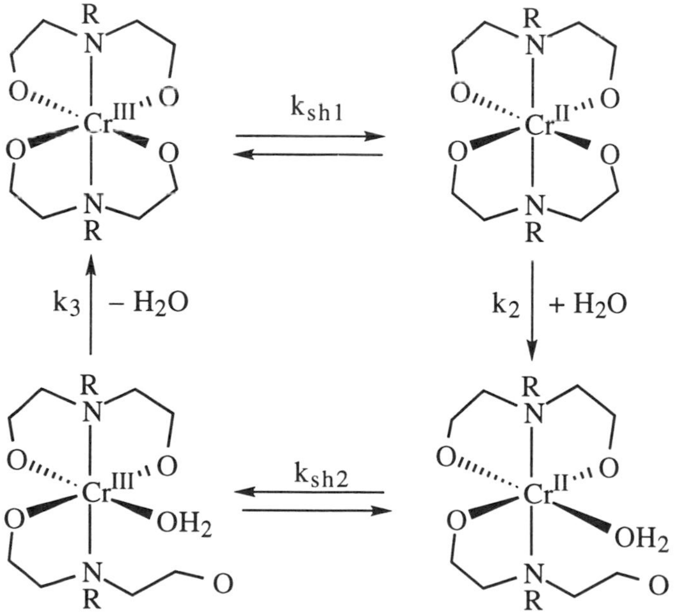

For such systems, initial CV reduction of the Cr(III) complex shows a broad cathodic wave whose position shifts from about -1.4 to -1.6 V as the sweep rate is increased. This is typical of an irreversible reduction with sluggish electrode kinetics and is assigned to the k_{sh1} process. If the voltage sweep is reversed after the irreversible reduction and the CV continued, then anodic (−1.18 V) and cathodic (−1.25 V) peaks appear, typical of a reversible process. These were assigned to k_{sh2}, the oxidation/reduction of the ring-opened species. Values of k_{sh1} and k_2 were determined by comparison of voltamograms at various sweep rates to digital simulations, and values of k_{sh2} and k_3 were obtained similarly from pre-electrolyzed solutions containing the Cr(II) complex. The results gave k_2 and k_3 values of 35 ± 6 and 1.3 ± 0.2 s^{-1}, respectively, for R = Me, and $7.3 \pm 3.4 \times 10^2$ and $2.2 \pm 0.2 \times 10^2$ s^{-1} for R = Et.

Hershberger, Klingler, and Kochi[40] studied the oxidation of $Mn(\eta^5\text{-}H_3CC_5H_4)(CO)_2(NCCH_3)$ by cyclic voltammetry with 0.1 M Et_4NClO_4 in acetonitrile at a platinum electrode. The acetonitrile complex has a reversible CV wave ($E^{o'} = 0.19$ V vs $Fe(Cp)_2^+/Fe(Cp)_2$), but when another ligand (PR$_3$) is added to the system, the observations can be understood in terms of Scheme 9.3.

Scheme 9.3

$$A \xrightarrow[-e]{k_{sh}} A^+$$

$$A^+ + PR_3 \xrightarrow{k_1} B^+ + CH_3CN$$

$$B^+ + A \xrightarrow{k_2} B + A^+$$

$$B^+ \xrightarrow{+e} B$$

$$A = Mn(\eta^5\text{-}H_3CC_5H_4)(CO)_2(NCCH_3)$$

$$B = Mn(\eta^5\text{-}H_3CC_5H_4)(CO)_2(PR_3)$$

	P(OPh)$_3$	P(Ph-4-Cl)$_3$	PPh$_3$	PMePh$_2$
k_1, M^{-1} s^{-1}	12	9.5×10^2	1.3×10^4	$>1 \times 10^5$

The reduction wave for the A$^+$ species is decreased to an extent dependent on the PR$_3$ concentration and the scan rate, and a new wave appears due to B/B$^+$. This is attributed to relatively rapid substitution of NCCH$_3$ by PR$_3$ in the 17-electron A$^+$ species, with a second-order rate constant k_1. By digital simulation of the voltamograms, the values of k_1 shown on Scheme 9.3 were determined for various PR$_3$ at 25°C. The span of these values provides an indication of the dynamic range of the method. This system is somewhat unusual because the k_2 reaction is thermodynamically favorable and provides the propagation step for a catalytic pathway for substitution of CH_3CN by PR_3. This work was extended[41] to substituted pyridines as leaving groups, and leaving and entering group effects on the kinetics were analyzed.

More recently, Sweigart and co-workers[42] have generated analogous anions $Mn(\eta^5-R_xC_5H_{5-x})(CO)_3^+$ (R = Me, Ph, ^nBu; x = 2, 3) by electrochemical oxidation and have studied their reactions with $P(OEt)_3$ in CH_2Cl_2 at 25°C. The rate constants for initial substitution to give $Mn(\eta^5-H_3CC_5H_4)(CO)_2(P(OEt)_3)^+$ and further substitution to form $Mn(\eta^5-H_3CC_5H_4)(CO)(P(OEt)_3)_2^+$ were determined to be 1×10^8 and 3.1×10^3 M^{-1} s^{-1}, respectively. The larger value was determined by the pre-wave method of Parker[43] under conditions of a deficiency of $P(OEt)_3$ (e.g., $[P(OEt)_3]/[Mn] = 0.55$). The rapid reaction of $P(OEt)_3$ depletes its concentration at the electrode surface so that the later part of the wave appears as a normal wave of the unreacted cation.

9.4 NUCLEAR MAGNETIC RESONANCE METHODS

There is a wide variety of applications of NMR to problems in inorganic kinetics. The time scale depends on the type of system and can vary from hours, when simply monitoring concentration changes of reactants and products, to microseconds, for exchange and fluxional processes on paramagnetic systems. One great advantage of NMR is that the temperature can be changed over a wide range, from about −200°C to +150°C without significant instrument modification. Another advantage is the molecular specificity of the NMR signal, which often permits an assignment of the composition and structure of stable intermediates and products. The specificity is augmented by the ability to detect a wide range of NMR active nuclei; ^1H, ^{13}C, ^{19}F, and ^{31}P are standard for most modern NMR instruments and many metals have NMR active isotopes that can be observed with appropriate modifications. A feature that is almost unique to NMR is the ability to measure rates of reactions in which there is no net chemical change, such as solvent exchange and ligand fluxionality. The major limitation of NMR is sensitivity, and concentrations must be typically about 0.01 M, unless signal averaging is possible. However, the small sample size of 0.5 to 2 mL allows for modest materials consumption.

Discussions of the theory and quantitative analysis in this area often use the lifetime, τ, of a nucleus in a particular site as the kinetic feature of interest. This lifetime has the conventional definition (see Section 1.1) of the concentration of nuclei in the site divided by their rate of disappearance from the site. To establish the relationship between the rate constant and the lifetime, it is necessary to define τ clearly because of ambiguities due to the number and populations of sites.

For example, the exchange of nuclei between a hydrated metal ion $M(OH_2)_n^{z+}$ and bulk solvent water can be represented by the reaction in

(9.17) with whole water molecule exchange, or it might just involve proton exchange as in (9.18).

$$M(OH_2)_n^{z+} + H_2O \rightleftharpoons M(OH_2)_{n-1}(OH_2)^{z+} + H_2O \qquad (9.17)$$

$$M(OH_2)_n^{z+} + H_2O \rightleftharpoons M(OH_2)_{n-1}(OH)^{(z-1)+} + HH_2O^+ \quad (9.18)$$

In either case there are two lifetimes, τ_m for the water ligands and τ_s for the bulk solvent. There is an ambiguity as to whether one is considering whole water molecule exchange with n coordinated sites as in reaction (9.17), or proton exchange with 2n such sites as in reaction (9.18). If ^{17}O NMR is used, then only whole molecule exchange will be observed and the definitions are straightforward and given by

$$\tau_m = \frac{n\,[M^{z+}]}{Rate} \qquad \tau_s = \frac{[H_2O]}{Rate} \qquad Rate = k\,[M^{z+}]^x\,[H_2O]^y \quad (9.19)$$

If ^1H NMR is used, the populations in each site are multiplied by 2 because there are two hydrogens per water molecule and the lifetimes are defined by

$$\tau'_m = \frac{2\times n\,[M^{z+}]}{Rate} \qquad \tau'_s = \frac{2\times[H_2O]}{Rate} \qquad Rate = k'\,[M^{z+}]^x\,[H_2O]^y \quad (9.20)$$

If one believes that the exchange involves a water molecule, then k in Eq. (9.19) is the rate constant for exchange of one water ligand. But k' in Eq. (9.20) is for exchange of one H and, since there are two H atoms per ^{17}O, then k = k' / 2. However, if H exchange occurs only by reaction (9.18), then k' is the rate constant as defined by Eq. (9.20). Therefore, ^{17}O NMR will give k for water molecule exchange, but the k' from ^1H NMR has an ambiguous assignment. For other solvents, such as acetonitrile, DMF and DMSO, where independent exchange of the methyl protons is very unlikely, the site population factor in Eq. (9.20) is often implicitly omitted and the τ definitions refer to whole solvent molecule exchange.

Fluxional processes present another example where it is important to define the rate and to understand the relationship between the rate constant from the NMR measurement and that for the chemical event. The latter aspect has been discussed in detail by Johnson and Moreland[44] and more recently by Green, Wong, and Sella.[45]

A simple example is the H exchange in an η^2-alkene metal hydride. The NMR experiment may be done by labeling either the hydride or alkene hydrogens, and this leads to different relationships between the NMR rate constant and the rate constant for the H shift. The labeling might be done by isotope substitution or by selective spin saturation or inversion in a pulsed NMR experiment.

The hydride labeling experiment is proposed to proceed through an η^1-alkyl intermediate (**I**), as shown in Scheme 9.4.

Scheme 9.4

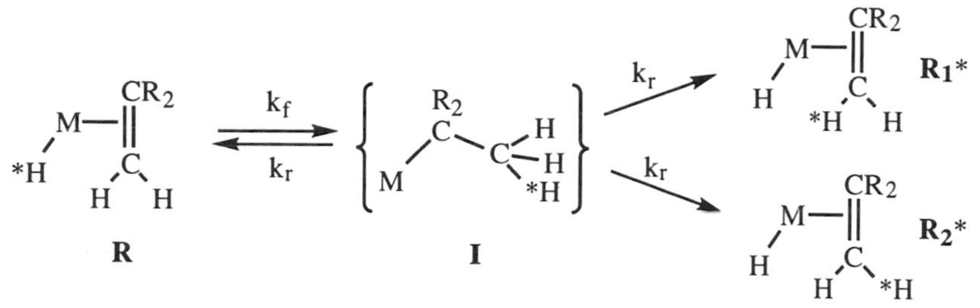

The general principle can be illustrated with the simplifying assumption that the H atoms in the η^2-alkene (**R**) are magnetically equivalent and then the products (**R$_1$*** and **R$_2$***) are identical in the NMR spectrum. The NMR experiment monitors the conversion of **R** to **R$_1$*** + **R$_2$***, and the problem is to determine how the rate of this process is related to k_f. First, one can make a steady-state assumption for the intermediate **I**

$$\frac{d\,[\mathbf{I}]}{d\,t} = k_f\,[\mathbf{R}] - 3\,k_r\,[\mathbf{I}] = 0 \tag{9.21}$$

and obtain the steady-state concentration, given by

$$[\mathbf{I}] = \frac{k_f\,[\mathbf{R}]}{3\,k_r} \tag{9.22}$$

Then, the rate of loss of **R** that is measured by NMR is given by

$$-\frac{d\,[\mathbf{R}]}{d\,t} = k_f\,[\mathbf{R}] - k_r\,[\mathbf{I}] \tag{9.23}$$

Substitution for **I** from Eq. (9.22) and rearrangement gives

$$-\frac{d\,[\mathbf{R}]}{d\,t} = \frac{2}{3}\,k_f\,[\mathbf{R}] \tag{9.24}$$

This is the rate measured by NMR and the observed rate constant is $k_{obsd} = (2/3)k_f$.

If the experiment is done by labeling the CH_2 protons, then the system is represented by Scheme 9.5.

Scheme 9.5

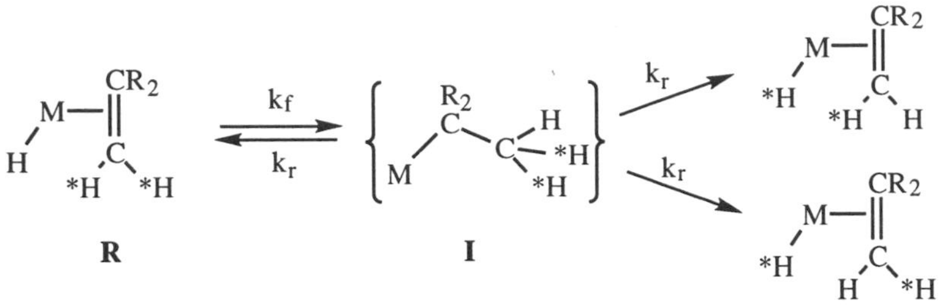

The steady-state concentration of **I** is given by Eq. (9.22), but the two reactions at the right return half of the magnetization or label to the original site **R** so that the rate of disappearance from **R** is given by

$$-\frac{d\,[\mathbf{R}]}{d\,t} = k_f\,[\mathbf{R}] - k_r\,[\mathbf{I}] - 0.5\,k_r\,[\mathbf{I}] - 0.5\,k_r\,[\mathbf{I}] \qquad (9.25)$$

Substitution from Eq. (9.22) and rearrangement gives

$$-\frac{d\,[\mathbf{R}]}{d\,t} = \frac{1}{3}k_f\,[\mathbf{R}] \qquad (9.26)$$

The NMR measured rate constant is $k_{obsd} = (1/3)k_f$ in this case. The difference between Eq. (9.24) and Eq. (9.26) results from the different site populations, and this depends on the assumption that the H atoms of the alkene are magnetically equivalent. Further examples are given by Green and co-workers.[45]

In the following discussion, the NMR methods are separated into four categories, roughly in the order of decreasing time scale of their applicability. However, the latter is quite dependent on the system and different methods might be used in different temperature ranges. The important variables of the chemical system are the correct assignment of the spectrum, the chemical shift differences of the species or sites involved in the reaction and the nuclear relaxation times of the nuclei being observed. There have been several recent reviews of dynamic NMR applications and the field is referred to as DNMR.[46]

9.4.a Signal Monitoring

This method refers to the simple monitoring of the changes with time of the concentration of reactants and products, as determined from the integrated intensities of the appropriate peaks in the NMR spectrum. The short time scale is the few minutes required for temperature equilibration and instrument setup and the long time is limited only by sample stability. For pulsed Fourier transform instruments, it is important to remember that the repetition rate or relaxation delay must be 8 to 10 times longer than the nuclear longitudinal relaxation time(s), T_1, in order to obtain correct relative intensities. The T_1 values for ^1H and ^{13}C nuclei can be in the 1- to 10-s range, and this puts a limitation on how quickly the system can be sampled.

A special example of this type of application is the measurement of exchange reactions using appropriate isotopes. For example, the exchange between free CO and CO ligands in metal carbonyls can be measured using ^{13}C-enriched CO and the exchange between H_2O and oxo anions[47] or water ligands can be measured in suitably inert systems using ^{17}O enrichment.[48] Deuterium replacement of ^1H can be used to measure proton exchange between water and weakly acidic ligands, such as amines.[49]

9.4.b Magnetization Transfer

This method is simple to qualitatively envisage and interpret.[50] A selective pulse (or DANTE series of pulses[51]) is used to produce spin inversion or saturation at one site. After a variable waiting period, t_m, a 90° pulse is used to generate the normal spectrum of the system. As exchange proceeds, the inverted nuclei appear in other sites and the intensities of the sites involved in the exchange will decrease. The pattern of the intensity changes is indicative of the exchange pathways in multisite systems. As t_m is increased to the stage where $t_m > T_1$, the natural nuclear relaxation processes tend to restore the intensity and the intensity of the sites involved in the exchange will increase due to T_1 processes.

The accessible range of exchange lifetimes, τ, for this method is determined on the short end by t_m and on the long end by T_1. For typical spectrometers, t_m can be as short as ~0.01 s and T_1 for protons is often ~1 s, so that first-order rate constants of ~100 to 1 s^{-1} can be determined. Because T_1 usually has a lower activation energy (~5 to 10 kJ mol^{-1}) than τ, it is often possible to adjust the temperature to meet the requirement of this method that $T_1 \geq \tau$. The apparent necessity to selectively invert only one signal could be a problem if resonances are close, but this limitation can be overcome by the suggestion of Muhandiram and McClung[52] to treat both the initial and final intensities as variables in the analysis. Inversion of a multiplet due to spin–spin coupling can be achieved with a single pulse, broad enough to cover the multiplet for small coupling constants, or by pulses of different frequencies in the DANTE sequence for large coupling constants. It has been found[53] that spin–spin coupling does not adversely affect rate constant determinations by this method. Because of the competition between exchange and the T_1 relaxation processes, it is advantageous for quantitative analysis to measure the T_1 values independently under slow exchange conditions.

9.4.c Two-Dimensional Exchange Spectroscopy

This method is discussed in detail in the review by Perrin and Dwyer[46] and is called 2D EXSY. In the experiment, a 90° pulse is applied to rotate the magnetization from the +z to the −x axis. After a time t_1, called the evolution or labeling time, a second 90° pulse rotates the magnetization from the xy plane into the xz plane and a field gradient or homospoil pulse is applied to dephase the magnetization along the x axis. After a further time t_m, called the mixing time, a third 90° pulse rotates the magnetization to the y axis and the free induction decay, FID, is collected during a time t_2. The magnetization evolves in the xy plane during the two time periods t_1 and t_2. For a single site, the angular rate of precession is ω, and maxima will occur when $\cos(\omega t_1)$ and $\cos(\omega t_2)$ equal 1. Then, a three-dimensional plot of intensity versus t_1 and t_2 will show a maximum when this condition is satisfied for t_1 and t_2. The peak is often represented by intensity contours and is really a cone. For a multisite system, nuclei in different environments have different precessional rates, ω_i, but will give maxima when $\cos(\omega_i t_1)$ and $\cos(\omega_i t_2)$ equal 1 and will give peaks along the diagonal of the t_1–t_2 plane when $t_1 = t_2$.

When the system is undergoing chemical exchange, magnetization can transfer between the sites during the mixing time and this produces off-diagonal cross peaks in the final three-dimensional plot of the spectra. These cross peaks give a map of the sites that are undergoing exchange. The evaluation of rate constants from the information is based on the intensities or, more properly, the volumes of the cross peaks. This analysis is not trivial, especially for multi-site systems, and requires special care in the collection and processing of the data to ensure that the volumes of the peaks are properly evaluated.

There are several other sources of cross peaks in the 2D EXSY experiment. Dipolar coupling with nearby nuclei (nuclear Overhauser effects) produces cross peaks as observed in the standard NOESY experiment. These can be identified because exchange usually has a larger temperature dependence than dipolar coupling. Scalar coupling interferes with 2D EXSY by producing so-called J cross peaks that can be eliminated by phase cycling.[54]

The choice of the mixing time is crucial for this method because it is quite time consuming to do studies by varying t_m as well as t_1 and t_2. If t_m is too short, then little mixing will occur and the intensities of the cross peaks will be small; if it is too long, the intensities approach those of the diagonal peaks and become insensitive to the exchange rate. For a two-site (AB) system, Perrin and Dwyer have suggested that the optimum mixing time is given by

$$t_m(\text{opt}) = \frac{1}{T_1^{-1} + k_{AB} + k_{BA}} \tag{9.27}$$

Clearly, for a multi-site system with different exchange rates, there will not be a single optimum value. In any case, the accessible range of rate constants is limited by the nuclear relaxation rate T_1^{-1}, and therefore is similar to the range for magnetization transfer.

The 2D EXSY method has been applied[55] to the rearrangements of tris(dithiolene) complexes of the general type $M(S_2C_2R_1R_2)_3$, where M is W or Mo, R_1 is phenyl or substituted phenyl, and R_2 is H or phenyl. The structures in solution are believed to be trigonal prismatic, based on the crystal strucures of analogous complexes,[56] and the asymmetric substitution gives the possibility of cis and trans isomers. However, there is the possibility that these are fac and mer isomers. The authors' assignments are shown by the left-hand structures in Scheme 9.6.

Scheme 9.6

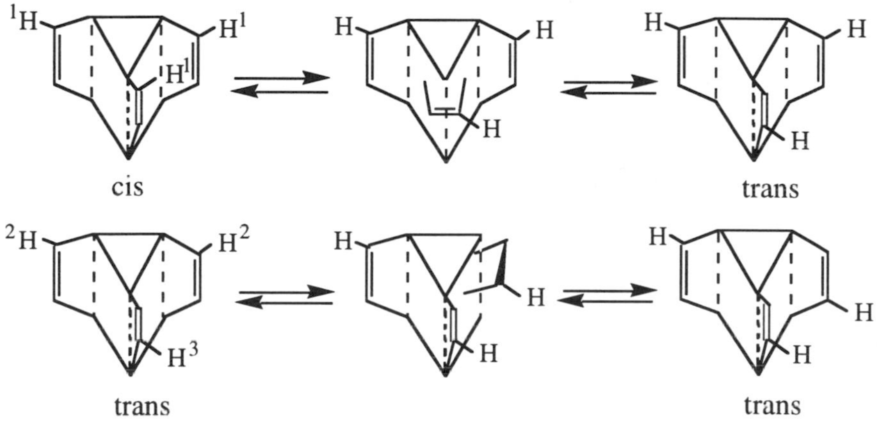

The low-temperature ^{1}H NMR spectrum has three peaks in the $R_2 = H$ region that were assigned to H^1, H^2, and H^3, as shown at the left in Scheme 9.6. The peaks have approximately a 1:2:1 intensity ratio, respectively, due to the ~3:1 equilibrium mixture of trans:cis isomers. The results of a 2D EXSY experiment on $W(S_2C_2H(p\text{-}CH_3OPh))_3$ are represented schematically in Figure 9.3. In addition to the diagonal peaks, cross peaks are observed for all the protons, indicating that all the sites are undergoing exchange with each other. The cross peaks for the H^1–H^3 protons are weakest and give less certain rate constants for this interchange.

Katakis and co-workers[55] suggested that the fluxionality is due to the rearrangment reactions shown in Scheme 9.6. The rotation of any ring in either isomer appears to have $\Delta H^* \approx 60$ kJ mol^{-1}. The relative rate constants for the interchange of the different proton types are reasonably consistent with the proposal. For the Mo analogue, bandshape analysis was found to be consistent with the same fluxionality mechanism.

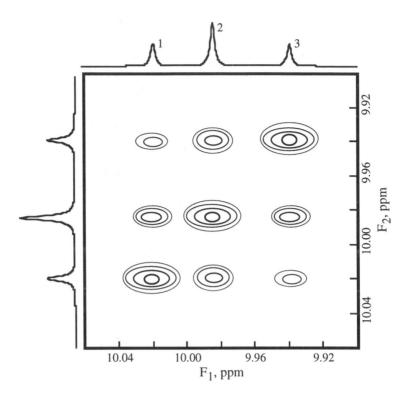

Figure 9.3. The 2D EXSY spectrum of $W(S_2C_2H(p\text{-}CH_3OPh))_3$ at $-40°C$ in the dithiolenic proton region with $t_m = 2$ s.

9.4.d Bandshape Analysis

This method was the first one used to show the applicability of NMR to dynamic processes. For a system with two sites, in the slow-exchange or low-temperature limit, one observes the normal spectrum with two peaks. If the temperature is raised and exchange starts to occur, then the two peaks begin to broaden and come together. The peaks will coalesce at some temperature, the coalescence temperature, which depends on the exchange rate and the chemical shift separation between the peaks in the absence of exchange. As the temperature is increased further, the signal changes to a single sharp resonance in the fast-exchange limit. For systems with several sites and therefore several peaks in the slow-exchange limit, it is possible to observe the broadening and coalescence of the exchanging sites. The quantitative analysis normally takes the form of calculating the spectrum for various exchange models and rate constants and then choosing the model that best fits the observed spectra and gives rate constants that have the normal temperature dependence. Computer programs are available to generate the calculated spectra.[57]

The time scale for this method depends on the chemical shift difference, Δv_0, between the exchanging sites. At the coalesence temperature, the rate constant is given by

$$k = \frac{\pi \Delta v_0}{\sqrt{2}} = 2.22 \times \Delta v_0 \qquad (9.28)$$

Below the coalesence temperature, in the slow-exchange region, two peaks are observed that are broadened over their natural full linewidth at half-height by δv, and $k \approx 2\pi\delta v$ in this region. Above the coalesence temperature, in the fast-exchange region, only one peak is observed with a linewidth of δv, and $k \approx 4\pi\Delta v_0^2 / \delta v$. For typical values for ^1H NMR of $\delta v \approx 5$ Hz and $\Delta v_0 = 100$ Hz, k can range from ~30 to ~3 x 10^4 s^{-1}. Since Δv_0 depends directly on the magnetic field strength, the range can be extended to larger k by working at higher fields. For other nuclei, such as ^{13}C, ^{19}F, and ^{31}P, and for paramagnetic systems, the Δv_0 can be much larger and the upper limit is greatly extended. The lower end of the range for bandshape analysis overlaps the upper range for the two methods discussed above, but the upper limit is extended by ~10^2 by bandshape analysis.

A limitation of bandshape analysis is that one needs the chemical shifts and line widths for the nonexchanging system. When possible, this is done by cooling the sample to well below the slow-exchange limit, but the temperature dependence of the shifts and line widths is rarely determined, and they are treated as constants in the analysis. Another problem is that the exchange pathways are not always clearly delineated by bandshape analysis, especially in multisite systems. A model is chosen and fitted to the data, but the initial choice is somewhat subjective, and some pathways may be missed. A problem can arise in the data collection on pulsed Fourier transform instruments. The pulse repetition rate must be substantially longer than the T_1 of any nuclei of interest to ensure that there are no intensity distortions, but there is always the temptation to shorten the repetition time in order to shorten the data collection time.

A fairly typical application of bandshape analysis to an inorganic mechanism problem is the recent study of Raymond and co-workers[58] on the fluxionality of tris-catecholate complexes of Ga(III), where the ligands are 2,3-dihydroxy-N,N'-substituted-terephthalamides. Under slow-exchange conditions, the ^1H NMR at 300 MHz of the isopropyl derivative shows two methyl resonances due to the chirality, as discussed in Section 4.2.c. In D_2O, as the temperature is raised these two peaks merge and coalesce at ~57°C. Further increase in temperature produces the expected sharpening to one methyl signal. The spectra over the temperature range of 20°C to 95°C were fitted by bandshape simulations to obtain $\Delta H^* = 55.2$ kJ mol^{-1} and $\Delta S^* = -39$ J mol^{-1} K^{-1}

(pD 9.8). The fluxional process was assigned to an intra- rather than an inter-molecular ligand rearrangement because the rate was independent of the concentration of free ligand added. From parallel observations in DMSO, the authors suggested that the solvent effect is minor and therefore also consistent with an intramolecular process. However, the solvent effect analysis was based on calculations of ΔG^* at the respective coalescence temperatures[59] of 57°C in D_2O and 87°C in DMSO. The resulting 7 kJ mol^{-1} difference was considered minor. This analysis represents an example of the all too common practice in this area of comparing two kinetic parameters at different coalescence temperatures. At 57°C, the calculated rate constants are 15.6 s^{-1} in DMSO and 126 s^{-1} in D_2O, and the eightfold difference in rate constants might not seem so minor.

Similar studies were reported[58] on the unsymmetrical amide ligand with a benzyl group on one nitrogen and a tertiary butyl on the other. This system has cis and trans isomers and the types of interconversion have been discussed in Section 4.2.d. The trans isomer undergoes inversion without isomerization with a coalescence temperature of 22°C, but a full analysis was not possible because the low-temperature limiting spectrum was not reached at 0°C. The cis–trans isomerization was observed at higher temperatures with coalescence at ~67°C. The inversion without isomerization of the trans isomer indicates a trigonal twist mechanism for the rearrangement.

9.4.e Relaxation Rate Measurements

This type of application is a specialized extension of bandshape analysis in which the temperature dependence of the transverse nuclear relaxation time, T_2, is used to measure rates of exchange. The T_2 can be determined from the line width of the NMR peak, or more accurately by special pulse sequences. The method is generally applied to simple systems with well-separated peaks in the NMR spectrum. It has been especially useful for measurements of solvent exchange rates from paramagnetic metal ions, such as the example in reaction (9.17). This special application assumes a two-site system with the NMR spectrum dominated by one peak, that of the bulk solvent. Swift and Connick[60] first published the basic equations that are a solution of the Bloch equations modified for chemical exchange by McConnell.[61] Analogous equations have been given for T_1, for the three-site problem[62] and for the rotating-frame relaxation time[63] in such systems. The method and results are the subject of several reviews.[64]

An idealized temperature dependence of the relaxation rates is shown in Figure 9.4. The parameter plotted, T_{ip}^{-1}, is the difference between the relaxation rate in the presence of the exchanging species and the rate for the pure solvent, divided by the metal ion concentration. In the high-temperature limit at the left of Figure 9.4, exchange is fast and the

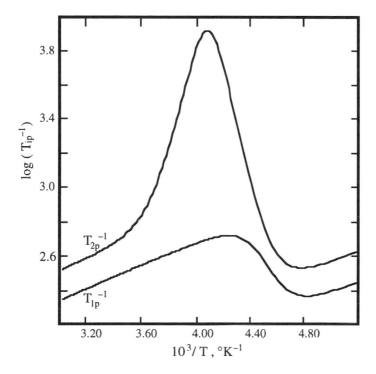

Figure 9.4. Representation of the temperature dependence of T_{2p}^{-1} and T_{1p}^{-1} for an idealized system in which all the regions are observable.

relaxation is controlled by the nuclear relaxation rate in the inner coordination sphere of the metal ion, T_{2m}^{-1}. As the temperature is lowered, exchange becomes slower and relaxation is controlled by dephasing of the nuclear precession frequency due to the difference in chemical shift between the bulk and coordinated nucleus, $\Delta\omega_m$. The slower the exchange, the more effective is the dephasing, so that the measured relaxation rate increases with decreasing temperature in what is called the non-Arrhenius region. At still lower temperature, the exchange becomes slow enough so that the dephasing is controlled by the exchange lifetime, τ_m, and the relaxation rate decreases with decreasing temperature in the Arrhenius region. Finally, at low temperature, inner-sphere solvent exchange is so slow that only relaxation due to outer-sphere interactions, T_{2o}^{-1}, is observed. The latter effect actually occurs at all temperatures but is generally obscured by the more effective relaxation processes. At the maximum between the Arrhenius and non-Arrhenius regions, $\tau_m^{-1} = \Delta\omega_m$. Since $\Delta\omega_m$ can be in the range of 10^3 to 10^5 s^{-1} for paramagnetic systems, this gives some indication of the range of applicability of this method.

In practice, a particular system often will show only two or three of these specific regions. The problem is to fit this temperature dependence to the known functions primarily to determine ΔH^* and ΔS^* for the exchange process, but this also requires some knowledge or estimates of the activation energies E_m and E_o for T_{2m} and T_{2o}, respectively. Measurements of T_1^{-1} can be helpful in this regard because T_1 is not affected by dephasing and exchange is apparent only in the Arrhenius region, as shown in Figure 9.4. The main difficulty is separating the various factors that affect the temperature dependence of T_2^{-1} when the limiting regions are not well defined. Least-squares analysis is far more satisfactory than the early graphical methods.

9.5 ELECTRON PARAMAGNETIC RESONANCE METHODS

This is a powerful technique for the detection and monitoring of species with unpaired electrons. This type of spectroscopy is designated by several acronyms: EPR, electron paramagnetic resonance; ESR, electron spin resonance; EMR, electron magnetic resonance. EPR is quite sensitive, with detection limits in the range of 10^{-6} M in favorable cases. It also can be quite informative as to structure because of electron–nuclear hyperfine coupling to metal and ligand nuclei. The main disadvantage is that many species with unpaired electrons do not give a useful EPR signal in solution because efficient electron spin relaxation leads to broad or undetectable signals. Signals are more generally detectable in the crystalline or frozen glassy state. For the first-row transition metals in their common oxidation states, solution EPR is useful for complexes of V(IV), Mn(II), and Cu(II), while Cr(III) and Fe(III) often give broad spectra in solution. Most organic radicals give EPR signals that are quite useful for detection and identification of such species as reaction intermediates.

Most EPR spectra are run at X-band frequency of 9.4 GHz in the microwave region of the electromagnetic spectrum, and the magnetic field of ~0.3 T is changed to give the resonance condition for signal detection. The sample tube should be quartz, in order to avoid impurity signals found in Pyrex, and the tube should be flattened for solvents with high dielectric constants, such as water, to minimize dielectric loss in the microwave cavity. The concentrations of paramagnetic species should be $<10^{-3}$ M, in order to minimize signal broadening due to intermolecular relaxation interactions. EPR spectra are usually displayed as plots of the derivative of signal intensity versus magnetic field. Double integration of such data is necessary to get proper integrated signal intensities. With proper calibration, the signal intensity can give a direct measure of the concentration of the EPR active species.

Pulsed EPR is becoming more widely available. A 90° pulse is typically in the range of 10 to 30 ns and the FID after such a pulse can be used to measure the electron spin relaxation time or to monitor the decay of radicals[65] that have been produced by some fast method, such as flash photolysis or pulse radiolysis. Electron spin-echo envelope modulation, ESEEM, spectroscopy[66] is a pulse method used primarily for detecting weak hyperfine coulping.

Most applications use EPR as a detection method for reactants or products. Various flow methods can be coupled with EPR to monitor the time dependence of EPR active species. It is also possible to use EPR line broadening to measure exchange rates, and in this area the time scale is in the range of 10^{-7} to 10^{-9} s.

Wang, Kumar, and Margerum[67] used EPR coupled to stopped flow to measure the electron exchange rate between complexes of Ni(II) and Ni(III) with half-times of ~1 s. The Ni(III) species is EPR active and was enriched with ^{61}Ni, which broadens the EPR signal due to hyperfine coupling. As electron exchange proceeds, the ^{61}Ni(III) is exchanged for the more abundant ^{58}Ni(III) and ^{60}Ni(III), which have no nuclear spin. As a result, the EPR signal sharpens and provides a measure of the extent of exchange. It was suggested that this method might be applied to other metals using isotopes such as ^{57}Fe, ^{99}Ru, and ^{53}Cr.

Spin trapping can be used to convert radical intermediates into more stable species that can be detected by EPR. For example, fumarate ion was used[68] to trap the aryl radicals, $^{\bullet}$Ar, formed in the oxidation of Fe(II) by benzenediazonium ions, ArN_2^+, in a stopped-flow EPR study. The $^-O_2C(C_6H_5)CH-^{\bullet}CHCO_2^-$ radical is sufficiently stable relative to dimerization so that its concentration after the 35-ms mixing time could be used to determine the rate of the initial oxidation reaction.

The much studied reaction of $Fe^{II}EDTA$ with H_2O_2 has been investigated[69] in a stopped-flow system with a deadtime of 18 ms using 5,5-dimethyl-1-pyrroline N-oxide, DMPO, to trap the $^{\bullet}OII$ radicals. The general reactions are shown in Scheme 9.7.

Scheme 9.7

$$Fe^{II}EDTA + H_2O_2 \longrightarrow Fe^{III}EDTA + HO^- + {^{\bullet}OH}$$

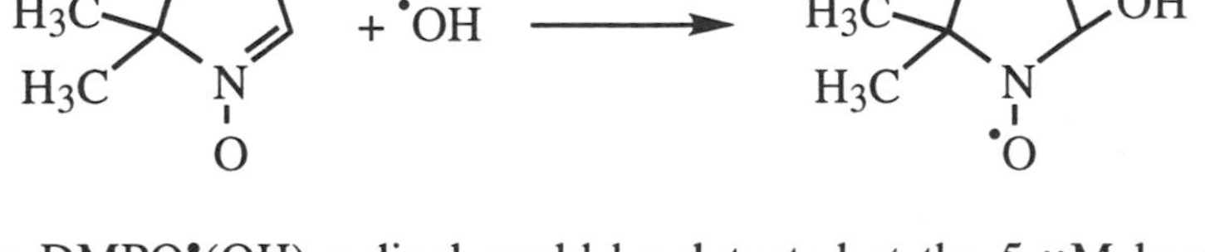

The DMPO$^{\bullet}$(OH) radical could be detected at the 5-μM level, and the time dependence of its formation was studied as a function of reagent concentrations. It was concluded from initial rates that the reaction is first order in [$Fe^{II}EDTA$] and [H_2O_2], but that only ~20 percent of the expected amount of DMPO$^{\bullet}$(OH) is formed with initial concentrations of 100 μM Fe(II), 200 μM EDTA, 600 μM H_2O_2, and 20 mM DMPO at pH 7.4. The amount of DMPO$^\bullet$(OH) decreased as the $[Fe(II)]/[H_2O_2]$ ratio increased. This indicates that $^\bullet$OH is reacting by other pathways in addition to being trapped by DMPO. One known reaction is that of Fe^{II}EDTA with $^\bullet$OH.[70]

9.6 PULSE RADIOLYSIS METHODS

It is possible to quickly generate reactive species and solvated electrons by passing a high-energy pulse of electrons through a solution. Results have been summarized and discussed in recent reviews.[71] The pulse is typically 5 to 100 ns long with energies in the range of 2 to 20 MeV, depending on the source apparatus. The high-energy electrons initially are present in hot spots or spurs and the thermalized species are present after $\sim 10^{-7}$ s. In water, the species and number produced per 100 eV of energy absorbed, in brackets, are: e_{aq}^- (2.65), $^\bullet$OH (2.65), $^\bullet$H (0.65), H_2O_2 (0.72), H_2 (0.45). In most applications, the initial radiolysis products are scavenged by additives to remove undesired species or to produce a new reactive species. For example, water saturated with N_2O (~ 0.022 M under 1 atm of N_2O) converts e_{aq}^- to $^\bullet$OH in ~ 50 ns by the following reaction:

$$e_{aq}^- + N_2O + H_2O \longrightarrow \ ^\bullet OH + N_2 + OH^- \quad (9.29)$$

The $^\bullet$OH radicals can be scavenged by t-butanol through the reactions in (9.30), but the t-butanol radical can be intercepted by sufficiently reactive substrates. Other alcohols react similarly, but their radicals are generally 5 to 10 times more reactive with substrates than the t-butanol radical.

$$^\bullet OH + (H_3C)_3COH \longrightarrow H_2O + (H_2C^\bullet)(H_3C)_2COH$$

$$2\ (H_2C^\bullet)(H_3C)_2COH \longrightarrow \begin{array}{c} HOC(CH_3)_2H_2C \\ | \\ HOC(CH_3)_2H_2C \end{array} \quad (9.30)$$

Sodium formate can be used to form the $^\bullet CO_2^-$ radical by the following reaction:

$$^\bullet OH + HCO_2^- \longrightarrow \ ^\bullet CO_2^- + H_2O \quad (9.31)$$

If O_2 is present, $^\bullet CO_2^-$ will react to give CO_2 and the superoxide radical $^\bullet O_2^-$. Methyl radicals can be produced from DMSO by the following sequence[72]:

$$\bullet OH + (H_3C)_2SO \longrightarrow (H_3C)_2S^\bullet OOH$$

$$(9.32)$$

$$(H_3C)_2S^\bullet OOH \longrightarrow \bullet CH_3 + (H_3C)SOOH$$

Bromide and thiocyanate ions can produce the corresponding $\bullet(X)_2^-$ radical anion by the following reactions:

$$\bullet OH + X^- \longrightarrow OH^- + \bullet X$$

$$(9.33)$$

$$\bullet X + X^- \longrightarrow \bullet(X)_2^-$$

The $\bullet(SCN)_2^-$ radical is moderately stable with an absorbance maximum at 478 nm. The amount of $\bullet(SCN)_2^-$ produced during radiolysis of a 0.01 M SCN^- solution is commonly used as a dosimeter[73] to determine the number of radicals produced per electron pulse. With azide ion, the reaction stops at the first stage in (9.33) to give predominantly $\bullet N_3$, but $\bullet(N_3)_2$ may be produced at high N_3^- concentrations.[74]

Combinations of reagents are used to produce the dominant radiolysis product of interest. For example, to study reactions of $\bullet OH$, the sample solution would be saturated with N_2O, but to study $\bullet CO_2^-$, the solution would also contain ~0.1 M HCO_2^-. The radicals produced in the above reactions are generally quite reactive: e_{aq}^- and $\bullet CO_2^-$ are strong reducing agents; $\bullet OH$, $\bullet N_3$, and $\bullet Br_2^-$ are strong oxidizing agents. These species also tend to decay by dimerization or disproportionation, but their concentrations are sufficiently low (1 to 10 μM) so that these second-order reactions are often slow compared to reactions with other substrates that are added at the mM level.

Most studies in this area use spectrophotometric detection and the time scale can be from microseconds to seconds. Because the products are produced at low concentrations, it is often possible to do multiple radiation pulses on the same sample. The main problem is the lack of molecular specificity of the spectrophotometric method, so that the nature of the reaction and the products are often inferred by analogy and by the concentration dependence of the reaction rate.

Pulse radiolysis was used[75] to rapidly generate $Co(NH_3)_6^{2+}$ and then follow the aquation of the NH_3 ligands. The radiolysis was done on solutions of $Co(NH_3)_6^{3+}$ at pH < 4.5 in the presence of t-butanol to scavenge $\bullet OH$ so that e_{aq}^- is the reducing agent. Detection was by conductivity change due to the loss of H_3O^+ by formation of NH_4^+. The overall observations are summarized in Scheme 9.8, where the times are half-times for the various steps. Similar rates were observed when several pentaammine and tetraammine cobalt(III) complexes were studied.

Scheme 9.8

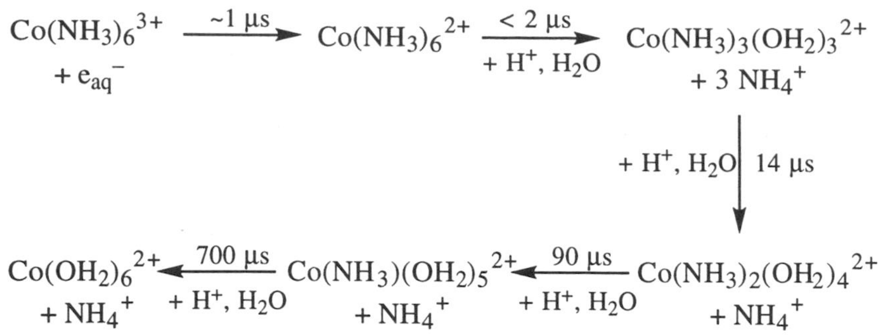

$$Co(NH_3)_6^{3+} \xrightarrow{\sim 1\ \mu s} Co(NH_3)_6^{2+} \xrightarrow[+\ H^+,\ H_2O]{< 2\ \mu s} Co(NH_3)_3(OH_2)_3^{2+}$$
$$+\ e_{aq}^- \qquad\qquad\qquad\qquad\qquad +\ 3\ NH_4^+$$

$$+\ H^+,\ H_2O \Big|\ 14\ \mu s$$

$$Co(OH_2)_6^{2+} \xleftarrow[+\ H^+,\ H_2O]{700\ \mu s} Co(NH_3)(OH_2)_5^{2+} \xleftarrow[+\ H^+,\ H_2O]{90\ \mu s} Co(NH_3)_2(OH_2)_4^{2+}$$
$$+\ NH_4^+ \qquad\qquad\quad +\ NH_4^+ \qquad\qquad\qquad +\ NH_4^+$$

The rate of aquation of $Co(en)_3^{2+}$, produced from $Co(en)_3^{3+} + e_{aq}^-$, is slower than in the ammine systems in Scheme 9.8. The half-time for loss of the first ethylenediamine is pH dependent and varies from 320 to 910 μs between pH 2.5 and 4.6. This effect was attributed to a protonated monodendate intermediate.

There have been several pulse radiolysis studies of the oxidation of aquo and other complexes of the M(II) ions of the first transition series. In several cases, the pulse radiolysis observations indicate that the reactions can be more complex than simple oxidation to the M(III) ion. For example, pulse radiolysis of N_2O saturated solutions of $Fe(OH_2)_6^{2+}$ and N_3^- involves the following initial reactions and equilibria[76] (rate constants at 25°C in $M^{-1}\ s^{-1}$):

$$e_{aq}^- + N_2O + H_2O \xrightarrow[9 \times 10^9]{k =} {}^\bullet OH + N_2 + OH^- \qquad (9.34)$$

$$^\bullet OH + N_3^- \xrightarrow[1.2 \times 10^{10}]{k =} OH^- + {}^\bullet N_3 \qquad (9.35)$$

$$Fe(OH_2)_6^{2+} + {}^\bullet OH \xrightarrow[2.3 \times 10^8]{k =} Fe(OH_2)_6^{3+} + OH^- \qquad (9.36)$$

$$Fe(OH_2)_6^{2+} + N_3^- \underset{245\ M^{-1}}{\overset{K =}{\rightleftharpoons}} Fe(OH_2)_5(N_3)^+ + H_2O \qquad (9.37)$$

The pH was 5.4 to 6.2, so that HN_3 is fully dissociated to N_3^-. When the Fe(II) concentration (~0.01 M) is 10 to 15 times larger than that of N_3^-, the major products after 0.26 μs are Fe(III) and $^\bullet N_3$ from the reactions (9.35) and (9.36), respectively. However, when the N_3^- concentration is the same or larger than that of Fe(II), then $^\bullet OH$ reacts mainly by (9.35) and the product after 28 μs has absorbance maxima at 300 and 419 nm. This spectrum does not correspond either to $^\bullet N_3$ or to the oxidized complex $Fe^{III}(OH_2)_5(N_3)^{2+}$, and was assigned to aqueous $Fe^{II}(^\bullet N_3)^{2+}$.

This intermediate decays to the final Fe(III) products, and the conclusions and rate constants (M^{-1} s^{-1} or s^{-1}) of Parsons et al.[76] are summarized in Scheme 9.9, where H_2O ligands are omitted for clarity.

Scheme 9.9

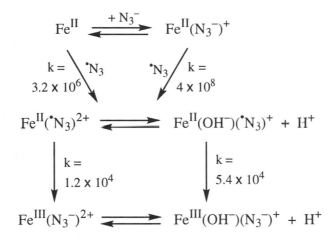

Reactions of $Fe(OH_2)_6^{2+}$ with $^\bullet N_3$, $^\bullet Br_2$, and $^\bullet O_2H$ have rate constants in the narrow range of 1.2 to 3.6 x 10^6 M^{-1} s^{-1}, and it appears that the reaction is limited by water ligand dissociation from $Fe(OH_2)_6^{2+}$. Similar observations have been reported for analogous reactions at other metal centers by van Eldik, Cohen, and Meyerstein.[77] The larger rate constant for $Fe(OH_2)_6^{2+}$ + $^\bullet OH$ in (9.36) indicates that this reaction proceeds by H atom abstraction.

With regard to the intermediate assigned as $Fe^{II}(^\bullet N_3)^{2+}$, it was noted by Parsons et al.[76] that the spectrum of this species is similar to that observed in the reaction of $Mn^{II}(NTA)$ with $^\bullet OH$[70] and $Fe^{II}(NTA)$ with $^\bullet CO_2^-$.[78] The former was assigned to a product of H abstraction from the ligand in $Mn^{II}(NTA)$, and the latter to $Fe^{III}(NTA)(OH_2)(CO_2)^{2-}$ with an Fe—C bond. Presumably the correspondence in these spectra is accidental, but it illustrates the difficulty in assigning species from their electronic spectra alone.

9.7 FLASH PHOTOLYSIS METHODS

This technique is somewhat analogous to pulse radiolysis in that the system is subjected to a short high-energy pulse and then subsequent events are monitored. In flash photolysis the pulse usually is provided by a laser beam of photons and the immediate product is some photoexcited state of the absorbing reactant(s). Subsequent events are monitored on the nanosecond or longer time scale, most commonly by

Fourier transform IR or UV-visible spectrophotometry. Flash photolysis is much cleaner than pulse radiolysis in that there is not the multiplicity of initial reactants or the need to add reagents to remove undesired reactants. In both methods, there is the problem of identifying the reactive intermediates from the often limited spectroscopic signatures that they provide.

There have been many studies of the activation of C—H bonds by the coordinatively unsaturated species that can be generated by flash photolysis.[79] In a recent study, Harris and co-workers[80] have observed the flash photolysis of $Rh(Tp^*)(CO)_2$ in pentane with a 295 nm laser pulse (Tp^* = hydridotris(3,5-dimethylpyrazolyl)borate). They observed reformation of $Rh(Tp^*)(CO)_2$ at 2054 cm^{-1} with $\tau \approx 70$ ps and cooling of a vibrationally excited intermediate with $\tau \approx 23$ ps, to give a vibrational ground state absorbing at 1972 cm^{-1}. This intermediate decays with $\tau \approx 200$ ps and seems to convert with the same rate to another intermediate absorbing at 1992 cm^{-1}. The final product, $Rh(Tp^*)(CO)(R)H$, appears on a longer time scale of ~500 ns. The intermediates were suggested to be weakly and more strongly solvated species, but their 20 cm^{-1} difference in CO stretching frequencies was acknowledged to be unexpected for such a model. It should be noted that the Tp* ligand in this system is known to undergo a rapid fluxional η^3 to η^2 change, but the η^3 form is dominant.[81] The wavelength dependence of the quantum yields and other chemistry of this system have been reported by Purwoko and Lees[82], and some synthetic and structural aspects are given by Chauby et al.[83]

Flash photolysis also provides access to electronic excited states whose photochemistry, energy transfer, and electron transfer properties can be observed after the flash. One of the most widely studied such systems is $Ru(bpy)_3^{2+}$, in which the photochemically generated excited state is both a good oxidizing and reducing agent.[84] Binding of analogues of this complex to proteins and then photoactivating the system has been exploited to study electron transfer in proteins.[85] Barton and co-workers[86] have used Ru and Os analogues to study electron and energy transfer for such complexes intercalated in DNA.

References

1. Pilling, M. J.; Seakins, P. W. *Reaction Kinetics*; Oxford University Press: Oxford, 1995; Wilkins, R. G. *Kinetics and Mechanism of Reactions of Transition Metal Complexes*, 2nd ed.; VCH: Weinheim, 1991; Bernasconi, C. F. *Relaxation Kinetics*; Academic Press: New York, 1976; Bradley, J. N. *Fast Reactions*; Oxford University Press: Oxford, 1975; Hague, D. N. *Fast Reactions*; Wiley-Interscience: London, 1971; Caldin, E. F. *Fast Reactions in Solution*; Blackwell: Oxford, 1964.

2. Wilkins, R. G. *Adv. Inorg. Bioinorg. Mech.* **1983**, *2*, 139.

3. Applied Photophysics (http://www.apltd.co.uk/); Hi-Tech Scientific (http://www.hi-techsci.co.uk/); OLIS, On-Line Instrument Systems Inc. (http://www.olisweb.com).

4. Funahashi, S.; Ishihara, K.; Aizawa, S.; Sugata, T.; Ishi, M.; Inada, Y.; Tanaka, M. *Rev. Sci. Instr.* **1993**, *64*, 130.

5. Sienkiewicz, A.; Qu, K. B.; Scholes, C. P. *Rev. Sci. Instr.* **1994**, *65*, 68.

6. Inada, Y.; Funahashi, S. *Rev. Sci. Instr.* **1994**, *65*, 18.

7. Gerhard, A.; Gaede, W.; Neubrand, A.; Zang, W.; van Eldik, R.; Stanitzek, P. *Instrum. Sci. Technol.* **1994**, *22*, 1.

8. van Eldik, R.; Palmer, D. A.; Schmidt, R.; Kelm, H. *Inorg. Chim. Acta* **1981**, *50*, 131; Nichols, P. J.; Ducommun, Y.; Merbach, A. E. *Inorg. Chem.* **1983**, *22*, 3993. Ishihara, K.; Miura, H.; Funahashi, S.; Tanaka, M. *Inorg. Chem.* **1988**, *27*, 1706.

9. Balny, C.; Saldana, T.-L.; Dahan, N. *Anal. Biochem.* **1987**, *163*, 309.

10. Bourke, G. C. M.; Thompson, R. C. *Inorg. Chem.* **1987**, *26*, 903.

11. Dickson, P. N.; Margerum, D. W. *Anal. Chem.* **1986**, *58*, 3153.

12. Beckwith, R. C.; Cooper, J. N.; Margerum, D. W. *Inorg. Chem.* **1994**, *33*, 5144.

13. Meagher, N. E.; Rorabacher, D. B. *J. Phys. Chem.* **1994**, *98*, 12590.

14. Massoud, S.; Jordan, R. B. *Inorg. Chim. Acta* **1994**, *221*, 9; Palmer, D. A.; van Eldik, R. *Chem. Rev.* **1983**, *83*, 651.

15. Buckingham, D. A.; Clark, C. R. *Inorg. Chem.* **1994**, *33*, 6171.

16. Brower, K. R. *J. Am. Chem. Soc.* **1968**, *90*, 5401; Jost, A. *Ber. Bunsenges. Phys. Chem.* **1976**, *80*, 316; Doss, R.; van Eldik, R.; Kelm. H. *Ber. Bunsenges. Phys. Chem.* **1982**, *86*, 825; Martinez, P.; Mohr, R.; van Eldik, R. *Ber. Bunsenges. Phys. Chem.* **1986**, *90*, 609.

17. Capitan, M. J.; Munoz, E.; Graciani, M. M.; Jiminez, R.; Tejera, I.; Sanchez, F. *J. Chem. Soc., Faraday Trans. I* **1989**, *85*, 4193.

18. Funahashi, S.; Ishihara, K.; Tanaka, M. *Inorg. Chem.* **1983**, *22*, 2070.

19. Grace, M. R.; Swaddle, T. W. *Inorg. Chem.* **1992**, *31*, 4674.

20. Betts, R. H.; Dainton, F. S. *J. Am. Chem. Soc.* **1953**, *75*, 5721.

21. Gerischer, H.; Heim, W. *Ber. Bunsenges. Phys. Chem.* **1967**, *71*, 1040.

22. Gerischer, H.; Holzwarth, J.; Seifert, D.; Strohmaier, L. *Ber. Bunsenges. Phys. Chem.* **1969**, *73*, 952; Bruhn, H.; Nigam, S.; Holzwarth, J. F. *Disc. Faraday Soc.* **1982**, *74*, 129; Eck, V.; Marcus, M.; Stange, G.; Westerhausen, J.; Holzwarth, J. F. *Ber. Bunsenges. Phys. Chem.* **1981**, *85*, 869.

23. Beckwith, R. C.; Wang, T. X.; Margerum, D. W. *Inorg. Chem.* **1996**, *35*, 995; Fogelman, K. D.; Walker, D. M.; Margerum, D. W. *Inorg. Chem.* **1989**, *28*, 986.

24. Eigen, M. *Pure Appl. Chem.* **1963**, *6*, 97.

25. Marcandalli, B.; Winzek, C.; Holzwarth, J. F. *Ber. Bunsenges. Phys. Chem.* **1984**, *88*, 368.

26. Ertl, G.; Gerischer H. *Z. Electrochem.* **1962**, *66*, 560.

27. Aubard, J.; Nozeran, J. M.; Levoir, P.; Meyer, J. J.; Dubois, J. E. *Rev. Sci. Instrum.* **1979**, *50*, 52.

28. Dawson, A.; Gormally, J.; Wyn-Jones, E.; Holzwarth, J. F. *J. Chem. Soc., Chem. Commun.* **1981**, 386; Bannister, J. J.; Gormally, J.; Holzwarth, J. F.; King, T. A. *Chem. Br.* **1984**, *20*, 227, 232; Fletcher, P. D. I.; Holzwarth, J. F. *J. Phys. Chem.* **1991**, *95*, 2550; Alexandridis, P.; Holzwarth, J. F.; Hatton, A. T. *Langmuir* **1993**, *9*, 2045.
29. Hague, D. N.; Martin, S. R. *J. Chem Soc., Dalton Trans.* **1974**, 254.
30. Doss, R.; van Eldik, R. *Inorg. Chem.* **1982**, *21*, 4108.
31. Winkler-Oswatitsch, R.; Eigen, M. *Angew. Chem., Int. Ed. Engl.* **1979**, *18*, 20.
32. Citi, M.; Secco, F.; Venturi, M. *J. Phys. Chem.* **1988**, *92*, 6399.
33. Stanley, B. J.; Marshall, D. B.; *Rev. Sci. Instr.* **1994**, *65*, 199.
34. Rieger, P. H. *Electrochemistry*, 2nd ed.; Chapman & Hall: New York, 1994; Faulkner, L. R.; Bard, A. J. *Electrochemical Methods*; Wiley: New York, 1980; Osteryoung, J. *Acc. Chem. Res.* **1993**, *26*, 77.
35. Rudolph, M. *J. Electrochem. Interfacial Electrochem.* **1991**, *314*, 13; Rudolph, M.; Reddy, D. P.; Feldberg, S. W. *Anal. Chem.* **1994**, *66*, 589A. DigiSim 1.0 program; Bioanalytical Systems Inc., West Lafayette, Indiana (http://www.bioanalytical.com:80/); O'Dea, J. J.; Osteryoung, J. G.; Lane, T. *J. Phys. Chem.* **1986**, *90*, 2761.
36. Connelly, N. G.; Orpen, A. G.; Rieger, A. L.; Rieger, P. H. *J. Chem. Soc., Chem. Commun.* **1992**, 1293.
37. Shaw, M. J.; Geiger, W. E. *Organometallics*, **1996**, *15*, 13; Zhang, Y.; Gosser, D. K.; Rieger, P. H.; Sweigart, D. A. *J. Am. Chem. Soc.* **1991**, *113*, 4062; Pike, R. D.; Alavosus, T. J.; Camaioni-Neto, C. A.; Williams, J. C.; Sweigart, D. A. *Organometallics* **1989**, *8*, 2631.
38. Lei, Y.; Anson, F. C. *Inorg. Chem.* **1995**, *34*, 1083.
39. Hecht, M.; Schultz, F. A.; Speiser, B. *Inorg. Chem.* **1996**, *35*, 555.
40. Hershberger, J. W.; Klingler, R. J.; Kochi, J. K. *J. Am. Chem. Soc.* **1983**, *105*, 61.
41. Zizelman, P. M.; Amatore, C.; Kochi, J. K. *J. Am. Chem. Soc.* **1984**, *106*, 3771.
42. Huang, Y.; Carpenter, G. B.; Sweigart, D. A.; Chung, Y. K.; Lee, B. Y. *Organometallics* **1995**, *14*, 1423.
43. Parker, V. D.; Tilset, M. J. *J. Am. Chem. Soc.* **1987**, *109*, 2521.
44. Johnson, C. S., Jr.; Moreland, C. G. *J. Chem. Educ.* **1973**, *50*, 477.
45. Green, M. L. H.; Wong, L.-L.; Sella, A. *Organometallics* **1992**, *11*, 2660.
46. Perrin, C. L.; Dwyer, T. J. *Chem. Rev.* **1990**, *90*, 935; Orrell, K. G.; Sik, V. *Ann. Rep. NMR Spectrosc.* **1993**, *27*, 103; Ibid. **1987**, *19*, 79.
47. Brasch, N. E.; Buckingham, D. A.; Evans, A. B.; Clark, C. R. *J. Am. Chem. Soc.* **1996**, *118*, 7969.
48. Aygen, S.; Hanssum, H.; van Eldik, R. *Inorg. Chem.* **1985**, *24*, 2853; Gonzalez, G.; Moullet, B.; Martinez, M.; Merbach, A. E. *Inorg. Chem.* **1994**, *33*, 2330; Galsbol, F.; Mønsted, L.; Mønsted, O. *Acta Chem. Scand.* **1992**, *47*, 43; Cusanelli, A.; Frey, U.; Richens, D. T.; Merbach, A. E. *J. Am. Chem. Soc.* **1996**, *118*, 5265.
49. Jackson, W. G. *Inorg. Chim. Acta* **1987**, *131*, 105; Buckingham, D. A.;

Marzilli, L. G.; Sargeson, A. M. *J. Am. Chem. Soc.* **1968**, *90*, 6028.

50. Gesmar, H.; Led, J. J. *J. Magn. Reson.* **1986**, *68*, 95; Grassi, M.; Mann, B. E.; Pickup, B. T.; Spencer, C. M. *J. Magn. Reson.* **1986**, *69*, 92.
51. Freeman, R. *Chem. Rev.* **1991**, *91*, 1397.
52. Muhandiram, D. R.; McClung, R. E. D. *J. Magn. Reson.* **1987**, *71*, 187; Ibid. **1988**, *76*, 121.
53. McClung, R. E. D.; Aarts, G. H. M. *J. Magn. Reson.* **1995**, *115A*, 145.
54. Johnson, E. R.; Dellwo, M. J.; Hendrix, J. *J. Magn. Reson.* **1986**, *66*, 399.
55. Argyropoulos, D.; Mitsopoulou, C.-A.; Katakis, D. *Inorg. Chem.* **1996**, *35*, 5549.
56. Smith, A. E.; Schrauzer, G. N.; Mayweg, V. P.; Heinrich, W. *J. Am. Chem. Soc.* **1965**, *87*, 5798.
57. Binsch, G. *J. Am. Chem. Soc.* **1969**, *91*, 1304; Binsch, G.; Stephenson, D. *J. Magn. Reson.* **1978**, *32*, 145; Binsch, G.; Kleier, D. A. *DNMR3, Program 165, Quantum Chem. Prog. Exchange*, Indiana University, **1970**. Szymanski, S. *J. Magn. Reson.* **1988**, *77*, 320; Bain, A. D.; Duns, G. J. *J. Magn. Reson.* **1995**, *112A*, 258.
58. Kersting, B.; Telford, J. R.; Meyer, M.; Raymond, K. N. *J. Am. Chem. Soc.* **1996**, *118*, 5712.
59. The relationship between ΔG^* and the rate constant k_C at the coalescence temperature T_C given in reference 58 contains minor errors and should be $\Delta G^*(\text{kJ mol}^{-1}) = 19.145 \times 10^{-3}(10.319 + \log(T_c/k_c))$.
60. Swift, T. J.; Connick, R. E. *J. Chem. Phys.* **1962**, *37*, 307.
61. McConnell, H. *J. Chem. Phys.* **1958**, *28*, 430.
62. Angerman, N. S.; Jordan, R. B. *Inorg. Chem.* **1969**, *8*, 1824; Led, J. S.; Grant, D. M. *J. Am. Chem. Soc.* **1977**, *99*, 5845; Jen, J. *J. Magn. Reson.* **1978**, *30*, 111.
63. Chopra, S.; McClung, R. E. D.; Jordan, R. B. *J. Magn. Reson.* **1984**, *59*, 361.
64. Merbach, A. E. *Pure Appl. Chem.* **1987**, *59*, 161; Merbach, A. E.; Akitt, J. W. *NMR Basic Princ. Prog.* **1990**, *24*, 189; van Eldik, R.; Merbach, A. E. *Comments Inorg. Chem.* **1992**, *12*, 341.
65. Mezyk, S.; Bartels, D. A. *J. Chem. Soc., Faraday Trans.* **1995**, *91*, 3127.
66. Kang, P. C.; Eaton, G. R.; Eaton, S. S. *Inorg. Chem.* **1994**, *33*, 3660; Wirt, M. D.; Bender, C. J.; Peisach, J. *Inorg. Chem.* **1995**, *34*, 1663.
67. Wang, J.-F.; Kumar, K.; Margerum, D. W. *Inorg. Chem.* **1989**, *28*, 3481.
68. Gilbert, B. C.; Hanson, P.; Jones, J. R.; Whitwood, A. C.; Timms, A. W. *J. Chem Soc., Perkin Trans. 2* **1992**, 629.
69. Jiang, J.; Bank, J. F.; Scholes, C. P. *J. Am. Chem. Soc.* **1993**, *115*, 4742.
70. Lati, J.; Meyerstein, D. *J. Chem. Soc., Dalton Trans.* **1978**, 1105.
71. Buxton, G. V.; Mulazzani, Q. G.; Ross, A. B. *J. Phys. Chem. Ref. Data* **1995**, *24*, 1055; van Eldik, R. *Pure and Appl. Chem.* **1993**, *65*, 2603.
72. Veitwisch, D.; Janata, E.; Asmus, K. D. *J. Chem. Soc., Perkin Trans.* **1980**, 146.
73. Buxton, G. V.; Stuart, C. R. *J. Chem. Soc., Faraday Trans.* **1995**, *91*, 279.

74. Butler, J.; Land, E. J.; Swallow, A. J.; Prutz, W. A. *Radiat. Phys. Chem.* **1984**, *23*, 265.
75. Lillie, J.; Shinohara, N.; Simic, M. G. *J. Am. Chem. Soc.* **1976**, *98*, 6516.
76. Parsons, B. J.; Zhao, Z.; Navaratnam, S. *J. Chem. Soc., Faraday Trans.* **1995**, *91*, 3133.
77. van Eldik, R.; Cohen, H.; Meyerstein, D. *Inorg. Chem.* **1994**, *33*, 1566.
78. Goldstein, S.; Czapski, G.; Cohen, H.; Meyerstein, D. *J. Am. Chem. Soc.* **1988**, *110*, 3903.
79. Arndtsen, B. A.; Bergman, R. G.; Mobley, T. A.; Peterson, T. H. *Acc. Chem. Res.* **1995**, 28, 154; Perutz, R. N. *Chem. Soc. Rev.* **1993**, 361.
80. Lian, T.; Bromberg, S. E.; Yang, H.; Proulx, G.; Bergman, R. G.; Harris, C. B. *J. Am. Chem. Soc.* **1996**, *118*, 3769.
81. Ghosh, C. K.; Graham, W. A. G. *J. Am. Chem. Soc.* **1987**, *109*, 4726; Ghosh, C. K. Ph.D. Dissertation, University of Alberta, Edmonton, Canada, 1988.
82. Purwoko, A. A.; Lees, A. J. *Inorg. Chem.* **1996**, *35*, 675.
83. Chauby, V.; Serra Le Berre, C.; Kalck, Ph.; Daran, J.-C.; Commenges, G. *Inorg. Chem.* **1996**, *35*, 6354.
84. Sykora, J.; Sima, J. *Coord. Chem. Rev.* **1990**, *107*, 1; Juris, A.; Balzani, V.; Barigelletti, F.; Campagna, S.; Belser, P.; von Zelewsky, A. *Coord. Chem. Rev.* **1988**, *84*, 85.
85. Gray, H. B. *Chem. Soc. Rev.* **1986**, *15*, 17.
86. Holmlin, R. E.; Stemp, E. D. A.; Barton, J. K. *J. Am. Chem. Soc.* **1996**, *118*, 5236; Murphy, C. J.; Arkin, M. R.; Ghatalia, N. D.; Bossmann, S.; Turro, N. J.; Barton, J. K. *Proc. Natl. Acad. Sci. U.S.A.* **1994**, *91*, 5315.

Problems

CHAPTER 1

1. The following reaction is studied in t-butanol at 35°C by measuring the integrated peak intensity, I, of the ^{1}H NMR of the product:

$$cis\text{-}M(NH_3)_2(Cl)_2 + Br^- \longrightarrow cis\text{-}M(NH_3)_2(Cl)Br + Cl^-$$

The experimental conditions for several kinetic runs and the values of I at various times are given in the table below.

(a) Determine the experimental rate constant for each of the runs.

(b) Plot the experimental rate constants versus the bromide ion concentration and calculate the specific rate constant for the reaction, if possible.

| $M(NH_3)_2(Cl)_2$ (M) | 3.0×10^{-3} | | 5.0×10^{-3} | | 8.0×10^{-3} | |
Br^- (M)	0.060		0.085		0.110	
	Time (s)	I	Time (s)	I	Time (s)	I
	50	0.30	50	0.68	50	1.4
	100	0.58	100	1.30	100	2.56
	150	0.80	150	1.80	150	3.50
	200	1.02	200	2.24	200	4.30
	300	1.40	300	2.94	250	4.94
	400	1.70	400	3.48	300	5.48
	600	2.16	500	3.88	350	5.92
	800	2.44	600	4.16	400	6.28
	1000	2.64	800	4.54	450	6.60
	1300	2.80	1000	4.74	500	6.82
	1600	2.90	1300	4.90	600	7.21
	2000	2.96	1600	4.96	800	7.64
	3500	3.00	2500	5.00	2000	8.00

2. (a) Develop an equation that can be used to determine the experimental rate constant from measurements of absorbance versus time for the following system:

$$A \underset{k_r}{\overset{k_f}{\rightleftharpoons}} B$$

Assume that both A and B are absorbing at the observation wavelength and that both species obey Beer's law and have molar absorbancies ε_A and ε_B, respectively.

(b) Show how the final absorbance can be used to determine the equilibrium constant for the reaction if ε_A and ε_B are known.

3. Develop an expression that could be used to determine the second-order rate constant for the following system. (See Section 1.2.d.)

$$A + B \underset{k_{-2}}{\overset{k_2}{\rightleftharpoons}} 2\,C$$

4. (a) There have been two studies of the following isomerization reaction:

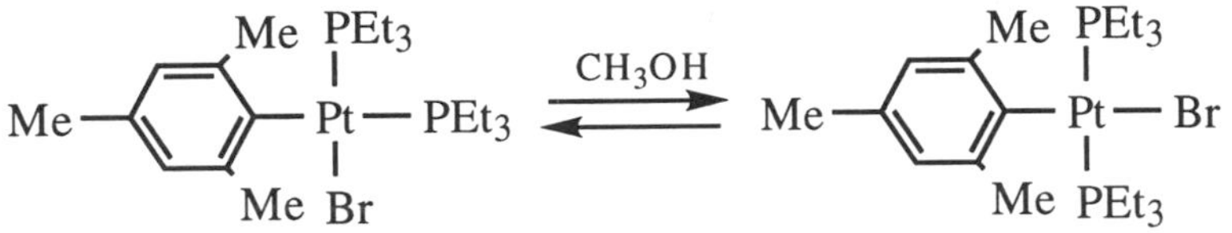

The results of the studies of the variation of the rate constant with temperature are given in the following table. Use each set of data to calculate the activation enthalpy and entropy for the reaction.

t (°C)	$10^4 \times k_{exp}$ (s^{-1})[a]	t (°C)	$10^4 \times k_{exp}$ (s^{-1})[b]
25	1.0	17.0	0.45
30	1.7	23.5	0.95
35	2.7	30.0	1.9
40	4.0	36.2	3.9
45	6.0	44.6	9.4

[a] Romeo, R.; Minniti, D.; Trozzi, M. *Inorg. Chem.* **1976**, *15*, 1134, using 5 x 10^{-5} M Pt(II) in 0.01 M LiClO$_4$.

[b] van Eldik, R.; Palmer, D. A.; Kelm, H. *Inorg. Chem.* **1979**, *18*, 572, using 5 x 10^{-4} M Pt(II) without added inert salt.

(b) Romeo et al.[1] found that the rate of the reaction in part (a) is affected by the bromide ion concentration. The results at 30°C are given in the following table. Devise an empirical expression that will describe the dependence of k_{exp} on $[Br^-]$.

$10^4 \times [Br^-]$ (M)	0.0	2.0	4.0	6.0	8.0
$10^4 \times k_{exp}$ (s^{-1})	2.13	1.05	0.69	0.53	0.43

(c) van Eldik et al.[2] also studied the pressure dependence of the following substitution reaction:

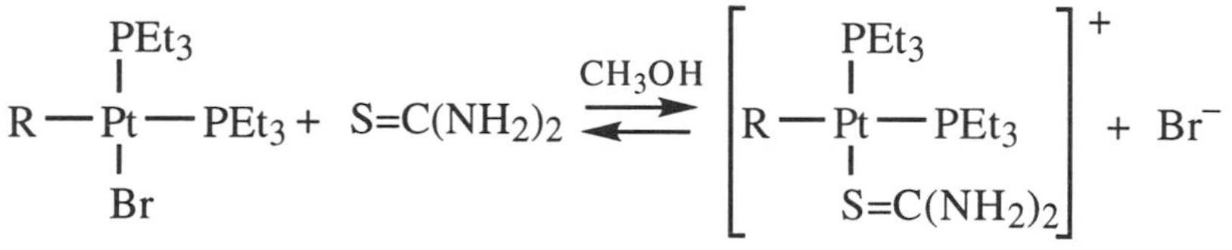

$$
\begin{array}{c}
\quad PEt_3 \\
\quad | \\
R-Pt-PEt_3 + S{=}C(NH_2)_2 \\
\quad | \\
\quad Br
\end{array}
\xrightleftharpoons{CH_3OH}
\left[
\begin{array}{c}
PEt_3 \\
| \\
R-Pt-PEt_3 \\
| \\
S{=}C(NH_2)_2
\end{array}
\right]^{+}
+ Br^-
$$

R ≡ 2,4,6 trimethylbenzene

Under pseudo-first-order conditions (1.0×10^{-4} M Pt(II), 0.01 to 0.1 M S=C(NH$_2$)$_2$) they found that the rate constant is given by the following two-term expression:

$$k_{exp} = k_1 + k_2 [S{=}C(NH_2)_2]$$

Use the pressure dependencies of k_1 and k_2 in the following table to calculate the volume of activation for each rate constant.

P (bar)	$10^4 \times k_1$ (s^{-1})	$10^3 \times k_2$ (s^{-1})
1	2.25	3.25
250	2.53	4.27
500	3.08	4.82
750	3.72	5.54
1000	4.36	5.56

CHAPTER 2

1. Develop the expression for the pseudo-first-order rate constant for the following mechanism. Assume that K_i is a rapidly maintained

equilibrium and that $[SO_4^{2-}] >> [Co(III)]_{total}$.

$$Co(NH_3)_5(OH_2)^{3+} + SO_4^{2-} \underset{}{\overset{K_i}{\rightleftharpoons}} \overset{\text{Ion pair}}{[Co(NH_3)_5(OH_2)\bullet(SO_4)]^+}$$

$$\downarrow k_1$$

$$Co(NH_3)_5(OSO_3)^+ + H_2O$$

2. Develop the expression for the pseudo-first-order rate constant for the following mechanism. Assume that [Y] and [X] >> $[M]_{total}$, the first step is a rapidly maintained equilibrium, and a steady state for the intermediate $\{(M)\bullet Y\}$.

$$M—X + Y \overset{K}{\rightleftharpoons} [(M—X)\bullet Y] \underset{k_2}{\overset{k_1}{\rightleftharpoons}} \{(M)\bullet Y\} + X$$

$$\downarrow k_3$$

$$M—Y$$

3. The mechanism shown below might be suggested as an explanation for the effect of bromide ion on the isomerization reaction in Problem 4(b) of Chapter 1:

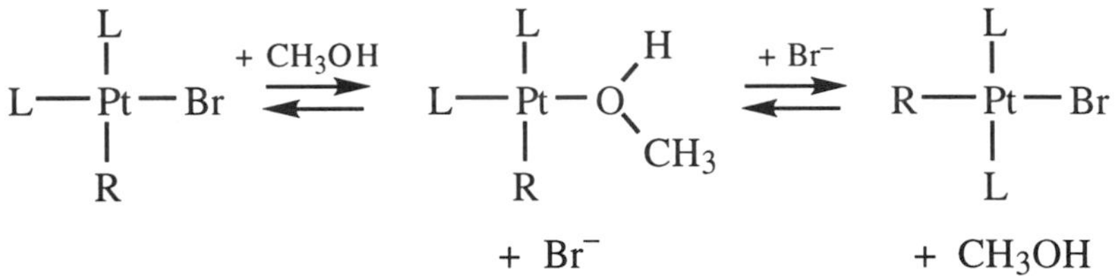

(a) Develop the rate law for this mechanism by assuming a steady state for the intermediate methanol complex.

(b) Does the predicted pseudo-first-order rate constant have a dependence on [Br⁻] consistent with the data in Problem 4(b) of Chapter 1?

(c) Is the mechanism as shown consistent with microscopic reversibility?

4. The rate law for the following reaction has been determined in aqueous 0.10 M $HClO_4$ with Fe^{3+} in excess. The rate is first order in the Cr–benzyl complex concentration and independent of the Fe^{3+} concentration.

$$(H_2O)_5Cr\text{—}CH_2C_6H_5{}^{2+} \qquad Cr(OH_2)_6{}^{3+} + 2\ Fe^{2+}$$
$$+\ 2\ Fe^{3+} + 2\ H_2O \qquad\longrightarrow\qquad +\ HOCH_2C_6H_5 + H^+$$

Suggest a mechanism consistent with this rate law. (Hint: See Zhang and Jordan.[3])

5. The reaction of $Ru(Cp)(AN)_3$ with benzene ($Cp = \eta^5\text{-}C_5H_5$, $AN = CH_3CN$) has been studied in acetone by Koefod and Mann.[4] They find that the rate is given by

$$Rate = \left(\frac{a}{[AN]^3} + \frac{b}{[AN]^2} \right)[C_6H_6][Ru(Cp)(AN)_3]$$

under conditions where [AN] and $[C_6H_6] \gg [Ru]$. The authors interpret these observations on the basis of the following rapidly maintained equilibria and the reactions given by k_1 and k_2:

$$Ru(Cp)(AN)_3{}^+ \rightleftharpoons Ru(Cp)(AN)_2{}^+ \rightleftharpoons Ru(Cp)(AN)^+ \rightleftharpoons Ru(Cp)^+$$
$$+\ AN \qquad\qquad +\ AN \qquad\qquad +\ AN$$

$$Ru(Cp)^+ + C_6H_6 \xrightarrow{\ k_1\ } Ru(Cp)(C_6H_6)^+$$

$$Ru(Cp)(AN)^+ + C_6H_6 \xrightarrow{\ k_2\ } Ru(Cp)(C_6H_6)^+ + AN$$

(a) Develop the rate law for this system. Assume that the reactant species are related by the relationships

$$\beta_1 = \frac{[Ru(Cp)(AN)^+]}{[Ru(Cp)^+]\,[AN]}; \beta_2 = \frac{[Ru(Cp)(AN)_2^+]}{[Ru(Cp)^+]\,[AN]^2}; \beta_3 = \frac{[Ru(Cp)(AN)_3^+]}{[Ru(Cp)^+]\,[AN]^3}$$

and the total reactant concentration is given by

$$[R]_{tot} = [Ru(Cp)^+] + [Ru(Cp)(AN)^+] + [Ru(Cp)(AN)_2^+] + [Ru(Cp)(AN)_3^+]$$

(b) Compare your result in (a) to the experimental rate law and describe under what conditions your result will correspond to the observations.

CHAPTER 3

1. The substitution reactions of (tetraphenylporphinato)chromium(III) chloride have been studied with several different leaving and entering groups.[5] The kinetic results have been interpreted in terms of a

limiting **D** mechanism as described by the following sequence:

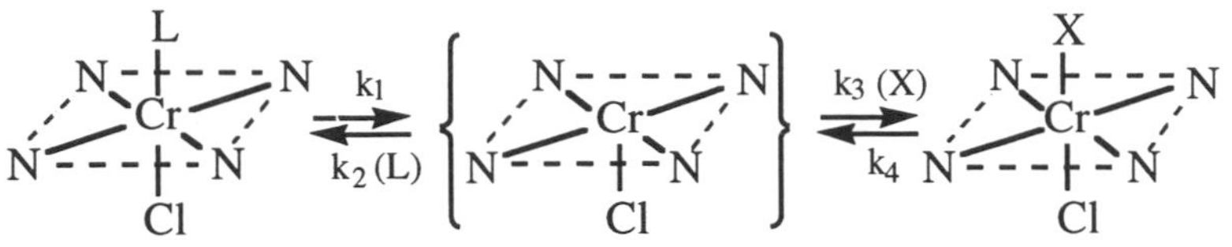

Some of the kinetic results (25°C in toluene) are summarized in the following table:

L	X	k_1 (s^{-1})^a	k_3/k_2 ^{a}	log K^a
PPh$_3$	MeIm^b	4.6	1.03 x 10^3	4.8
P(OPr)$_3$	MeIm^b	95	24	4.1
P(C$_2$H$_4$CN)$_3$	MeIm^b	80	~2 x 10^2	(5.3)^c
py	MeIm^b	5.0	1.7	2.5
PPh$_3$	py	3.6		2.6

^{a} O'Brien, P.; Sweigart, D. A. *Inorg. Chem.* **1982**, *21*, 2094.
^{b} MeIm is N-methylimidazole.
^{c} Calculated from data in the table.

(a) Use other data in the table to predict k_3/k_2 for the last system listed.

(b) Determine the order of nucleophilicity of the L and X ligands based on the k_3/k_2 values. Analyze this ordering on the basis of the Hard–Soft theory of acids and bases and other factors thought to affect nucleophilicity.

(c) Calculate k_4 from the overall equilibrium constant and the kinetic data for the three systems with X = MeIm and L = PPh$_3$, P(OPr)$_3$, and py. Are these values of k_4 consistent with the mechanistic proposal? Suggest how the value of log K for P(C$_2$H$_4$CN)$_3$ was calculated.

(d) Calculate k_4 for the system with L = PPh$_3$ and X = py. Is the result consistent with other results in the table?

2. Substitution reactions on Pt(II) are thought to proceed with associative activation. The following reaction has been studied by Romeo et al.[6] using various amine, am, entering groups:

cis-Pt(Ph)$_2$(CO)(SEt$_2$) + am $\longrightarrow$ *cis*-Pt(Ph)$_2$(CO)(am) + SEt$_2$

The pseudo-first-order rate constant is given by $k_{obsd} = k_1 + k_2[am]$. Representative results at 25°C in CH_2Cl_2 and some extrakinetic parameters are given in the following table:

am^a	pK_a	Cone Angle (deg)	E_r (kcal M^{-1})	10^4 x k_1 (s^{-1})	10^2 x k_2 (M^{-1} s^{-1})
$NH_2{}^nPr$	10.56	106	31	1.83	8.39
NH_2Cy	10.62	115	41	1.64	6.82
$NH_2{}^nBu$	10.64	113	43	1.97	4.66
$NH_2{}^tBu$	10.65	123	53	1.72	2.18
Pip	11.12	121	61	1.91	3.26
$NHEt_2$	10.92	125	73	1.74	0.669
NH^iPr_2	11.09	137	105	1.72	0.039
NEt_3	10.76	150	109	1.69	
$NHCy_2$		133	113	1.71	

^{a} Abreviations are: ^nPr = n-propyl; Cy = cyclohexyl; ^nBu = n-butyl; ^tBu = t-butyl; Pip = piperidine; ^iPr = iso-propyl.

(a) Suggest a mechanism for the k_1 pathway that is consistent with the data.

(b) Discuss the variation of k_2 with cone angle. Does this pathway exhibit a steric threshold? Estimate values of k_2 for NEt_3 and $NHCy_2$. Do these estimates seem consistent with the apparent inability to evaluate k_2 relative to k_1 for [am] = 1.0 M?

(c) Discuss the variation of k_2 with E_r by considering the same factors described in (b).

3. Will the following complexes be labile or inert as defined by Taube? $Cr^{II}(L)_6$, octahedral low spin; $V^{III}(L)_6$, octahedral; $Rh^{II}(L)_6$, octahedral.

4. Various types of evidence indicate that the substitution reactions on $Co(OH_2)_6{}^{2+}$ in water have a **D** mechanism. If this is correct, explain why the reaction rates are observed to be first order in the entering ligand.

5. (a) Use the d orbital energies in Table 3.14 to calculate the crystal field activation energies for an octahedral d^3 system undergoing substitution by a trigonal pyramidal and by a pentagonal bipyramidal transition state. Use these results and the data in Table 3.15 to predict the most favorable transition state of regular geometry and the mechanism for such a system.

(b) Calculate the crystal field activation energy for a square planar d^8 system undergoing associative substitution by square pyramidal and trigonal bipyramidal transition states.

6. The reaction of aqueous vanadium(III) with salicylic acid has the following rate law:

$$\text{Rate} = \left(6.0 + \frac{2.6}{[H^+]} \right) [V(III)]_{total} \, [\text{salicylic acid}]_{total}$$

The relevant ionization reactions in this system are the following:

$$V(OH_2)_6^{3+} \; \underset{1.4 \times 10^{-3} \text{ M}}{\overset{K_m =}{\rightleftharpoons}} \; V(OH_2)_5(OH)^{2+} + H^+$$

$$C_6H_4(OH)(CO_2H) \; \underset{1.6 \times 10^{-3} \text{ M}}{\overset{K_{a1} =}{\rightleftharpoons}} \; C_6H_4(OH)(CO_2)^- + H^+$$

$$C_6H_4(OH)(CO_2)^- \; \underset{1 \times 10^{-12} \text{ M}}{\overset{K_{a2} =}{\rightleftharpoons}} \; C_6H_4(O)(CO_2)^{2-} + H^+$$

(a) Propose two reaction schemes that are consistent with the rate law.

(b) Calculate the specific rate constants for each scheme in (a) and determine which, if any, is more probable on the basis of other data in Table 3.23.

7. Several studies have shown that the replacement of L in $Fe^{II}(CN)_5L$ complexes has kinetic properties typical of a **D** mechanism. The rate shows a saturation effect at high entering group concentrations, consistent with Eq. (3.3). Reddy and van Eldik[7] have determined the ΔV^* for k_1 for such reactions with L = various amines as the entering group. The partial molar volume, $\bar{V}^\circ$, of the amines was varied to determine the effect on ΔV^*. The results are tabulated below:

	NH_3	NH_2CH_3	$NH_2C_2H_5$	$NH_2^iC_3H_7$	NH_2CH_2Ph
ΔV^* (cm^3 M^{-1})	16.6	24.0	16.3	18.5	17.4
$\bar{V}^\circ$ (cm^3 M^{-1})	24.8	41.7	58.4	85.6	109

(a) What is unusual about these results, since, for a **D** mechanism

$$\Delta V^* = \bar{V}^\circ(Fe(CN)_5) + \bar{V}^\circ(L) - \bar{V}^\circ(Fe(CN)_5L)$$

(b) Suggest an explanation for these observations. (See Jordan.[8])

CHAPTER 4

1. For the following complex, draw diagrams describing a process that:

    (a) Causes racemization but no isomerization.

    (b) Causes cis–trans isomerization.

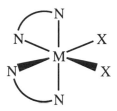

2. For each of the isomers of the following asymmetric chelate:

    (a) Predict the products if the reaction proceeds by one-ended dissociation of A' to give a trigonal bipyramidal intermediate with the monodentate ligand in the trigonal plane. Assume that ring closing occurs with equal probability along the trigonal edges.

    (b) Indicate if a new structural isomer or stereoisomer of the reactant has formed in each case.

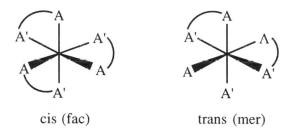

cis (fac)   trans (mer)

3. Stereochemical mobility in a square-based pyramidal molecule may be due to an "anti-Berry" pseudorotation in which the intermediate in the normal pseudorotation is actually the reactant. Sketch an anti-Berry process for a square-based pyramidal molecule.

4. The fluxional behavior of the Ru(III) acetylacetonate derivatives in the following scheme have been studied by Hoshino et al.[9] The authors isolated the isomer in the lower left of the scheme and deduced that, as it isomerizes in DMF, it does not convert directly to the isomer on the lower right, i.e., k_{13} and k_{31} are much smaller than the other rate constants. However, the isomer at the top of the scheme can interconvert to both of the other isomers. This behavior places restrictions of the types of rearrangement that occur.

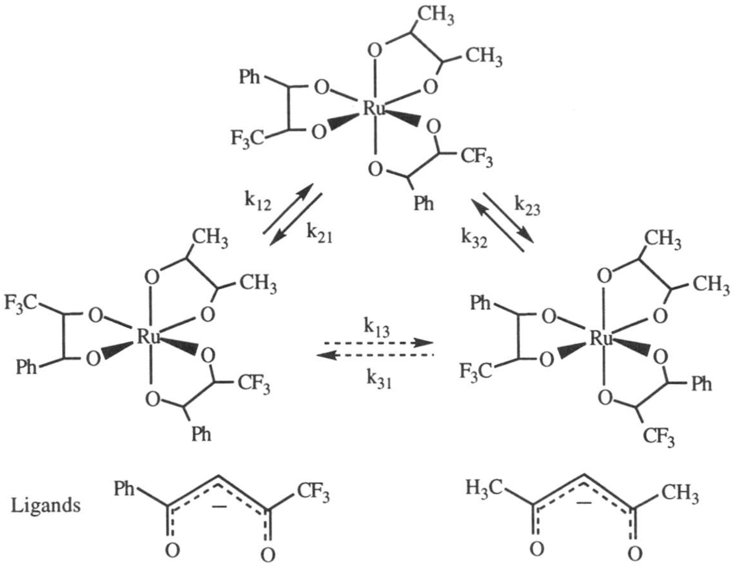

(a) Show two intramolecular twist mechanisms that will bring about the k_{13} reaction. These are presumably elimated as mechanistic possibilities because this change is not observed.

(b) Show two mechanisms that proceed by trigonal bipyramidal intermediates and give the k_{21} and k_{23} process. Assume that the bond broken is always that of an O next to CF_3.

(c) Determine if the process described in (b) gives the k_{13} process.

5. Show with clear diagrams why an η^2-C_6H_6 system should be more readily fluxional than an η^4-C_6H_6 system.

6. For $Rh(\eta^5$-$C_5H_5)(C_2F_4)(C_2H_4)$, the pressure dependence of the rotation of ethylene has been measured by Peng and Jonas[10] using ^1H NMR line shape analysis. Representative results in pentane at 0°C are given below. Note that the solvent viscosity, η, increases with increasing pressure.

P (MPa)	0.10	105	202	308	401	495
k (s^{-1})	235	295	340	359	346	313
η (cP)	0.448	0.762	1.21	1.87	2.60	3.47

The observations are unusual because the rate constant goes through a maximum. The results were fitted to a polynomial to obtain

$$\ln k = 5.461 + 2.508 \times 10^{-3}\, P - 3.585 \times 10^{-6}\, P^2$$

(a) Since $(\delta(\ln k)/\delta P)_T = -\Delta V^*/RT$, the derivative of the above polynomial can be used to calculate the value of ΔV^* at any pressure. Calculate ΔV^* at pressures of 0, 100, 200, 300, 400, and 500 MPa.

(b) The authors have ascribed the pressure-dependent component of ΔV^* to a frictional effect of the solvent. Intuitively, one might expect such an effect to correlate with the solvent viscosity. Plot the ΔV^* values in (a) versus η to determine if this supposition is reasonable.

CHAPTER 5

1. Explain the trends in the rate constants given in the table below for the following reaction:

$$Cr(CO)_5L + {}^*CO \longrightarrow Cr(CO)_4({}^*CO)L + CO$$

Compound	k $(s^{-1}, 30°C)^a$
$Cr(CO)_6$	1.0×10^{-12}
$Cr(CO)_5(PMe_2Ph)$	1.5×10^{-10}
$Cr(CO)_5(PPh_3)$	3.0×10^{-10}
$Cr(CO)_5Br^-$	2.0×10^{-5}
$Cr(CO)_5Cl^-$	1.5×10^{-4}

[a] Atwood, J. D.; Brown, T. L. *J. Am. Chem. Soc.* **1976,** *98*, 3160, and references therein.

2. Substitution reactions on some metal carbonyls proceed by parallel dissociative and associative pathways and the reaction may have significant reversibility. The situation has been considered recently by Schneider and van Eldik.[11] If the entering group is Y and the leaving group is CO, then Eq. (3.17) can be readily adapted to give the dissociative contribution to k_{exp} and k_5 and k_{-5} can be defined as the forward and reverse rate constants for the parallel associative pathway. Then, the pseudo-first-order rate constant is given by

$$k_{exp} = \frac{k_1 k_3 [Y] + k_2 [CO] k_4}{k_2 [CO] + k_3 [Y]} + k_5 [Y] + k_{-5} [CO]$$

(a) Apply the principle of detailed balancing to this cyclic system by noting that the equilibrium constant $K = k_1 k_3/k_2 k_4 = k_5/k_{-5}$. Use these relationships to eliminate the reverse rate constants, k_4 and k_{-5}, and express k_{exp} in terms of the other rate constants and K.

(b) For the limiting conditions that $k_3[Y] \gg k_2[CO]$ and $k_1k_3[Y] \gg k_2k_4[CO]$, the expression for k_{exp} appears to reduce to

$$k_{exp} = k_1 + k_5[Y] + k_{-5}[CO]$$

Show that this equation must be simplified further to be consistent with the assumptions and the principle of detailed balancing.

3. For the following reaction, the rate constant in toluene has been found to be independent of the nature and concentration of L.

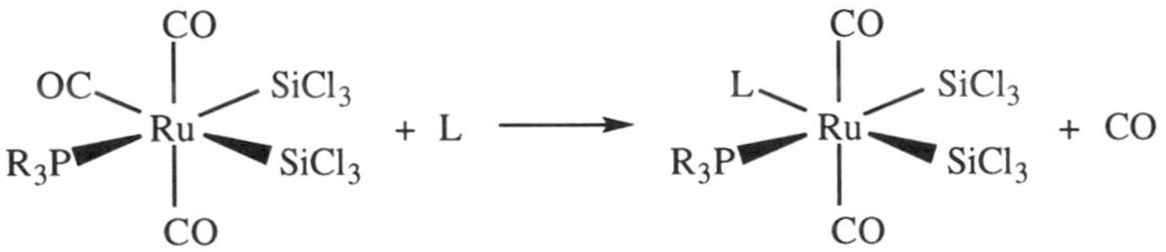

The kinetics have been studied for various PR_3 ligands and the rate constants are given in the following table:

Phosphine	pK_a	Cone Angle	$- \log k^{\,a}$
$P(OCH_2)_3CEt$	1.74	101	5.96
$P(OMe)_3$	2.6	107	5.14
$PPh(OMe)_2$	2.64	120	4.95
$PPhMe_2$	6.50	122	4.72
$P(OPh)_3$	-2.0	128	4.70
$PPh_2(OMe)$	2.69	132	4.64
PPh_2Me	4.57	136	4.15
$P(p\text{-}ClC_6H_4)_3$	1.03	145	3.17
$P(p\text{-}FC_6H_4)_3$	1.97	145	3.12
PPh_3	2.73	145	3.05
$P(p\text{-}MeC_6H_4)_3$	3.84	145	3.08
$P(m\text{-}MeC_6H_4)_3$	3.3	165	2.89
$P(m\text{-}ClC_6H_4)_3$	1.03	165	3.02

[a] Chalk, K. L.; Pomeroy, R. K. *Inorg. Chem.* **1984**, *23*, 444; at 40°C in toluene, k in s^{-1}.

(a) Suggest the reaction mechanism, based on the kinetic results with varying L.

(b) The relative constancy of the rate constants for the series of PR_3 with a cone angle of 145° but different pK_a indicates that the reaction is relatively insensitive to σ-bonding effects. However, the rate constant seems to vary with the cone angle. Make a plot of log k versus cone angle and analyze the steric effects in terms of their influence on the ground state and transition state for the reaction. (*Hint*: See Giering et al.[12])

4. Nineteen-electron organometallic complexes are rare and generally quite reactive. Recently, the ligand substitution kinetics have been studied for the following system:

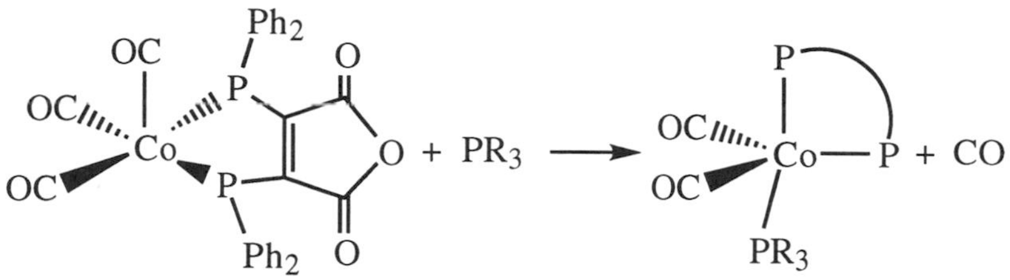

Although formally 19-electron systems, these complexes may be viewed as Co(I) with a reduced ligand radical, P_2^-. In any case, the mechanism of substitution is of interest, and the following data have been obtained. All the data are for 1.4×10^{-3} M Co(CO)$_3$(P-P).

Table A. Rate Constants (25°C) for Co(CO)$_3$(P-P) + PPh$_3$ in CH$_2$Cl$_2$

[PPh$_3$] (M)	0.148	0.296	0.371	0.445	0.556	0.704
10^3 x k (s^{-1})	5.48	5.45	5.47	5.52	5.42	5.50

Table B. Rate Constants for Co(CO)$_3$(P-P) + PPh$_3$ in CH$_2$Cl$_2$

t (°C)	10.0	15.0	20.0	25.0	30.0
10^3 x k (s^{-1})	0.598	1.26	2.67	5.47	10.2

Table C. Rate Constants (25°C) for Co(CO)$_3$(P-P) + PR$_3$ in CH$_2$Cl$_2$

PR$_3$	PPh$_3$	P(OPh)$_3$	PMePh$_2$	PBu$_3$	P(OMe)$_3$
10^3 x k (s^{-1})	5.47	5.18	5.20	5.58	5.03

(a) Are the kinetic results consistent with an I_a mechanism?

(b) Are the kinetic results consistent with an **A** mechanism?

(c) Are the kinetic results consistent with rate-controlling chelate ring opening followed by substitution? (See Scheme 5.2.)

(d) Are the kinetic results and activation parameters for PPh₃ consistent with an **I_d** mechanism?

5. Substitution on 17-electron systems usually proceeds with associative activation. $Mo(Cp)(PMe_3)_2(I)_2$ is unusual for a 17-electron complex because it is more readily reduced than oxidized. This suggests that it is not a willing electron acceptor and therefore might undergo substitution by an associative pathway. Poli, Owens, and Linck[13] have studied the following system in dichloromethane:

$$Mo(Cp)(PMe_3)_2(I)_2 + (PPN)Cl \longrightarrow Mo(Cp)(PMe_3)_2(I)Cl + (PPN)I$$

$$(PPN)^+ \bullet Cl^- \underset{}{\overset{K_d}{\rightleftharpoons}} (PPN)^+ + Cl^-$$
$$\text{ion pair}$$

(PPN)Cl = (triphenylphosphinium)iminium chloride

The study is complicated by the ion pairing, and the general problem is to determine if the active nucleophile is free Cl^- or the ion pair, $(PPN)^+ \bullet Cl^-$, and if there is a dissociative component to the rate. The ion pair formation constants were determined from conductivity measurements by Algra and Balt.[14] Therefore, it is possible to calculate the amount of free Cl^- and ion pair, $(PPN)^+ \bullet Cl^-$, for a particular total (PPN)Cl concentration. Typical results at 30°C and $K_d = 4 \times 10^{-4}$ M are tabulated below:

$10^2 \times$ (PPN)Cl (M)	0.546	1.224	2.372	3.400	6.490
$10^2 \times [Cl^-]$ (M)	0.139	0.219	0.313	0.379	0.532
$10^2 \times (PPN)^+ \bullet Cl^-$ (M)	0.407	1.005	2.059	3.021	5.958
$10^5 \times k_{obsd}$ (s^{-1})	0.60	0.92	1.48	1.83	3.00

(a) If $[Cl^-]$ is assumed to be the only nucleophile, then least-squares analysis gives

$$k_{obsd} = -(1.86 \pm 0.95) \times 10^{-6} + (5.43 \pm 0.40) \times 10^{-3} [Cl^-]$$

Why does this assumption appear unreasonable?

(b) If the ion pair, $(PPN)^+ \bullet Cl^-$, is assumed to be the only nucleophile, then least-squares analysis gives

$$k_{obsd} = (4.35 \pm 0.36) \times 10^{-6} + (4.58 \pm 0.24) \times 10^{-4} [(PPN)^+ \bullet Cl^-]$$

What could be the significance of the term that is independent of $[(PPN)^+ \bullet Cl^-]$ in this equation?

(c) If one assumes that both Cl⁻ and the ion pair are reactive, then

$$k_{obsd} = (3.64 \pm 0.22) \times 10^{-3} [Cl^-]$$
$$+ (1.64 \pm 0.34) \times 10^{-4} [(PPN)^+ \cdot Cl^-]$$

Do the relative values of the rate constants for these two nucleophiles seem reasonable?

(d) The ambiguities in this rate law analysis might be resolved if the rate constant with Cl⁻ could be determined. Suggest how this might be done by using a chloride salt with a quite different value of K_d.

6. It has been observed that *cis*-Pt(PEt₃)₂(H)(Solv)⁺ (Solv = methanol or acetone) reacts with 1-hexene but only brings about the isomerization of the 1-hexene to 2-hexene and 3-hexene. Suggest a mechanism for this process.

7. Assume that the reaction of Rh(Cp*)(CO)Kr with cyclohexane involves rate-controlling dissociation of Kr. Verify that the k_{exp} has the form of Eq. (5.40) and the conclusion of Bergman et al. that α should be independent of the nature of the alkane.

8. The following reaction is of possible relevance to the mechanism of the hydroformylation reaction:

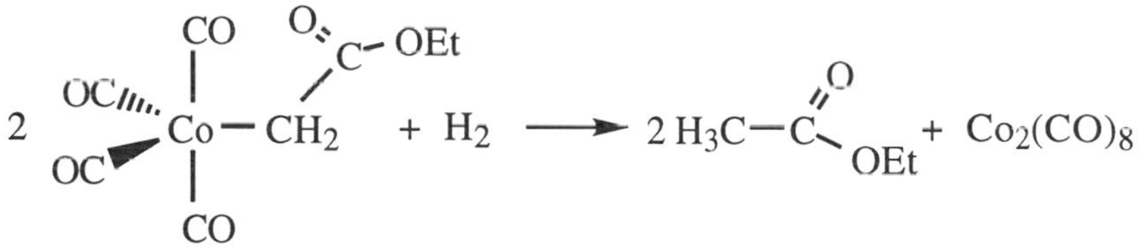

The dependence of the reaction rate on H₂ and CO concentrations has been determined at 35°C in *n*-heptane, and the results are given in the following table:

$10^3 \times [H_2]$ (M)	3.68	3.68	9.94	3.30	2.59	2.13
$10^3 \times [CO]$ (M)	1.31	1.31	3.56	2.30	4.16	5.35
$10^5 \times k_{exp}$ (s⁻¹)	1.89	1.93	1.60	1.03	0.43	0.25

(a) Deduce the rate law that seems to describe the results best.

(b) Suggest a mechanism that is consistent with the rate law in (a).

(c) Suggest a further type of experiment or kinetic study that would help to confirm your suggestion in (b).

9. In a study of the substitution reactions of $Re_3(\mu\text{-}H)_3(CO)_{11}(NCCH_3)$, Beringhelli et al.[15] have proposed the dissociative mechanism in the following scheme:

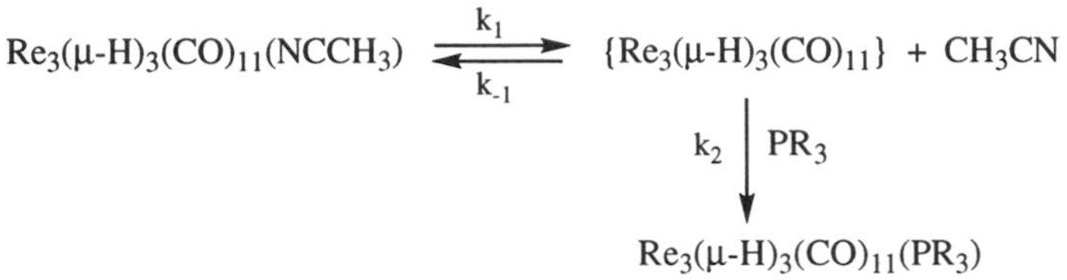

$$Re_3(\mu\text{-}H)_3(CO)_{11}(NCCH_3) \underset{k_{-1}}{\overset{k_1}{\rightleftharpoons}} \{Re_3(\mu\text{-}H)_3(CO)_{11}\} + CH_3CN$$

$$\downarrow k_2 \quad PR_3$$

$$Re_3(\mu\text{-}H)_3(CO)_{11}(PR_3)$$

(a) Determine if this mechanism is consistent with the dependence of k_{obsd} on reagent concentrations given in the following table for the reaction at 300 K in $CDCl_3$.

$[Re_3]_{tot}$ (M)	0.019	0.014	0.0141	0.0121
$[PPh_3]$ (M)	0.219	0.143	0.290	0.143
$[CH_3CN]$ (M)	0.262	0.367	0.152	0.265
$10^5 \times k_{obsd}$ (s^{-1})	1.71	0.89	3.68	1.22

(b) For several phosphines, the analysis gave values of k_2/k_{-1}, as tabulated below along with some extrakinetic factors. Discuss the trends in this ratio with respect to the basicity of the phosphine and steric factors, as measured by the cone angle or the repulsion parameter, E_r.

Phosphine	pK_a	Cone Angle	E_r	$\log k_2/k_{-1}$
PMe_3	8.65	118	39	−0.04
$PPhMe_2$	6.50	122	44	−0.21
$P(OMe)_3$	2.6	107	52	−0.45
PPh_2Me	4.59	136	57	−0.42
P^nBu_3	8.43	132	64	−0.68
PPh_3	2.73	145	75	−0.92

(c) Are the data in (a) consistent with a mechanism involving pre-equilibrium hydride-bridge opening followed by phosphine substitution on the unsaturated metal center and then rapid bridge closing and CH_3CN elimination?

(d) If the reaction were done in the presence of CD_3CN, would the observations in (a) predict appearance of CD_3CN in the reactant?

CHAPTER 6

1. The oxidation of iron(II) by lead(IV) might proceed by either of the following mechanisms:

Mechanism A	Mechanism B

$$Fe^{II} + Pb^{IV} \rightleftharpoons Fe^{III} + Pb^{III} \qquad Fe^{II} + Pb^{IV} \rightleftharpoons Fe^{IV} + Pb^{II}$$

$$Fe^{II} + Pb^{III} \longrightarrow Fe^{III} + Pb^{II} \qquad Fe^{II} + Fe^{IV} \longrightarrow 2\ Fe^{III}$$

(a) Assume a steady state for Pb^{III} and Fe^{IV} and develop the rate law for each mechanism.

(b) Could these alternatives be distinguished experimentally?

2. Use the data in Table 6.2 and the Marcus relationship, Eq. (6.33), to predict the rate constants for the following reactions:

$$Co(sep)^{3+} + Cr(OH_2)_6^{2+} \longrightarrow Co(sep)^{2+} + Cr(OH_2)_6^{3+}$$

$$Fe(OH_2)_6^{3+} + Ru(NH_3)_6^{2+} \longrightarrow Fe(OH_2)_6^{2+} + Ru(NH_3)_6^{3+}$$

3. For the following reaction, $k = 4.2 \times 10^2$ M^{-1} s^{-1} and $K = 1.3$ at 25°C:

$$V(OH_2)_6^{3+} + Cr(bpy)_3^{2+} \longrightarrow V(OH_2)_6^{2+} + Cr(bpy)_3^{3+}$$

Use additional data in Table 6.2 to estimate the self-exchange rate constant for $Cr(bpy)_3^{2/3+}$.

4. The Ru(IV) and Ru(III) complexes of sarcophagine, sar, undergo deprotonation to give $Ru^{IV}(sar(-H))^{3+}$ and $Ru^{III}(sar(-H))^{2+}$, and this couple has a reduction potential of 0.05 V. Bernhard and Sargeson[16] have reported the rate constant of 7×10^7 M^{-1} s^{-1} for the following reaction:

$$Ru(sar)^{3+} + Ru(sar(-H))^{2+} \longrightarrow Ru(sar)^{2+} + Ru(sar(-H))^{3+}$$

(a) Use the simple Marcus relationship to estimate the self-exchange rate constant for $Ru^{IV/III}(sar(-H))$, given that the $Ru^{III/II}(sar)$ couple has a self-exchange rate constant of 1.2×10^5 M^{-1} s^{-1} and a reduction potential of 0.29 V.

(b) For the reaction of $Ru^{II}(sar)^{2+}$ with O_2, $k = 1.4$ M^{-1} s^{-1} and the

O_2/O_2^- couple has a reduction potential of -0.15 V. Use this and other information in (a) to estimate the self-exchange rate constant for the O_2/O_2^- couple. Compare your answer to other results in Table 6.5.

5. Predict the products of the following reactions:

(a) $Cr(OH_2)_6^{2+} + Co (NH_3)_5(CN)^{2+} \longrightarrow$

(b) $Ru(OH_2)_6^{2+} + Co(NH_3)_5Cl^{2+} \longrightarrow$

6. Based on the reduction potentials and self-exchange rate constants given in the following scheme, estimate whether either Ce^{IV} or MnO_4^- would be suitable analytical reagents for $Fe(CN)_6^{4-}$.

	$E°$ (V)	k $(M^{-1} s^{-1})$
$*Ce^{IV} + Ce^{III} \rightleftharpoons$	$+1.44$	4.6
$*Fe(CN)_6^{3-} + Fe(CN)_6^{4-} \rightleftharpoons$	$+0.36$	3×10^2
$*MnO_4^- + MnO_4^{2-} \rightleftharpoons$	$+0.56$	3.6×10^3

CHAPTER 7

1. In Figure 7.1, assume that the photoactive state A can react by two pathways to give products P_1 and P_2 with rate constants k_1 and k_2, respectively. Develop the expression for the quantum yields of the two products.

2. The energy levels for the $d–d$ transitions in an octahedral chromium(III) complex are given in Figure 7.3. Suggest two explanations for the observation that the quantum yield for photosubstitution on such complexes is independent of the irradiation wavelength over the region from 350 to 700 nm.

3. Use Adamson's rules to:

(a) Predict the product of the photolysis of *cis*-$Cr(NH_3)_4(Cl)_2^+$.

(b) Explain the very low quantum yield for the photoaquation of *trans*-$Cr(en)_2(OH_2)_2^{3+}$.

4. For Scheme 7.12, develop the kinetic expression that can be used to determine the various rate constants from the dependence of the rate on [CO] and [THF].

5. The activation volumes (cm^3 M^{-1}) for the photolysis of two chloro complexes of chromium(III) and rhodium(III) are given in the table below. Suggest rationalizations for the similarities and differences.

	$\Delta V^*(\Phi_{Cl})$	$\Delta V^*(\Phi_{NH_3})$
Cr(NH$_3$)$_5$Cl^{2+}	−13	−9.4
Rh(NH$_3$)$_5$Cl^{2+}	−8.6	+9.3

6. The photolysis of *trans*-Rh(PMe$_3$)$_2$(CO)Cl has been studied in benzene/THF solution and the products and proposed reaction scheme are shown below:

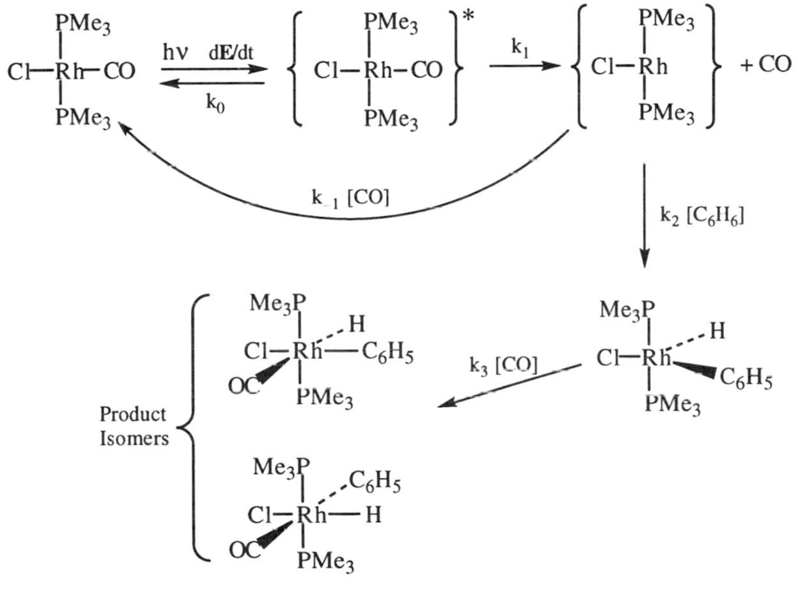

(a) Develop the expression for the quantum yield of the products, assuming a steady state for all intermediate species and that excess CO and benzene are present.

(b) Would a study of the variation of product quantum yield with the concentration of benzene and CO help to establish the validity of the proposed scheme?

(c) The interconversion of the product isomers has been proposed to proceed through an η^2-benzene intermediate. Show, with clear diagrams, how this might occur.

7. The photolysis of $Re_2(CO)_{10}$ in acetonitrile produces a mixture of $Re_2(CO)_9(NCCH_3)$ and $Re(CO)_5$. In the absence of other reagents, two $Re(CO)_5$ radicals recombine with $k = 1 \times 10^{10}$ M^{-1} s^{-1}. Sarakha and Ferraudi[17] observed that the addition of various macrocyclic complexes of Cu(II) of the general form $Cu(N)_4^{2+}$ had no effect on the decay of $Re(CO)_5$ unless halide ions (Cl^-, Br^-, or I^-) were added to the acetonitrile solution. It also was observed that NaCl or NaBr alone did not affect the spectrum or decay of $Re(CO)_5$.

   (a) Suggest a mechanism for the reaction. (*Hint*: See Section 6.4.)

   (b) On the basis of your mechanism, suggest other species that might be added to promote the reaction of $Re(CO)_5$ and $Cu(N)_4^{2+}$.

CHAPTER 8

1. Describe in detail the complete mechanistic sequence proposed by Finke and co-workers for the reaction of diol dehydrase with ethylene glycol.

2. (a) For a system with competitive inhibition, develop expressions for the slope, intercept when $[S]^{-1} = 0$ and intercept when $v_i^{-1} = 0$ for Lineweaver–Burke plots, by rearrangement of Eq. (8.16).

   (b) For a system showing uncompetitive inhibition, develop expressions for the slope, intercept when $[S]^{-1} = 0$ and intercept when $v_i^{-1} = 0$ for Lineweaver–Burke plots, by rearrangement of Eq. (8.18).

   (c) Draw sketches to show how the Lineweaver–Burke plots would vary with [I] for competitive and uncompetitive inhibition.

3. Assume that the hydrolysis of phenyl acetate by carbonic anhydrase proceeds by a mechanism analogous to that for CO_2 hydration. Describe the sequence of reactions that would be involved in the ester hydrolysis.

4. Show that the steady-state solution for Scheme 8.6 yields Eq. (8.28).

5. Consider the electronic structure of the reactants and suggest why the rate-controlling step for CO binding to myoglobin may be formation of the Fe—CO bond, while this step is not rate-controlling for O_2 binding.

6. Show that the mechanism in Scheme 8.10 gives the experimental rate law for NO oxidation. Make any steady-state approximations that are necessary.

7. If reactants are involved in protolytic equilibria, there is often a proton ambiguity as to the state of protonation of the actual reacting species. In the case of the reaction of peroxynitrite ion with carbon dioxide, Lymar and Hurst[18] were able to show that the reactants are CO_2 and ONO_2^-.

(a) Develop two rate expressions in terms of total carbonate species and total peroxynitrite species for the following scheme, assuming the reaction proceeds only by the k_2 path and only by the k_2' path. Assume that the equilibria are rapidly maintained.

$$CO_2 + H_2O \underset{1.1 \times 10^{-6}}{\overset{K_a' =}{\rightleftharpoons}} H^+ + HCO_3^-$$

$$ONO_2H \underset{2.5 \times 10^{-7}}{\overset{K_a =}{\rightleftharpoons}} H^+ + ONO_2^-$$

$$ONO_2^- + CO_2 \xrightarrow{k_2} ONO_2CO_2^-$$

$$ONO_2H + HCO_3^- \xrightarrow{k_2'} ONO_2CO_2 + OH^-$$

(b) In the above development, the assumption that the CO_2 hydration is rapid is not really true. It has a half-time of ~25 s at pH 7. Explain how this would allow the actual nature of the reacting species to be determined.

References

1. Romeo, R.; Minniti, D.; Trozzi, M. *Inorg. Chem.* **1976**, *15*, 1134.
2. van Eldik, R.; Palmer, D. A.; Kelm, H. *Inorg. Chem.* **1979**, *18*, 572.
3. Zhang, Z.; Jordan, R. B. *Inorg. Chem.* **1994**, *33*, 680.
4. Koefod, R. S.; Mann, K. R. *J. Am. Chem. Soc.* **1990**, *112*, 7287.
5. O'Brien, P.; Sweigart, D. A. *Inorg. Chem.* **1982**, *21*, 2094.
6. Romeo, R.; Arena, G.; Scolaro, L. M.; Plutino, M. R. *Inorg. Chem.* **1994**, *33*, 4029.
7. Reddy, K. B.; van Eldik, R. *Inorg. Chem.* **1991**, *30*, 596.
8. Jordan, R. B. *Inorg. Chem.* **1996**, *35*, 3725.
9. Hoshino, Y.; Takahashi, R.; Shimizu, K.; Sato, G. P.; Aoki, K. *Inorg. Chem.* **1990**, *29*, 4816.
10. Peng, X.; Jonas, J. *J. Chem. Phys.* **1990**, *93*, 2192.
11. Schneider, K. J.; van Eldik, R. *Organometallics* **1990**, *9*, 92, 1235.
12. Rahman, Md. M.; Liu, H.-Y.; Prock, A.; Giering, W. P. *Organometallics* **1987**, *6*, 650; Lorsbach, B. A.; Bennett, D. M.; Prock, A.; Giering, W. P. *Organometallics* **1995**, *14*, 869.

13. Poli, R.; Owens, B. E.; Linck, R. G. *Inorg. Chem.* **1992**, *31*, 662.
14. Algra, G. P.; Balt, S. *Inorg. Chem.* **1981**, *20*, 1102.
15. Beringhelli, T.; D'Alfonso, G.; Minoja, A. P.; Freni, M. *Inorg. Chem.* **1991**, *30*, 2757.
16. Bernhard, P.; Sargeson, A. M. *Inorg. Chem.* **1988**, *27*, 2754; *J. Am. Chem. Soc.* **1989**, *111*, 597.
17. Sarakha, M.; Ferraudi, G. *Inorg. Chem.* **1996**, *35*, 313.
18. Lymar, S. V.; Hurst, J. K. *J. Am. Chem. Soc.* **1995**, *117*, 8867.

Chemical Abbreviations

Ar	aryl
azacapten	1-methyl-3,13,16-trithia-6,8,10,19-tetraazabicyclo[6.6.6]eicosane
bpy	2,2'-bipyridyl
Bu	butyl
Cp	cyclopentadienide
Cp*	pentamethylcyclopentadienide
Cy	cyclohexyl
cyclam	1,4,8,11-tetraazacyclotetradecane
dien	diethylenetriamine
DMA	*N,N*-dimethylacetamide
DMF	*N,N*-dimethylformamide
DMP	dimethylphenanthroline
DMSO	dimethylsulfoxide
EDTA	ethylenediaminetetraacetate
en	ethylenediamine; 1,2-diaminoethane
Et	ethyl
Et$_4$dien	tetraethyldiethylenetriamine
gly	glycine
hfac	hexafluoroacetylacetonate
Im	imidazole
isn	isonicotinamide
Mb	myoglobin
Me	methyl
nic	nicotinamide
NTA	nitrilotriacetic acid
ox	oxalate
Ph	phenyl
phen	1,10-phenanthroline
Pr	propyl
Pro	proline
Prop carb	propylene carbonate; 1,2-propanediol cyclic carbonate
py	pyridine

sar	sarcophagine; 3,6,10,13,16,19-hexaazabicyclo[6.6.6]eicosane
sep	sepulchrate; 1,3,6,8,10,13,16,19-octaazabicyclo[6.6.6]eicosane
tacn	1,4,7 triazacyclononane
THF	tetrahydrofuran
tn	trimethylenediamine; 1,3-diaminopropane
TPP	tetraphenylporphinate; 5,10,15,20-tetraphenyl-21H,23H-porphine
tren	triethylenetetramine
ttacn	1,4,7-trithiacyclononane
[9]aneS$_3$	1,4,7-trithiacyclononane

Index